Robert Winston is one of the  scientists. As Emeritus Professor Imperial College, London, and reproductive physiology, he has made advances in fertility medicine and is a leading voice in the debate on embryo research and genetic engineering. His television series, including *Your Life in Their Hands*, *Making Babies*, *The Human Body*, *Child of Our Time*, *Human Instinct*, *Human Mind* and *The Story of God*, have made him a household name across Britain. He became a life peer in 1995.

Also by Robert Winston

INFERTILITY: A SYMPATHETIC APPROACH
GETTING PREGNANT
MAKING BABIES
SUPERHUMAN (with LORI OLIWENSTEIN)
GENETIC MANIPULATION
THE IVF REVOLUTION
HUMAN INSTINCT
THE HUMAN MIND
WHAT MAKES ME ME
THE STORY OF GOD

A Child Against All Odds

ROBERT WINSTON

BANTAM BOOKS

LONDON • TORONTO • SYDNEY • AUCKLAND • JOHANNESBURG

TRANSWORLD PUBLISHERS
61-63 Uxbridge Road, London W5 5SA
a Random House Group Company
www.rbooks.co.uk

A CHILD AGAINST ALL ODDS
A BANTAM BOOK: 9780553817447

First published in Great Britain
in 2006 by Bantam Press
a division of Transworld Publishers
Bantam edition published 2007

Typeset in 11/13pt Sabon by
Falcon Oast Graphic Art Ltd.
Printed and bound in Great Britain by
CPI Cox & Wyman, Reading, RG1 8EX

2 4 6 8 10 9 7 5 3 1

To Stephen Hillier, Alan Handyside,
Kate Hardy and Carol Readhead

Invaluable scientific colleagues whose intelligence,
knowledge and friendship constantly informed me,
whose commitment to young scientists has inspired so many,
and whose selfless dedication to improving reproductive
treatment continues to be of immeasurable value to
thousands of people.

Contents

Acknowledgements   9
Prologue: Visions of Mortality   13

1   The Beginnings of Life   21
2   IVF: A 'blind assertion of will against our bodily
            nature'?   77
3   IVF: When, Why and How   131
4   Freezing: The Refrigeration Game   189
5   Donated Eggs and Sperm: The Most
            Generous Gift?   235
6   Womb to Rent   296
7   Pre-implantation Diagnosis   322
8   Cloning: A Feat of Clay   374
9   Stem Cells: Nature's Magic   405
10   The Trojan Horse   447

Notes   501
Glossary   529
Index   541

Acknowledgements

Unquestionably my greatest debt is to the many men and women who came as patients hoping that my team and I at Hammersmith could help them. These people, coming from many parts of the world, bared their pain and their inmost fears, their aspirations, and often the most private part of their lives. I learnt hugely from them. Fertility treatment is life-changing when it works, but it is accompanied more often by failure than by success. Though I have too frequently dealt in failure, my hope is that sometimes even this will have brought resolution for those undergoing treatment.

So this book, which I have written and rewritten in my mind over many years, is a somewhat autobiographical account of some of my medical and scientific work. Although it has been many years in the making, it has been put down on paper quite rapidly to coincide with a new television series. There was a need to research some of the material in a short space of time and write extremely quickly when I was often occupied elsewhere. Rapid writing may have ensured that the book is up-to-date, but there may be mistakes or omissions of recall; for any such, I alone am to blame.

I am grateful to various people who have made my job

much easier. My old friend Matt Baylis, with his usual insight into scientific and ethical problems, helped me with a number of areas of research and avenues of interest. As in the past with other projects, his wide knowledge and enthusiasm for interesting themes have been immensely valuable. He helped me at ridiculously short notice to evaluate some of the issues raised by reproductive manipulation and also checked references – and this book is much the better for his advice. I am also most grateful to John Oates, who helpfully collated much historical material, not all of which have I had space to use. His was a difficult task that added considerably to my thinking about the subject-matter.

Lira, my wife, was as supportive as ever, lovingly and assiduously reading through the entire manuscript, making valuable criticisms and correcting numerous imperfections. As usual, she has tolerated my preoccupied behaviour which is notable whenever I am engaged on projects of this kind. I am also very grateful to Rachel Ward, my most willing PA, who typed part of the manuscript.

Once again I have been truly blessed with the good fortune of having Gillian Somerscales as my copy-editor. She has improved this book hugely with the insight of her questions and her suggestions to advance both its clarity and my expression. She has worked faster than ever with this manuscript, but never losing her sensitivity and wisdom. I am deeply in her debt and greatly touched by her enthusiasm for the material. As usual, I am immensely grateful to Sally Gaminara at Bantam Press, whose continued interest and confidence in what I write always surprise me. She and her team at my publishers, Transworld, have been immensely supportive. Thanks also to Katrina Whone, Managing Editor at Bantam Press, and

Transworld's Managing Director, Larry Finlay, who has always encouraged me in my writing.

As usual, I have used very little of the material in the BBC TV series that accompanies this book. But the input of the BBC production team – a gifted and stimulating group of people – has been a great help. First, the Series Producer, Kim Shillinglaw, has been continuously engaged by this fascinating subject, and her enthusiasm for reading my meanderings about these complex themes has been of the greatest benefit to me. I am also very grateful to Sadie Holland, producer, for her enthusiastic comments on the text, and to the other members of the production team who have been so sensitive with the people we have filmed. Thanks too, to my Executive Producer, Jessica Cecil. This has been the third or fourth project on which we have worked together, and as always it has been a highly positive experience.

Finally, as ever, I am particularly grateful to my old friend Maggie Pearlstine and to her colleague, Jamie Crawford. They have excelled as my literary agents, as they always do. It still mystifies me why they are prepared to be so positive and encouraging. I am, as always, in their debt.

Prologue: Visions of Mortality

In the Old Testament, one of the earliest of known written works, there is a gripping narrative that, irrespective of the reader's background, gives universal insights into the nature of human impulses, desires, instincts and behaviour. Its opening volume, Genesis, is an account of human nature with all its flaws, illustrated by the story of Abraham and his family. This story contains one the most puzzling events in the entire Bible: the account of Abraham's binding of Isaac.

After years of agonizing childlessness, the hopes of Abraham and his wife Sarah are finally fulfilled when they produce Isaac. Once Isaac reaches maturity, the Bible recounts how Abraham is commanded by God to sacrifice him ('your son, your only son, the one you love, even Isaac': Gen. 22: 2) as a burnt offering. Abraham does not protest at all. Yet Abraham is the first person in the Bible to converse repeatedly with God, and has previously disagreed vigorously with Him about other, less personal, matters – such as God's apparently unjust decision to destroy the entire population of Sodom and Gomorrah. Nevertheless, on this occasion there is no argument: Abraham meekly obeys. Accompanied by two servants, he journeys with Isaac for three days. Leaving the servants to await his return at the bottom of Mount Moriah, he starts the long climb with his much loved son.

On the ascent to find a suitable spot to build an altar, Isaac carries the wood for the impending sacrifice on his back,[1] while Abraham carries a knife and the fire. Once a site is found, stones are placed together and the wood is placed on top of the altar. Abraham binds Isaac, lays him on the brushwood and raises his knife, ready to cut his throat. Throughout this terrifying episode, Isaac too submits meekly, without the slightest struggle.

The account is highly enigmatic. First, human sacrifice – widely practised in Canaan at the time – was always absolute anathema to Jews. Second, and more relevant to my present purpose, Isaac was most deeply loved. We know how much he was desired because the Bible is explicit about Abraham's years of infertility. When God promises that Abraham will become father of a great nation, Abraham simply replies, 'God, what will you give me, seeing that I go childless?' The Hebrew word that Abraham uses for 'childless' is stark: *ariri*, a term that does not just imply 'sterile', but has the connotations 'desolate', 'abandoned', even 'destroyed'. In his barren situation, Abraham sees his own mortality all too clearly.

Irrespective of whether one believes in the Bible as historical record or divine revelation, this ancient text shows the deepest perception. It is no accident that, when God answers Abraham in his distress at being childless, the Hebrew construction reporting this conversation is unusual. At this point, instead of the direct dialogue Abraham has always been privileged to have with God, the text says, 'And, behold, the word of the Lord came unto him saying . . . "He that cometh forth from out of thine own bowels shall be thine heir."' This is not a conversation – these words reflect a deep primeval instinct, Abraham's instinct that he needs his own genetically related child, 'out of his own bowels'.

A remarkable painting by Caravaggio in the Uffizi Gallery in Florence depicts the binding of Isaac. A moment in time is frozen. Abraham is about to plunge the knife into his son's neck. An angel has appeared at Abraham's side and points across the altar to a ram: the animal is to be the substitute victim in Isaac's place. Oddly, the face of the angel on the left of the picture and that of Isaac on the right show the closest resemblance to each other – are they the same person? But the most striking element of the image is the rictus of terror on Isaac's face. This seems to conflict with the biblical story, where Isaac follows his father passively up the mountain and allows himself to be bound without a struggle. Caravaggio paints Isaac's face contorted with fear, a visible representation of what is going on inside Isaac's mind. Perhaps this vision of abject terror reflects the most puzzling aspect of the whole story. Once Abraham has sacrificed the ram, Isaac virtually disappears from the narrative. It is almost as if he *had* been sacrificed. Abraham continues his journey with the servants towards Beer-Sheva, but Isaac is not even mentioned. Thereafter, Isaac is a shadowy figure. In spite of his inheritance from a rich, powerful father, his days lack achievement and he ends life blind and impotent, deceived by his wife, Rebecca, and his son, Jacob. Isaac bears the scars of the trauma caused by his father's actions to his deathbed.

The focus on fertility and family at the beginning of the Bible is no accident. The very first commandment, after man is created, contains four verbal decrees: 'Be fruitful and multiply, fill the Earth and subdue it.' This commandment, which the childless cannot naturally fulfil, comes before any of the conventional decrees about preserving life, submitting to God, moral behaviour or justice. Each of these four decrees has special relevance for us, to which we shall return.

How striking that all four biblical matriarchs, mothers of the nation, suffer with infertility! Sarah, as we have seen, is the first. Getting old, she urges Abraham to bed her maid Hagar, hoping she can adopt any child that results from the union. Poignantly she says, 'Perhaps I can be builded up through her,' asserting that her status as a wife and a woman is reduced by being childless. Thousands of years after these words were written, this sadness, this lack of self-esteem and these feelings of human inadequacy are expressed daily, in different ways, by women in every infertility clinic.

The arrangement with Hagar is the earliest example of surrogate motherhood. Even today some infertile couples still consider surrogacy. Experience suggests that, like Sarah, they frequently fail to understand its emotional pitfalls and the psychological risks run by all involved. Hagar despises Sarah for her infertility; but she is threatened by Sarah the moment her pregnancy becomes visible. Hagar flees the household, temporarily preferring to die in the desert, before being persuaded to return to give birth to her son, Ishmael. After Ishmael's birth, the tension between Sarah and Hagar continues for years. A decade later, Sarah eventually conceives and Isaac is finally born. Then Sarah throws Hagar out, unconcerned that Ishmael, now a teenager, may die of thirst or starvation with his mother in the wilderness.

Pregnancy in infertile women can produce serious medical problems. Sarah's daughter-in-law, Rebecca, is also infertile. When she eventually conceives twins, the pregnancy is so difficult that she says she would prefer death. As we shall see, even now pregnancy in infertile women raises particular problems. First, miscarriage is more common. Moreover, the babies of infertile mothers are more likely to be of low birth-weight and more likely to die in infancy. Multiple pregnancy (for example, twins or triplets), so common after in vitro fertilization (IVF), presents substantial difficulties.

Jacob, one of Rebecca's twin boys, marries in turn the two daughters of Laban, Leah and Rachel. Although he falls in love with Rachel, Jacob is tricked by their father into marriage with Leah. She, the older sister, is unattractive but initially fertile. Seven years later, he is allowed to marry Rachel, who turns out to be infertile. Rachel is deeply threatened by her sister's fecundity. Her agonized cry, 'Give me children, or else I die' (Gen. 30: 1), rings down the ages, and is still to be heard in the course of medical practice. Her husband's response seems almost indifferent. Perhaps it reflects the belief that infertility is a female problem and a kind of punishment. Jacob replies: 'Am I in God's stead, who hath withheld from thee the fruit of the womb?'

But now Leah, having had four sons, finds it difficult to conceive again – the first recorded case of secondary infertility.[2] In search of a remedy, she sends her eldest son, Reuben, into the fields to dig up mandrakes.[3] Rachel, believing like Leah in the plants' legendary medicinal powers in promoting conception, pleads with her sister to give her some of Reuben's mandrakes. In exchange, she offers Leah the opportunity to spend an extra night with Jacob. Thus Leah conceives Issachar – after the first documented fertility treatment – and the herbal remedy works also for Rachel, who finally conceives, giving birth to Joseph.

Genesis goes beyond mere descriptions of infertility and attempts to overcome it. It acknowledges that humans manipulate reproduction in other ways. Jacob is eventually enriched by his knowledge of genetic engineering. Having worked for Laban for fourteen years without payment, he is asked by his father-in-law to state what recompense he wants. Jacob says his wants are little, but suggests that Laban gives him the less valued sheep, the small minority of the flock that are speckled black and white. Laban agrees, but, always on the make, immediately arranges to have his other shepherds remove all the spotted, speckled sheep to a remote place, three days' journey away. Jacob is undaunted. He knows how to breed speckled sheep selectively from the white remainder of Laban's herd. The ancients believed that antenatal experiences, sights or sounds could influence the developing characteristics of a child, and Jacob exploits this belief. So, while Laban's animals are drinking or copulating, Jacob shows the sheep speckled, whitish sticks which he has whittled. These sheep then give birth to mottled offspring.

Jacob was no fool. Fourteen years breeding sheep for Laban must have told him that visual stimuli at the moment of conception do not affect fetal lambs. These speckled sticks were simply a smokescreen to deceive Laban's sons and divert their attention from his true methods. Speckling is a recessive genetic characteristic. Carriers of recessive genes for a particular trait do not show the features of that trait unless they inherit that recessive gene from both their father and mother. A single copy of the gene from one or other parent merely leaves them carriers of the trait, while their phenotype – the way they look – leaves them undistinguishable from normal white sheep. Jacob surely recognizes, from years of husbandry, that pure white sheep occasionally give birth to sheep with abnormal speckling. By observing which animals produce speckled offspring, he manages to identify – several millennia before Gregor Mendel[4] – the likely carriers of the recessive gene for the colouring. With selective breeding, he soon has a massive flock of speckled and entirely healthy sheep. Genesis records the commercial value of his knowledge and how his new flock soon makes him richer than Laban.

These and other stories in Genesis, eternal in their resonance, form the central themes and context of this book. The need to reproduce is one of the strongest instincts that humans possess, second perhaps only to the drive for personal survival – of which it is, indeed, one expression. Parenthood is the most important experience most humans will have, influencing our adult lives most profoundly. Yet we are capable of damaging the lives and development of our children in many different ways. In the case of Isaac, it seems that he never recovers from the mental trauma inflicted upon him during the binding. Rebecca suffers in carrying her twins, and multiple pregnancy remains the biggest concern in modern infertility medicine. Mandrakes were sought as a means of empirical fertility treatment; even today, qualified experts still give therapy whose efficacy is at best poorly supported by evidence. And although Jacob uses his skill in sheep genetics to take revenge on an unscrupulous and evil man, the misuse of genetic manipulation and the intrusion of commercial interests remain two great threats in human endeavours to exploit reproductive technology.

Infertility frequently has a severe impact on people who suffer from it. Although a few infertile people are not greatly affected by being childless, for the majority it causes serious anxiety. This anxiety is often followed by depression, loss of self-esteem, feelings of pointlessness and/or loss of libido. Some infertile people even think of suicide. For many others it represents a deep threat to their most precious human relationships. And, as our children are our immortality – for most of us, our most important achievement, irrespective of our success in various walks of life – so childlessness can cause utter despair.

The fourth of the verbal decrees at the beginning of Genesis – 'Be fruitful, multiply, fill the Earth and subdue it' – is, in a sense, the key to this book. The Hebrew verb used is wrongly translated as 'subdue'. It does not mean 'conquer' or 'overcome'. Rather, it implies the intelligent use of human power – including human technology – in recognition of humans' responsibility for the health of the Earth. This book is about how humans have used, still use and will in the future use technology to improve or manipulate reproduction. And a major issue that I wish to cover is how that technology can now, and must in the future, be used with responsibility and care. Reproductive manipulation is incredibly powerful. It has caused great happiness and can cause deep unhappiness; it can improve and destroy health; it can cement relationships and under-mine them – and it can leave offspring, like Isaac, deeply damaged. But above all, I contend that reproductive manipulation – and in particular genetic meddling – is so powerful that it could ultimately change our idea of what it is to be human.

I

The Beginnings of Life

> They are very short, the tallest being only two cubits in height, most of them only one and a half. Their hair is very long, going down to the knees and even lower, and their beards are larger than those of any other men. When their beards are full grown they leave off wearing clothes and let the hair of their head fall down behind far below the knees, while their beard trails down to the feet in front. When their body is thus entirely covered with hair, they fasten it round them with a girdle, so that it serves them for clothes. They are snubnosed and ugly.

So writes the Greek physician Ctesias of Cnida, who travelled the eastern world in his role of doctor to the King of the Persians in the fifth century BCE. Homer and Pliny also refer to this tribe – called the Pygmaioi – as being permanently plagued by giant storks, whose eggs they consumed rapaciously in order to prevent the birds from gaining the upper hand.

It is likely that these early accounts of 'pygmies', although radically embellished for the historians' audiences, testify to genuine encounters between Europeans and Central Asians and the darker-skinned tribes at the fringes of their empires. What intrigues me is that they, in turn, may have given rise to a myth which is still referred to in every maternity clinic and GP's surgery.

Until quite recently, some rather Victorian ideas of decency prevented adults from telling inquisitive children where babies came from. When I was growing up, it was not uncommon to meet small children who really did believe that their baby brothers or sisters were delivered down the chimney by a stork. My parents were not particularly prudish about sex – rather the contrary – but my father repeatedly told me, aged six, that I had been found under a hedge. It was only later that my mother said to me, 'Look at you – one moment of pleasure, and all I got was you!'

I do not know any friends who would still explain pregnancy to their children using the story of a stork; but the image still appears in occasional popular advertisements, and it is not uncommon for physicians, when confirming the joyful news of a positive pregnancy test, to make the odd quip about a visiting stork. It is just a part of our culture – as much as people saying 'touch wood' or throwing salt over their shoulders.

One possible source for this tale – perpetuated by an understandable reluctance to discuss the mechanics of love-making with one's small offspring – lies in the Greek myth of Oinoe, a beautiful Queen of the Pygmaioi, who was so beloved and treasured by her subjects that she forgot to pay tribute to the goddesses Artemis and Hera. In revenge, the snubbed goddesses turned Oinoe into a stork, preventing her from raising her newborn son Mopsos, and stirring up war between her and the Pygmaioi. Oinoe and the other storks began to plague the tribe, hovering over their houses with their menacing beaks and swooping on their heads.

The stork myth might also have crept into our popular

culture because of the way the bird behaves. It is monogamous, a bird that mates for life and has been observed to go to great lengths nurturing not only its own offspring but also sick and elderly relatives. In addition, European storks are fond of nesting close to chimneys – presumably for the simple reason that a chimney stack gives off plenty of warmth. It may well be that in former centuries people noticed that, wherever a baby was born, a stork seemed to be close by; for it is probable that any house containing a newborn infant and an exhausted mother was likely to have the fires on full blast, making its chimney an attractive resting spot for the elegant white bird.

But while the stork myth is a charming tale, it does not remotely compare in power or beauty to the remarkable ballet of physical and chemical processes that actually give rise to a human pregnancy. And, though I hesitate to describe it as a mystery, we still have a huge amount to learn about how human conception occurs and is maintained.

The egg: a waiting game

In humans, pregnancy is a result of sexual intercourse – unless fertility experts are helping out. In a nutshell, as everyone reading this book will know, a man's ejaculate contains millions of sperm, and just one of these, given a fair wind, will fertilize a woman's egg and occasionally lead to successful pregnancy. But the complex process of human reproduction begins much earlier. A woman's ability to achieve a pregnancy is decided when she herself is in her mother's uterus – when she is a tiny embryo less than 2 millimetres long. This is when her own eggs start to be created. She will carry these throughout her adult reproductive life, even though at this stage of development she has an unformed heart and no other recognizable organs. Astonishingly, her ability to create further life is in place well before her own ability to survive outside the womb has been secured.

Where do these eggs come from? They are created from the germ cells, a collection of tiny, primitive cells that grow outside the embryo, in the yolk sac. Around one hundred of these progress along the primitive highway that in later pregnancy becomes the umbilical cord and stream into the embryo itself, where they divide and multiply astonishingly. Nobody knows precisely how many eggs are made, but the best estimate is that by mid-pregnancy a female fetus may well have around seven million eggs in her ovaries.

Gloomy poets and novelists have pointed out that we begin to die almost from the moment we are born. The truth is, if anything, even darker. By the time she is born, the female will have lost most of these eggs – and, as far as we know, no more eggs are formed after birth. My colleagues at Hammersmith, Kate Hardy and Stephen Franks, estimate that probably only 600,000 are left when a girl is born. By the time she reaches puberty – the point at which her body is potentially capable of using these eggs to create and sustain life – she will have perhaps 100,000 or fewer in each ovary. In the western world, around two or three of these eggs will probably go on to become children.

It was William Harvey, in 1652, who first suggested that mammals produce eggs, though he declared that ovaries were a useless set of 'female testicles', with no role in the business of baby-making – it was not clear to him from where eggs actually came. In one way, he was thinking on the right lines, as the ovaries do rather little until puberty. But during childhood they have been gently stimulated by the pituitary, a gland buried inside the brain, which controls the production of hormones; and as puberty approaches, the pituitary starts to swing into more aggressive action, releasing increasingly regular amounts of follicle stimulating hormone (FSH).[1] This sparks a maturing process in a few of the follicles in the ovary,[2] which balloon to become so-called Graafian follicles. Every month, perhaps twenty follicles each containing a single egg will enter this balloon or 'blister' stage – but in humans, unlike most other mammals, just one of these, with its egg, eventually wins out at the expense of the others. In truly Darwinian style, this one dominant follicle absorbs the lion's share of available nutrients and, by a mechanism that still is not fully understood, actively prevents the other follicles from having a chance of developing further. It is because of this process that humans normally produce single births as opposed to the multiple litters of cats, dogs and rabbits. Anyone who has watched with mounting horror as a pet produces seemingly endless armies of mewling life-forms will be glad that this is a feature we do not share with our animal cousins.

Occasionally, the human mechanism does not function entirely smoothly: two follicles form and become dominant, both ovulate, both eggs fertilize, and both resulting embryos implant. Humans probably produce a twin pregnancy more often than we realize.[3] Recent evidence has shown that it is quite common for human pregnancy to start with two implantations, but for only one of the fetuses to survive. Around one in eighty pregnancies ends in a twin birth, and most of these twins are dissimilar (that is, not identical) because they come from two different eggs, carrying different genes. It is not known how often a single follicle may contain two eggs, but it is likely that some dissimilar twin pregnancies arise from just one follicle.

We also differ from other mammals (apart from a few members of the monkey family) in the way our bodies prepare themselves for reproduction. Most other mammals have an *oestrus cycle* whereby, in response to a hormone surge which can occur from once every four days to once a year, depending on the species, the uterus prepares itself to house babies. In addition, the female becomes interested in having sex with the male. In humans, the *menstrual cycle* means that the body prepares itself once a month – but human males and females are interested in sex most of the time, irrespective of any preparation for fertility. The reasons for why we evolved in this way, and what advantages it gave us, are still hotly debated.

In spite of its miraculous potential, the egg at this point is just one rather fragile cell. It depends entirely for its survival on a protective cushion of several million helper cells which surround it. These helper cells, the *granulosa*, ensure a steady supply of nutrients to the egg and produce the hormone oestrogen, which plays an important role in the next act of the reproductive ballet.

A steady pulse of FSH helps the follicle to grow to its maximum size before ovulation, which occurs roughly midway between one menstrual period and the next. Once the follicle is about 2 centimetres in diameter and the egg is close to maturity, the pituitary gland orders the release of luteinizing hormone (LH), which times and directs the release of the egg from its follicle. This surge of LH is critically important as it enables the egg to complete its final stages of maturation, and an incompletely matured egg may not have the capacity to undergo fertilization. It is prompted by the increasing amounts of oestrogen produced by the granulosa cells as the egg approaches maturity in the follicle: once the amount of this hormone in the follicle reaches a certain level, the brain realizes what is going on and instructs the pituitary gland to release a pulse of LH. Within approximately thirty-six hours this will cause the follicle to release the egg. This is the process called ovulation.

This process is fragile, depending as it does on so many different mechanisms working properly and at the right time. It is one reason why humans produce so few offspring compared to their mammalian cousins. But the chain of events sketched out above also ensures that only the most 'ripe' eggs go on to compete in the fertilization game. It is like selecting a team for the World Cup – you cannot determine whether your side will enter the finals, or whether they'll win it. But by entering the best possible side into the competition, you maximize your chances. Delivering up into the uterus an undeveloped or immature egg will be a waste of the body's resources, because it won't fertilize, or, if it does, it will produce an embryo with deep flaws that is unlikely to be viable.

The pregnancy cannot occur until the egg is in the right place, and those remarkable helper cells, the granulosa, play a further role in getting it there. Some of these leave the follicle with the egg, and their sticky surface assists the egg to find a resting place in one of the two fallopian tubes. The tubes themselves also help: they are lined with cells bearing hair-like processes, called *cilia*, which beat rhythmically, waving the egg towards the uterus. These beating hair-cells seem magical, their rhythmic pulsation, seen under a microscope, miraculous. In their absolutely regular pulses they resemble a myriad of wheat stalks repeatedly bending in response to rapid waves of wind across a field – and always the wind is in one direction, away from the ovary towards the uterus.

In the 1970s it seemed that these hair-cells, beating the way they did, were important in ensuring fertility. Inevitably the thought occurred that if the beat could be altered, or stopped, or its direction changed, we might have a perfect method of contraception, free of serious side-effects. If the egg did not reach the uterus, but was simply arrested in its passage down the tube – or not even wafted into it in the first place – this little cell would just die and disintegrate.

So in 1972 I, like a number of other scientists, tried a number of ways to assess how these beating, ciliated cells actually functioned. One of the delights was working with my rabbits, who became very friendly towards me. I visited them without fail every morning, and as my breeding experiments became increasingly successful, these ward rounds down the expanding lines of hutches got longer and longer. Two rabbits were particularly memorable at this stage of my career – one was Beatrice, the other Xanthippe. They use to lollop around my laboratory, much preferring this to being bundled back into their hutches for the night. The studies I did on their ciliated cells yielded important information about the cause of ectopic pregnancy – one of the main causes of maternal death in both developed and less developed countries.

In one group of experiments, I transplanted a fallopian tube from one rabbit to another to see what would happen to the control of this repetitive, rhythmical waving. This was a difficult procedure, requiring me not only to fashion a careful join between the new tube and the stump of the old, but to join their tiny blood vessels too. These vessels are only half a millimetre in diameter and such a joint must not leak, otherwise a clot will form and the blood vessel will block. At the time this was a revolutionary technique, largely devised by a friend, Robert Acland, a plastic surgeon! I used a needle and thread so fine that if I inadvertently dropped the steel needle it would not fall to the floor but would be wafted upwards by the gentle currents of air in my sub-basement laboratory. I also had to find an entirely new method for anaesthetizing the rabbits, because sometimes these operations took eight or ten hours. The operating table I built contained the first ever electric blanket made for bunnies, sealed inside the metal of the table so it could not get wet and electrocute either me or the rabbit.

That sub-basement room was literally a concrete cell deep under the remote, barely visited corridors of Hammersmith Hospital. Thirty years later, it has been filled in by contractors. It might have been an ideal refuge if there was a nuclear attack, but it was unheated and bitterly cold in winter. Sometimes, stitching with these fine instruments, I lost all sensation in my extremities and would drive home at two-thirty in the morning with numb toes and blue fingers. But the rabbits, being furry and having been kept warm with the electric blanket, made a very rapid recovery.

And after the transplants, what happened? Nothing – the beat could clearly be seen under the microscope, still rhythmic, still towards the uterus. It was clearly controlled not by the nerves supplying the tube – they had been severed – but by some remarkable inherent property in the cells themselves. In later experiments I placed tiny microspheres, the size of a rabbit's real eggs, in the tubes. These little spheres emitted faint radioactivity so I could monitor their progress down the tube with a Geiger counter. Nothing I did to the tube seemed to make any difference. These 'surrogate' eggs still entered the uterus ninety hours after ovulation, just as they did in the intact normal rabbit.

In another experiment I cut out a middle section of the fallopian tube and turned it through 180°, sewing it back into place with its blood vessels still working. Now finally the little hairs beat in the 'wrong' direction in the transplanted segment: towards the ovary and not the uterus. Any egg placed in the reversed segment moved back towards the ovary and stopped at the join; any egg placed in the tube before the join moved away from the ovary and stopped at the join. This was good evidence of the value of the cilia, but not much use – in itself – as an easily administered contraceptive.

So what might happen if we paralysed the beat of these little hair-cells using drugs to immobilize them? Colleagues around the world tried innumerable drugs. Many could be shown to slow the beat or stop it completely. Theoretically, any of them would have been ideal as a contraceptive as they had few if any side-effects. But what happened when we mated the animals whose cilia were paralysed by these drugs? They were fine – too fine: not only did they have no side-effects at all, but they got pregnant as normal, with the same number of babies in each litter as when their tubal cilia were working normally. So after months of careful surgery, deep underground at night in the cold on my own, I ended in a scientific blind alley.

What was happening in my rabbits? It turns out that – as so often in reproductive biology, it being too risky for most organisms to rely on one method of getting pregnant – there is a back-up system. The muscles of the tube help the process of transport towards the uterus, squeezing the tubal walls together and pushing the egg in the right direction. The egg's progress would be arrested only in the exceptionally rare event of the cilia and the muscles both being paralysed.

Even after ovulation, the empty follicle continues to play a part. The granulosa cells left in the follicle now produce progesterone, a hormone that will become vital if the egg goes on to be successfully fertilized by a sperm. Without progesterone, no developing pregnancy will survive. Because these cells also produce a yellow pigment they colour the ruptured follicle, which is why it is often called a *corpus luteum* or 'yellow body'. If no fertilization happens, the corpus luteum simply withers away.

Sperm and the reproductive marathon

It is remarkable to consider that, on any day of any month, millions of female bodies are readying themselves to create more humans, an army of potential Einsteins, Marie Curies and David Beckhams, the greater proportion of which will never be given the spark of life. For life cannot emerge without a contribution from the male. It is a curious irony of our biology that, whereas the female is born with a limited stock of eggs, a man produces new sperm throughout most of his life. It may be of interest to consider that if you, an averagely fast reader, are a woman of child-bearing age, you will have lost five or six eggs while reading this short chapter. In the same period of time, your male partner may have made 200,000 new sperm. If he is not a particularly quick reader, by the time he has got through four or five of the longer chapters, he will have made enough sperm to impregnate all the women of child-bearing age in the United Kingdom.

As a woman gets older, the potential of her eggs to create healthy babies decreases. It is not merely the number of eggs that declines rapidly before the menopause, but also their quality. It seems also to be the case – although this is less well documented – that, although there are many recorded instances of men fathering children into their nineties, the quality of sperm also decreases with age, albeit more gradually. My friend Dr Sherman J. Silber, from St Louis, Missouri, points out that men over the age of forty often produce sperm with somewhat reduced motility. That, and the fact that older men produce a lower volume of semen when they ejaculate, means that they may not produce enough good sperm to achieve conception. In addition, there is now increasing evidence that the sperm of older men more frequently show alteration in their genetic make-up, increasing the risk of genetic abnormality in any embryo that develops; and Professor Schill, from the Department of Dermatology and Andrology at Justus Liebig University in Giessen, Germany, points out that men much over forty also make many more malformed sperm,[4] making conception more difficult.

The sperm have a remarkably arduous job; one sperm does indeed produce a baby against all odds. Inside each testis is a miraculous network of fine plumbing, a series of tubes in the lining of which each sperm is created. This complex pipework leads to a Spaghetti Junction of more tubing and then eventually to one central tube, the *epididymis*. Even this is finer in diameter than a piece of cotton; it is heavily coiled, and if you could uncoil it and measure it end to end would stretch for about 12 metres. This tubing is really fragile – its wall has the consistency of wet tissue paper. Given its extreme complexity and delicacy, the potential for mishaps in this mechanism is considerable. A minute blockage at any point in the coiled lengths of the epididymis can scupper the whole reproductive process, not least because, although there is a back-up system – two testes, two epididymes, two sets of tubing – if there is damage on one side, there tends also to be symmetrical damage on the other side. The male reproductive apparatus is just as prone to problems as that of the female, but reproductive problems within women can more often be fixed by delicate surgery. Despite advances in microsurgery, it is not always possible to do the same with men, because of the tiny scale of these tubes and the complexity of the system.

The epididymis ends in connection with another length of human piping, the *vas deferens*. This is a thicker, more muscular but still narrow-bore tube that transports the sperm to the prostate gland. Here the tubing is connected to the piping from the bladder, the urethra, which transports seminal fluid through the penis at the moment of orgasm, with its muscle contractions. In some men, the muscle contractions may not be strong enough, or properly co-ordinated, to propel the seminal fluid properly around the circuitry, and in some cases this may result in the sperm being delivered into the bladder, where they will die or be lost in the urine. Provided no such mishap occurs, the sperm leave the penis as part of the ejaculate, roughly a teaspoonful of fluid also containing liquid from the tubes, the prostate gland and glands within the urethra. This fluid acts as fuel for the sperm, providing their energy for the onward journey. Interestingly, it seems most of the sperm are contained in the first part of the ejaculate – perhaps a case of nature trying to give them a head-start. At the moment of orgasm, between 100 million and 300 million sperm will be released, most of which possess not a cat in hell's chance of going on to create an embryo. It is like the London Marathon – a very few athletes are outnumbered by hordes of sorely unfit amateurs.

It is worth pausing in the account of their journey for a moment to consider just what proportion of these reproductive runners are not up to the challenge ahead. Human males produce more poor-quality semen than any other mammal that has been studied – with the possible exception of the gorilla.[5] Some studies show that as much as 40 per cent of human sperm in the average ejaculate are incapable of adequate movement. In many men it is not uncommon to find that 60–80 per cent of the sperm are abnormal in shape and size. So the unfit amateurs clearly predominate – to an increasing extent as we age – and even some of those that stay the course are not in very good shape. Older men, for example, are more likely to father children with the disease of achondroplasia – dwarfism – which may be attributable to mutations in the sperm DNA.

And their continuing journey *is* a challenge. In the first place, many millions of sperm simply fall out of the vagina. The seminal fluid is a thick, jelly-like substance when it emerges from the penis, but within half an hour it thins out and liquefies. Some animals have a thicker, more durable ejaculate which seals off the partner's vagina; for humans, chance plays a greater role. In addition, human sperm encounter a very hostile environment once they enter the vagina. To protect itself, and the deeper structures of the reproductive apparatus, from infection, the vagina is acidic – which kills off infection, but also kills off millions of sperm. Some 95 per cent of the sperm ejaculated will fall victim to one of these fates, so that only about 5 per cent get as far as the cervix.

Even those that do face more obstacles. Mingling with

the special mucus in the cervix at ovulation time, the fit sperm undergo changes, readying them for the business of finding and eventually penetrating the egg.[6] A certain proportion of them will then go on into the uterus and the fallopian tubes, reaching these within five to ten minutes of ejaculation. (We know this because a number of scientists have inseminated women, with their approval, some ten to fifteen minutes before abdominal operations – and sperm have been recovered from the ends of the tubes in many of these patients.) Given the speed at which sperm are capable of swimming, this is far too fast for the sperm to get there without help: their journey is aided by muscular contractions in the uterus, possibly initiated by prostaglandins, chemical substances in the seminal fluid that are powerful muscle stimulants. Some claim that rhythmic contractions experienced by a woman during orgasm may act to propel the sperm upwards. The evidence for this is not convincing; although it could be that female orgasm increases the chance of conception, having worked with couples with fertility problems for so long I think that gratuitous exhortations to strive for orgasm only add to the anxiety of people whose sex life has already become fraught with the pressure to make a baby. Whatever the truth, having an orgasm could improve a woman's chances of conceiving only marginally.

A second, reserve army of sperm remains in the cervical mucus for around seventy-two hours, detachments being continually being sent up into the uterus. Many couples feel certain that they can pinpoint the exact occasion on which a child was conceived. But since the process can take up to five days, possibly even longer, people having regular sex can be mistaken about this. A child named 'Florence' after a romantic weekend break might be more accurately called 'Neasden'.

Once again, the defences of the uterus, designed to protect the fragile tissues from infection, will see off hundreds of thousands of sperm – leaving perhaps only a few hundred with the potential to reach the fallopian tubes. The fallopian tubes also act as a filter, preventing most dead or abnormal sperm that have got this far from entering the upper reaches where the egg rests. There are, in short, so many checks and balances on the whole procedure to this point that it is surprising that anything other than the most viable sperm get as far as the egg. But, as the sperm reach the outer surface of a ready egg, problems can still occur. Defects in the sperm's chemistry may prevent it sticking. In general, though, most problems at this stage are related to the egg. If, as a result of hormonal deficiencies, it is not mature enough before it is released into the fallopian tubes, its surface may be unreceptive or too tough for the head of the sperm to penetrate. If, on the other hand, it is 'over-ripe', its surface may allow two or more sperm to get through. A properly mature egg instantaneously creates a barrier to other sperm once it has been penetrated. Multiple penetration is a problem because each individual sperm carries twenty-three chromosomes, the building blocks of our genetic make-up. For fertilization to take place, the twenty-three chromosomes of the sperm must line up alongside the twenty-three present inside the egg. These forty-six chromosomes are vital constituents of a successful pregnancy. If more than one sperm enters the egg, the total number of chromosomes is greater than forty-six and normal pregnancy cannot result.

The fertilization process continues over more than eighteen hours. First the sperm has to penetrate the outer coating of the egg. Then, once the head of the winning sperm and the egg are merging, the egg should eventually divide into two cells. Thereafter, cell divisions happen roughly every fifteen hours, so that by the time two days have passed there is usually an embryo of eight cells present in the fallopian tube. At this stage there is something about the embryo that, even after all these years, I still find remarkable. Each cell, visible only under a microscope, seems to be totipotential – that is, if we split it into eight separate cells, each of these could theoretically begin its own process of division and multiplication. In other words, it would be just possible to produce eight human beings from a single embryo. This, of course, is how twins, triplets and other multiple births occur naturally – occasionally an embryo may divide completely into two, or three, or more.

Sexless reproduction

It is also possible, rarely, for the process of cell division within an egg to begin without penetration by any sperm; this is called parthenogenesis,[7] and it occurs naturally in certain insects, reptiles and lower mammals. The phenomenon was first noted in the eighteenth century by the Swiss naturalist Charles Bonnet, who used it as the basis for some rather startling theories about evolution. Having noticed that aphids could reproduce without sex, Bonnet became a convinced ovist, concluding that the egg, rather than the sperm, was the source of life. Bonnet merely observed the process of parthenogenesis within the natural world. But at the close of the nineteenth century a German scientist, Jacques Loeb, achieved it for himself by exposing sea-urchin eggs to varying concentrations of salt water. His work was somewhat hampered by the fact that he was based at the University of Chicago, and had to keep travelling to California to obtain a year-round supply of suitable sea urchins. But these difficulties were nothing compared to what he faced when the public learned about his work.

By 1902 journalists had heard of Loeb and took to citing him as a revolutionary figure, whose work would enable people to reproduce without the need for sex. The shy and studious Loeb had never been so stupid as to claim any such thing. Other scientists were beginning to replicate his work with different species, and visitors flocked to Loeb's laboratory to learn directly from the master. One young evolutionist, Francis B. Sumner, was moved to describe Loeb's circle, rather disparagingly, as a 'cult'.

As so often happens when a scientific discovery is made public, the pioneering scientist's caution was lost in a tidal wave of speculation. The popular press claimed that women had stopped bathing in the sea for fear of getting pregnant. Feminists delighted in the idea that women were being released from their 'bondage' to men, and Loeb and other scientists were inundated with letters demanding treatment for a 'sexless child'.

Reporters doorstepped Loeb, stole his students' notebooks and eavesdropped on his lectures. Other aspects of his research were picked over and then amplified into tabloid theatrics. For example, Loeb once remarked that salt should be seen not as a food, but as a necessary ingredient for the electrochemical activity upon which life depended. In the hands of the press, this became a claim that electricity was the source of life. Quack doctors suggested that his comments were supportive of electric shock treatment for a range of illnesses and claimed that their devices were based on the wondrous discoveries of life-making Doctor Loeb.

Loeb himself, whose heavy German accent, Jewish

background and pronounced atheism all contributed to his deep-rooted feeling of being an outsider, deteriorated sharply under the pressure of this public attention. He felt that many of his colleagues shunned him because of claims he had never made. He suffered depression and heart palpitations, and had a historic falling-out with a student he had previously nurtured.

His contribution to science remained valuable, however, and was considered influential by Gregory Pincus, who claimed in 1936 that he had produced a live birth by artificially stimulating rabbit eggs by means of chemical compounds and temperature change. The resulting rabbit appeared on the cover of *Life* magazine. However, as far as I am aware, Pincus never published the scientific details of this experiment, and whether or not he really achieved a virgin birth remains highly dubious. Experiments with other mammals, particularly mouse embryos, have shown that nearly all attempts at activating an egg without sperm may result in cell division and the beginnings of organ growth, but after a while further development stops. It is only very recently, following experiments by researchers in Tokyo, that a truly parthenogenetic mouse has been born.[8]

Whatever the validity of this particular claim, like Loeb before him Pincus under-estimated the power of public opinion: an outcry following the appearance of the photo on the cover of *Life* led to his resigning from his post at Harvard University. Fortunately this furore did not stop Pincus from going on to make one of the great advances of twentieth-century science when in 1954, together with Dr M. C. Chang, he developed the first oral contraceptive.

Could parthenogenesis occur in humans? 'Virgin birth' is, of course, familiar to many of us as the means by which Jesus Christ is said to have arrived on earth. Nor is

Christianity the only religion in which such a phenomenon features. The ancient Greeks believed that the goddess Aphrodite emerged from a sea of foam; the early Persians that their prophet Zoroaster was conceived when his parents drank a sacred mixture of milk and herbs; and the Jains of India that their Saint Mahavira descended from heaven into his mother's womb. Many tribal peoples, too, believe in the possibility of virgin births. The Trobriand Islanders of the South Pacific, for example, seem to believe that the sun's rays can make a woman pregnant. But while ultraviolet rays can induce parthenogenesis, it seems this method only works in sea urchins; and although human eggs can be activated spontaneously, as is sometimes seen after IVF when sperm have not penetrated the egg successfully, the resultant embryos have serious defects and do not live for more than a few days.

The Roman Catholic idea of conception being the point at which a new life begins does not take note of current biology, certainly. Life can now be shown not to begin – or at least, not to begin only – with fertilization. An amazing experiment by Dr Tomohiro Kono from Tokyo has shown that it is possible to activate a mouse egg, get it to cleave repeatedly, and see it develop into a normal mouse that is entirely fertile.[9] This mouse, Kaguya, is the first mammalian parthenote to have survived. The egg that made her was simply activated by a minor mechanical injury – pricking its coat. The egg never saw a sperm. It is possible that this birth of a normal fertile mouse was just a random event, but it may have been due to a rearrangement of some of the imprinted genes I describe later (see page 396).

Finding a home: implantation in the uterus

Once the embryo has entered the uterus, it floats around for two or three days. Even though it has gone through an extensive process of cell division, it is no larger than the original egg cell from which it began and is still invisible to the naked eye – about one-tenth of the size of this full-stop. At this stage, the embryo forms a cavity and two types of cell can be identified. About one-tenth of the cells become the *inner cell mass* – which will end up as the person. The remaining cells, the *outer cell mass*, become the throw-away part of the pregnancy, the placenta and membranes. The embryo at this stage is called a *blastocyst*. About one week after fertilization, the blastocyst finally starts to increase in size and begins to implant in the lining of the womb, the *endometrium*.

At this point the embryo releases the pregnancy hormone, *human chorionic gonadotrophin* (HCG), which triggers the endometrium to thicken and helps the embryo to implant. This hormone is a very stable chemical and is not readily destroyed by the mother's body, so eventually it appears in the urine. It is this substance which, in sufficient concentration, yields a positive result in the over-the-counter pregnancy test. Before we could measure HCG, the only early way to detect pregnancy was by injecting a woman's urine into a rabbit or toad, and then inspecting the animal's ovaries. If she was pregnant, follicles developed on the ovarian surface. This is because HCG is very similar in chemical structure to LH and is capable of causing ovulation.

Many things can still prevent a successful pregnancy becoming established. The successful embedding of an embryo into the endometrium is a remarkable event. The uterine lining may grow abnormally. The embryo may not implant, or it may implant in the wrong place. Above all, successful implantation seems to be against the body's most basic nature. The human body is designed to repel invaders. Any foreign protein or tissue is attacked and rejected. If it were not, we would soon die from bacterial infection. After a kidney transplant, a patient must remain on immuno-suppressant drugs for ever to prevent the body's immune system from rejecting this genetically different chunk of tissue. And a developing embryo is just another foreign body, like a transplanted organ. It has a unique combination of maternal and paternal DNA, different from that of its mother alone, with its own configuration of genes and therefore with its own special proteins. Yet, except in a few rare cases, the body of the mother accepts this 'foreign' presence and nurtures it. This phenomenon is still not fully understood. Nevertheless, by the time it has reached an age of fourteen days, the embryo should be firmly implanted in the endometrium and beginning to develop what has been called the *primitive streak* – the first sign of any organ, the crude beginnings of a nervous system.

I describe the reproductive process in some depth for two reasons. First, familiarity with the basic workings and terminology will be valuable when we consider the causes of infertility and their treatment, as well as the ethical implications of such treatment. Second, I hope I may be forgiven a surfeit of detail in order to hammer home a vital point. Humans, compared to the rest of the animal kingdom, are a notoriously infertile species. The average fertile couple, engaging in regular unprotected sex, have no more than an 18 per cent chance of developing a pregnancy each month. From the intricate plumbing of a man's testes to the complex hormone surges within the female ovaries, from the range of factors that prevent a sperm from swimming up to the fallopian tubes to the fault-lines that can occur within a woman's egg, the business of being fruitful, multiplying and filling the earth is fraught with difficulty. Professionals working in other fields of medicine are sometimes moved to comment – a little uncharitably, in my view – that the subject of fertility receives a disproportionate amount of media attention and even of funding. But the science of reproduction has a vital impact on our understanding of all development, genetics, cell biology and many diseases such as cancer; and having a baby is a totally life-changing event.

Time's arrow: the problem of ageing

In January 2005 Adriana Iliescu, a former university professor of literature and author of children's books, gave birth to a baby girl weighing not much more than a bag of sugar. What interested the world's press in the birth was the fact that Ms Iliescu, from Romania, was sixty-six years old, and had undergone nine years of IVF treatment. She was able to conceive at all only because she had received an egg from an anonymous donor. The tone of the response to this event was redolent of both Phineas T. Barnum's travelling freakshow and the McCarthy witch-hunts of the 1950s. The Orthodox Church condemned Ms Iliescu's actions as selfish. The Romanian government declared a review of the ethics of assisted reproduction. Once again the world debated whether technology was taking us into some scary new territory where our traditional concepts of right and wrong dissolved in a Petri dish. My own straw poll of the opinions being put forward in internet chat-rooms at the time showed almost universal condemnation of the mother, with very few supporting the view that all women had a right to a baby, regardless of their age.

Though I did not really know Ms Iliescu, it seemed irresponsible for her body to have been coaxed into giving birth at such an advanced age. At the time, I believed that the condemnation – if there was any point in handing it out – should have been more rightfully directed not to her but to the clinicians who treated her. There was nothing particularly clever in helping this woman achieve a pregnancy. We have reached a stage in our understanding of reproduction where it is quite possible to give babies to women of sixty-six. In fact, with a donor egg and sufficient hormone injections, it would be quite easy to get an eighty-year-old pregnant – or for that matter an eight-year-old. But this infertile woman had undergone nine years of unsuccessful treatment. Nine years of medical expenses. Nine years of emotional turmoil. Nine years of physical toll upon a reproductive system weakened by age. There comes a point when, regardless of the acutely felt needs of the patient, it is no longer responsible for a clinician to continue to hold out the hope of successful IVF treatment. In my view, that point had almost certainly passed a long time before Ms Iliescu became pregnant. Moreover, the older the woman, the more risky her pregnancy will be, and the higher the chance that the baby will be born very small or premature – or, of course, dead.

I have just returned from Bucharest, having gone to Romania to meet Adriana Iliescu and the doctor who treated her, Dr Bogdan Mariescu. I wanted to understand for myself why she had been prepared to go through so much therapy – and why her doctor felt his treatment justified. In some ways, meeting her was quite a sad experience; in others, an uplifting one. Adriana lives in a small, crowded apartment on the eighth floor of a

Soviet-style concrete block in a leafy suburb of Bucharest. Although she claims to be very fit, she looks a great deal older than her sixty-six years. In the middle of the tiny sitting room where she served me tea and cake with shaking, gnarled hands, was her delightful little girl, Eliza, standing in her cot – eighteen months old. Almost the moment I entered the room, Eliza stretched her arms out to me, a total stranger, chuckling happily as I lifted her up and played with her on my knee.

The pregnancy, Adriana told me, had been twins. One of the babies, weighing 800 grams, died at birth. Eliza herself was very small – at thirty-four weeks, premature and weighing 1,400 grams (about 3 pounds 4 ounces) – but she survived. I put it to her that there were so many abandoned babies in Romania – indeed, many British women travel all the way to her country to find a baby to adopt – that, if she had wanted a baby so badly, it would surely have been better to care for one of these needy souls rather than put a new little life at risk. 'I didn't feel a proper woman – I needed to give birth to my own,' she told me. I asked her whether she felt there were limits to what we should do with science. What was so interesting, and for me disappointing, in her reply was her marked hostility to the idea of lesbian couples having children. This notion clearly shocked her deeply because it was 'unnatural' – even though she herself had conceived with donor egg and donor sperm, after lengthy treatment with hormones that she could no longer produce naturally. She will strike many people as single-minded to the point of obsession, yet there can be no doubt about the love that now suffuses her life. As to what would happen to Eliza if her mother died, Adriana argued that she felt God would see that the child would be looked after.

My meeting with Dr Mariescu was no more reassuring. He was amiable, brusque and quite businesslike. To him, the treatment seemed to be a matter of Adriana Iliescu's choice, provided she could pay for the cost of the drugs – the only charge made, it seems, in that clinic. She was physically fit, he constantly reasserted, and that seemed to be almost his sole responsibility. He showed no concern about the dead twin. To be fair to Dr Mariescu, I had a strong feeling that he had thought more deeply about his patient's treatment (he constantly repeated that he had never had doubts about it) than he stated. He has treated a few other women in their late forties, but seemed un-interested in his world record. Certainly he has never attempted this treatment on anybody nearly as old as Adriana.

But there can be no escaping the fact that, apart from the risks to Ms Iliescu and her living child – and the death of another – considerable harm was done by this case to the image of reproductive medicine, and especially IVF. Of course, the number of women even ten years younger than Ms Iliescu being given this treatment is really tiny. To be brutally frank, very few women will even countenance having a baby much beyond the age of fifty. But this story raises an important consideration. Although Ms Iliescu's was an extreme case, her advancing years bear witness to the over-riding reason why most women are now seeking treatment for infertility. It is because of ageing.

The fact that a woman's stock of eggs begins to decline before she is even born hangs over her entire reproductive life. Most women think they are fertile until they stop menstruating. Certainly, there is a potential for some of them to become pregnant up to that point, but the chances reduce rapidly from the mid-thirties. If the ovaries are a kind of larder, then the eggs that have been there longest seem to be of the poorest quality. Quite why they deteriorate is not fully understood, but female age is the most critical factor in a range of infertility problems. Older women produce eggs that have a lower chance of being fertilized in the normal way. Their embryos are more likely to have genetic or chromosomal problems. Older women who do get pregnant are more likely to miscarry, or to give birth to an infant that does not thrive.

As a woman ages, the chemical processes behind the scenes of the ballet of ovulation also begin to falter. The menstrual cycle can become more irregular, the neatly timed surges of the right hormones in the correct concentration less efficient. As a consequence, even if an embryo forms without flaws, the lining of the uterus may not be sufficiently receptive to allow it to implant successfully. By the time they reach forty, almost a third of all women have fibroids, benign growths of tissue inside the uterus which can impair its function.

Older couples also tend to have sex less frequently. To anyone of advancing years who wants to become pregnant, I would counsel regular sex before trying the difficult, costly, time-consuming process of IVF. Ensuring that a regular supply of sperm is present in the cervical mucus will improve the chances that one of them may lead to a successful pregnancy. Even if IVF is advisable, it is still less successful in older women. We can sometimes stimulate the ovaries to produce more eggs – but we cannot guarantee their quality. For women aged between forty and forty-four, success rates after IVF are under 5 per cent and two-thirds of pregnancies miscarry.

People age at different rates. One woman may have several thousand eggs in her ovaries when she reaches forty, another equally fit woman hardly any at all. Just as we all vary in height, so all of us will vary to some degree in our basic fertility. 'Unexplained infertility' in a woman in her late thirties or early forties is most likely to be due to her ovaries containing fewer eggs than average for her age group.

Infertile older women present a very modern problem. The average age at which women choose to start their families has increased substantially over the past fifteen years. In 1990 just 3 per cent of first-time mothers were over thirty-five; by 2002 the proportion had risen to 10 per cent. This trend has its roots partly in the social revolution of the Sixties – the widespread availability of efficient contraception through the pill. But economics and gender politics also play a part. If you scan through the advertisements of any popular magazine from the Fifties, you will see that 'Mother' is the figure to whom all vacuum cleaners, canned foods and detergents are being proffered. Now more girls and women gain skills and education, and increasingly go on to compete with men in the workplace. Although women are still dramatically under-represented in certain jobs – for example, in the world of the physical sciences and engineering science – there is every expectation that this too will change so that the gender ratio approaches equality across all sectors. As more women establish themselves in their chosen occupations, so more of them have also tended to delay the business of having children. Rising costs of living also mean that, especially in western Europe, many couples find it more difficult to rely on one salary alone – and that also affects the age at which people decide to start a family. People now tend to make more careful choices about lifestyle, and the quality of life they feel they can offer their offspring, before having children. Also, many women delay having children because they feel they haven't yet found the right partner.

But this is an area where, in spite of the enlightened attitudes professed on all sides, prejudice lurks close to the surface. There's an image of the typical woman seeking IVF treatment as a sort of 'Bridget Jones' character: someone in her mid- to late thirties who has spent her best reproductive years quaffing Chardonnay or rapaciously climbing up the career ladder – and now she expects public-sector medicine to pay for her to have a baby? The only possible reasons for her not having had children already are careerism, selfishness, greed, irresponsibility and immaturity. The frequency with which women are stigmatized in this way suggests that genuine equality is still a long way off.

Are we 'right' to help older people to conceive? Older parents may have less energy to deal with the demands of a growing child, or the rebellions of a teenager. There is a greater risk of the parents dying before the child is able to look after itself. An only child may grow up finding it difficult to form relations with others in its peer group. And if there are elder brothers or sisters, there may be such a 'generation gap' between this child and its siblings that it might as well be an only child.

In practice, this area is not so fraught as it seems. At such an advanced age as Ms Iliescu, a pregnant woman is more prone to high blood pressure, diabetes and potentially life-threatening conditions such as thrombosis and heart attacks. For women between twenty and twenty-four in Britain, the deaths in childbirth stand at just 7.3 per 100,000. In women over forty, the figure leaps fourfold to 35.5 – comparable to countries with less comprehensive health care, such as Iran and Armenia. This is still a small proportion; but is it responsible to provide the possibility of pregnancy, when there is such a risk that the patient or her baby may die?

For the most part, common sense prevails. No practitioner I respect would consider treating a post-menopausal woman much beyond the age of fifty, because of the physical risks and ethical difficulties. We use the normal menopause (the average age for which in Britain is now fifty-two) as an obvious threshold – because it is nature's own way of deciding when a woman's child-bearing years have passed. Some younger women experience a very premature menopause, and this is a strong argument for treatment; but these cases apart, there seems no good reason for interfering with natural processes to make a post-menopausal woman pregnant.

Treatments like that given to Ms Iliescu do considerable harm to our branch of health care, provoking widespread public distaste that – whether it be right or wrong – colours how reproductive technology as a whole is per-ceived. At the individual level, the prevailing attitude can determine whether a GP listens with sympathy and respect to a patient and refers her for IVF. At a broader level, it determines how governments legislate. We are not artists attempting to change the world by shocking people. As practitioners in a field replete with ethical dilemmas and strong emotions, we should ply our trade with responsi-bility and respect.

The ultimate decision must rest with physician and patient, reached by open dialogue illuminated by the right information, assisted by counselling and without conflicts of financial or other interests. A danger of the 'Iliescu phenomenon' is that it creates an impression that IVF is a simple matter of giving drugs to someone, fertilizing their eggs and producing a baby. As we shall see, IVF is certainly not the answer for everybody. Comprehensive testing is needed, long before anybody mentions IVF, to determine the root causes of infertility and to decide on the right treatment – if there is one. Many women, apprised of all the facts, go away and get on with their lives. Most weigh up the risks and probabilities and, if the physician is of a similar opinion about the length of the odds, take the treatment offered. This system isn't perfect. But it is, I believe, fairer than that which arbitrarily forbids or rations IVF for women over thirty-five, forty or any other age.

In families where conception takes place without medical intervention, nobody tries to step in and prevent a couple having a baby because they are too old or there would be too great an age gap between the child and its siblings. If a woman of forty-five conceived naturally, her GP would – and should – advise her of the risks. But there would be no question of him or her turning the woman away from the clinic, or of the local maternity unit refusing to treat her when she gave birth. Why do we scrutinize infertile people more closely?

Identifying and treating the causes of infertility

A very detailed description of all the causes of male or female infertility is inappropriate in this book.[10] But it will be useful, and germane to my purpose here, to sketch some broad outlines. Infertility in women tends to stem from problems in one or more of three main categories. These are: problems with the ovaries, and the cycle of ovulation; problems with the fallopian tubes; and problems within the uterus and cervix. I cannot emphasize enough that IVF is not a 'magic bullet' that can remedy all cases of infertility; very often there are cheaper and more effective treatments.

The need for careful testing and examination before any treatment, including IVF, is considered cannot be emphasized enough. Regrettably, infertile couples have always been prey to exploitation. Even today, treatments are given without first undertaking a sound assessment of what is wrong – and many couples are so desperate that they will try almost anything that is suggested, in blind faith that it may help them.

One of the earliest examples of this was the 'magical' use of electricity by one James Graham in the eighteenth century. Dr Graham (1745–94) claimed innumerable health benefits from the voguish science of electricity, which had been popularized somewhat earlier that century by Benjamin Franklin. After attracting society patients including Georgiana, Duchess of Devonshire, Graham apparently spent £10,000 building a Temple of Health at the Adelphi Theatre in London. Visitors to the Temple, which opened in 1779, found themselves in candle-lit rooms decorated with erotic art. The air was heavy with perfume and gentle music. In this environment Graham lectured his audience on sexual health and demonstrated his electrical devices. He was ably assisted by attractive young women who posed wearing virginal white gowns – among them Emma Lyon, who would go on to become famous as Lady Hamilton, the mistress of Lord Nelson. A book stall in the Temple sold Graham's tracts, and visitors were encouraged to buy a Nervous Ætherial Balsam which guaranteed fertility.

Graham's ingenuity did not stop there, however. If the balsam failed to achieve its supposedly guaranteed results, wealthier members of the public could pay £50 for a night frolicking in the Celestial Bed. According to the advertising,

The super-celestial dome of the bed, which contains the odoriferous, balmy and ethereal spices, odours and essences, which is the grand reservoir of those reviving invigorating influences which are exhaled by the breath of the music and by the exhilarating force of electrical fire, is covered on the other side with brilliant panes of looking-glass.

The biblical command to 'Be fruitful, and multiply, and replenish the Earth' was wreathed with 'electrical fire' on the bedhead. Electromagnets ensured that love-making occurred within a magnetic field, and the bed could also be tilted after sex to encourage semen to make its way into the womb.

The Temple was one of the top public attractions of the day, despite the barbs of critics like Horace Walpole, who sniffed that 'the apothecary, who comes up a trap-door (for no purpose, since he might as well come up the stairs), is a novelty'. But it was expensive to run; Graham was eventually forced to move to cheaper quarters, and finally closed the establishment in 1784. Mired in debt, he moved to Edinburgh and became increasingly eccentric, being arrested at one point for giving all his clothes to a beggar. As well as promoting the health-giving properties of mud baths, he founded his own church, but failed to inspire any followers. Graham's final book, *How to Live for Many Weeks or Months or Years Without Eating Anything Whatsoever*, is out of print.

Problems with ovulation

As we saw earlier, the monthly cycle of egg production depends on hormones controlled by the pituitary gland in the brain. A significant proportion of infertile women have problems within either the pituitary gland or the hypothalamus, a region of the brain that acts as the control centre for much of the body's hormonal activity.

But the most common cause of failure to ovulate is the condition *polycystic ovary syndrome* or PCO. This was first recognized in 1935 in the USA, when Dr Irving Stein and Dr Michael Leventhal observed that some women who had irregular periods also had enlarged ovaries containing a number of cysts and did not ovulate. For some of these women, the misery was compounded by being overweight and having excessive body hair, often on the face and chest. All ovaries contain cysts – this is what the follicles are – but polycystic ovaries contain an increased number arranged around the circumference of the ovary. Around one in five women have this formation and don't experience fertility problems, but a smaller number have difficulty conceiving. It is now known that PCO causing infertility tends to have a genetic origin and is associated with resistance to insulin, a hormone that helps regulate ovarian function. When someone is insulin resistant, this means that cells throughout the body do not readily respond to insulin circulating in the blood, so that the amount of insulin in the blood remains high. Associated with this are increased male hormone secretion and decreased ovulation. People with PCO syndrome are more likely to have miscarriages.

Irregular periods are more common in the late thirties and early forties. Rarely, periods may stop altogether, although this does not always indicate a failure to produce eggs.

The good news about ovarian infertility is that it can normally be successfully treated with drugs. Infrequent or absent periods, a lack of the usual feelings of tenderness in the breasts before or during menstruation, an increase in bodily or facial hair, or obesity may all indicate that the problem lies within the ovaries.

A number of tests can be used to find out whether this is the case. One simple, DIY investigation can be done at home with a thermometer. Usually, body temperature drops just before ovulation and rises sharply afterwards. But this is distinctly imperfect. Daily temperature-taking is awkward and can be a constant reminder of a perceived problem. Some women who are ovulating normally may not have any measurable temperature changes, while some who have ovulation problems may still see the classic changes of a normal cycle.

I am very dubious about the snazzily packaged home testing kits that ostensibly predict a rise of luteinizing hormone (LH) in the urine, indicating that 'ovulation is taking place normally'. The manufacturers claim to help women time intercourse to give the best possible chance of conceiving. In fact, timing intercourse is generally a bad idea, adding stress to an already difficult situation. Frequent, unhurried sex is far more likely to help and far more likely to sustain relationships that may be under strain. Moreover, a rise in LH does not show if there are problems with the rest of the cycle. Women with PCO, for example, tend to have high levels of LH all the time.

Blood tests provide somewhat better information. The level of progesterone is significant: it rises sharply a few hours after ovulation, and stays high for several days, but is seldom raised in women who have not ovulated. Progesterone produces thickening of the endometrium in readiness for a fertilized embryo. Without pregnancy, progesterone levels drop, and the endometrium shrivels and is shed by menstruation.

The level of follicle stimulating hormone (FSH) is also important. As we saw, the pituitary releases it to start follicle development. If the ovaries do not respond promptly, or are producing fewer eggs, the FSH bombardment increases in a bid to coax the ovaries into contributing the maximum possible effort to the fertility process. Essentially, one can imagine a telephone. The FSH is the ring. The pituitary rings the ovary, but if the ovary doesn't or can't answer the call, the phone continues to ring. So prolonged, high levels of FSH often indicate that the ovaries are working inefficiently or not at all. FSH rises increasingly as women get older and the ovaries work less well. Unfortunately, normal FSH levels do not necessarily mean the ovaries are working normally.

Ultrasound has been the single biggest advance in recording ovarian function. One of the pioneers of this technology, Dr Karl Theodore Dussik, produced his first paper on the possibilities of ultrasound in medicine in Nazi-occupied Austria in 1942. He carried out research on Luftwaffe airmen with head and spine injuries – a reminder, pertinent to the debates surrounding reproductive medicine, that technology is neither good nor bad; it depends who is using it for what purpose. Another pioneer was my predecessor at Hammersmith Hospital, Professor Ian Donald. He started his researches after the Second World War, saying that he 'arrived with . . . a rudimentary knowledge of radar from my days in the RAF and a continuing childish interest in machines, electronic and otherwise'. In my view, his paper on ultrasound in the *Lancet* in 1958 deserved the Nobel Prize. Developed and applied using his techniques, ultrasound eventually became one of the most important diagnostic tools in medicine. In reproductive medicine, ultrasound now enables us to look at the surface and shape of the ovaries, measure the follicles, record ovulation and detect the very earliest stages of a pregnancy in the uterus.

One test that went out of fashion for a time but is now

regaining favour is an endometrial biopsy. Taking a small sample of tissue from the womb lining can help determine whether the progesterone is doing its job. More importantly, we now know that certain genes work to help the embryo implant, and examination of the lining of the uterus using specialized microscopic techniques can sometimes give a clue to genes that may not be functioning normally.

One important tool in investigation of the ovaries is the laparoscope. A small incision is made in the navel under general anaesthetic, and a telescope about as thick as a fountain-pen is passed into the abdomen. The laparoscope enables us to see the ovaries close up and determine whether ovulation has taken place and whether the ovaries have an excessive number of cysts, scarring or other disorders. Laparoscopy also gives information about the state of the fallopian tubes, and the shape and size of the uterus. This procedure should never be done without the surgeon making a photographic record of what he has seen, for this clear snapshot of how the ovaries looked at a particular moment in time is priceless information to be archived. Laparoscopy may be combined with hysteroscopy (the passage of a fine telescope into the inside of the uterus) to seek and examine possible defects or polyps.

Laparoscopy requires a general anaesthetic, but it is far better than undergoing IVF treatment when there might be more appropriate alternatives, or when there is no hope of pregnancy. Many treatments for ovarian problems bypass the costly, draining business of IVF. Possibly the most taxing aspect of my career as a fertility specialist has been meeting women who have come to me after years of failed IVF treatment, only to discover within a few minutes that their problem could have been remedied after a few simple tests. In some cases, it is by this time too late to offer them successful treatment. This is one of the principal issues regarding assisted fertility treatment: not one of ethics, but one of inefficiency. Too often, IVF is seen as a cure-all by patients and practitioners alike. There should always be proper attempts to determine any underlying causes of infertility, and to consider other remedies, before IVF is even considered.

Drugs frequently help. The classic fertility pill is clomiphene, developed originally – and paradoxically – in attempts to find a contraceptive pill. In early trials, it was noted that some of the rats that had been given the drug became pregnant very easily – which rather scuppered the drug's future as an agent of birth control. Clomiphene is a relatively gentle stimulant which encourages production of FSH from the pituitary gland, causing the ovaries to work harder to produce follicles. It is cheap and safe, but some of its side-effects can hamper treatment by creating cysts within the ovaries, thickening the mucus in the cervix or interfering with the growth of the endometrium.[11]

Nun's urine sounds like a preparation to be found in a medieval apothecary's handbook, but it formed a major component of one of the leading fertility drugs until recently. Menopausal women secrete large amounts of LH and FSH as the pituitary tries, in vain, to force the ovaries to thrust out more eggs. In the 1960s an Italian laboratory hit upon the idea that convents, based as they are on communal female living, would be an ideal source for large quantities of hormone-heavy urine. This was collected and made into human menopausal gonadotrophin (HMG). What wonderful symbolism that these devout Catholic women, their lives dedicated to God, should give others the chance to create life! But the outcome was less appealing. There tended to be problems purifying the material, as a result of which people injected with drugs made in this way occasionally developed painful reactions. Also, as the urine was a naturally derived material, the strength of the hormone content varied from batch to batch. So a race began to develop a purer, synthetic form of FSH, and in the 1990s quantities of the pure hormone were successfully produced from genetically engineered hamster cells. Pure FSH does not necessarily result in more live births than HMG, but it comes with fewer side-effects. Unfortunately, because it is genetically engineered, it is very expensive. In an average course of treatment, FSH is injected daily until ovulation begins; then an injection of chorionic gonadotrophin prepares the egg for fertilization.

These powerful drugs can be dangerous without careful monitoring, and regular ultrasound scans of the ovaries and hormonal tests are needed. The main risk is hyperstimulation – producing ovaries that are over-active and swollen with many follicles. Hyperstimulation can cause abdominal pain and swelling, fluid retention, fatigue and breathlessness. In very rare, extreme cases it can be life-threatening, which is why monitoring is so important. Because the follicles are being stimulated, too many eggs and therefore multiple embryos may be produced. Because multiple pregnancy is very undesirable, treatment may need to be temporarily halted if many ovulations are detected.

Some drugs work further up the chain of command, targeting the communication between the pituitary gland and the hypothalamus, the brain's hormone-control centre. In a normal cycle, the hypothalamus sends a hormonal signal to the pituitary to start releasing FSH and LH. But in a few cases, the signal doesn't get through, so injections of LH-releasing hormone (LHRH) can be effective.[12] Because LHRH targets only the beginning stages of the process, it does not carry the same risk of hyperstimulation or multiple pregnancies.

A treatment for ovarian problems caused by PCO that is less commonly used – though still often more effective than IVF – is follicular puncture, done during laparoscopy using an electrical needle. Mention should also be made of ovarian endometriosis. For reasons that are not entirely clear, the lining of the uterus may grow outside the cavity. When this happens, there is commonly an island of endometrium in the ovary itself from which menstruation occurs. Over a period of time a cyst full of old menstrual blood may develop, and ovarian function is compromised. In some cases these cysts require surgical removal, but this must be done by an experienced surgeon, because the removal of too much ovarian tissue can hasten the menopause.

Problems of the fallopian tubes
Many women have damaged fallopian tubes and yet are completely unaware of this until they try to conceive, for blocked or scarred tubes usually produce no symptoms. One of the most common causes of damage is infection. This can prevent the embryo from being picked up by the tubes or passing through them to the uterus. The degree of damage may vary from a slight narrowing in the tubes, or loss of the hair-like cilia, to total blockage. Usually, if one tube is damaged, the other will be as well. Occasionally, all or part of the tubes can stick to some other part of the anatomy, such as an ovary, the uterus or the bowel – a so-called adhesion; or areas of the lining of the tube itself can adhere to other parts of the lining, causing partial blockage.

Sexual activity is one cause of infection that can lead to

this kind of damage. Common organisms include gonorrhoea and chlamydia. But in my experience, this connection is over-diagnosed. Many women go through agonizing guilt over whether an unwise sexual encounter could have ruined their chances to conceive, when there are plenty of other bacteria, not associated with sexual intercourse, that can invade a woman's body and lead to infection – as can appendicitis, peritonitis (a burst appendix) and some bowel disorders.

The widespread availability of the contraceptive pill means that the 'coil' or intra-uterine device is less frequently chosen as a method of birth control these days. As well as being less effective than the pill, coils can sometimes be associated with inflammation or infections leading to damage of the fallopian tubes.

Miscarriage or abortion may also cause infection in the fallopian tubes. Even giving birth – especially if delivery is complicated – can cause this damage. Therapeutic abortion nowadays is done so carefully that few women are likely to emerge from it with reproductive problems. This is one reason why I regret the decision taken in March 2006 by the American state of South Dakota to outlaw abortions. If more states follow its lead, many women – probably the youngest, poorest and most vulnerable – may well be compelled to seek risky alternative methods to end their pregnancies. A law supposed to defend 'life' could actually end lives, or prevent the creation of life in future.

Roughly one in 100 pregnancies may be ectopic, where the embryo implants outside the uterus, usually within one of the fallopian tubes. Tubal damage following infection or surgery is the usual cause, though occasionally a congenital abnormality of one tube or the uterus may be implicated. If an ectopic pregnancy goes unnoticed, the embryo may grow to the point where it ruptures the tubes, possibly causing severe internal haemorrhage. Ectopic pregnancy is one of the commonest causes of maternal death, and prompt action is always taken if it is diagnosed – which has fortunately become much easier to do with the use of ultrasound scans and laparoscopy. The problem can then be treated with surgery, either through the laparoscope or through a small incision in the abdomen.

Laparoscopy applied to diagnose tubal damage generally uses a coloured dye to show up any problem areas in the tubes, and can be done in around forty minutes. In some cases, it may be possible for the surgeon to introduce fine instruments into the abdomen during the laparoscopy to cut adhesions around the tubes; alternatively, once a diagnosis is made by laparoscopy an appropriately trained surgeon may offer microsurgery. In carefully selected cases, microsurgery offers better results than IVF and is less stressful, allowing natural conception to take place later. But it does require a few days in hospital and recuperation afterwards. If it does not work, IVF can still be an option later.

Hysterosalpingogram or HSG is a neglected but really valuable investigation. A dye, colourless but opaque to X-rays, is squirted into the uterus. In skilled hands, the resulting X-ray photographs provide invaluable information about scarring, blockages or narrowing in the tubes, congenital abnormalities of the uterus or tubes, and the condition of the uterine cavity. In my view, all infertile women – particularly those contemplating IVF – should first have an HSG. Apart from its value in diagnosing tubal damage, it has other irreplaceable benefits. Each year at Hammersmith numerous women are seen who have repeatedly undergone IVF treatment in other clinics with no success, but have never had a hystero-salpingogram. So often, when the HSG is finally done, we find an undiagnosed, correctable abnormality. Thereafter, with a simple, easy and quick treatment, natural conception becomes a possibility.

Problems of the uterus and cervix

Abnormalities in the uterus and the structure at its lower end, the cervix, cause perhaps 10 per cent of female infertility problems. Unlike abnormalities of the ovaries or fallopian tubes, they do not always result in infertility; but certain malformations can contribute to infertility. When a female embryo develops, her uterus is formed from two tubes, which eventually fuse together to form the single cavity. Sometimes this fusion may be incomplete, so that there remain two cavities – the so-called double or *bicornuate* uterus. Alternatively, one tube may develop more than another, creating a *unicornuate* uterus. These conditions result in a distorted, small uterine cavity that is frequently associated with problems in embryonic implantation, or with miscarriages and premature birth. There are various congenital uterine abnormalities, many of which can be corrected or improved by simple surgery.

The commonest uterine disease is fibroids. These are benign growths that may grow into the cavity and distort it; they may also cause abnormalities of blood flow that seem to reduce the chances of implantation. Occasionally they can grow to such a size that the uterus bulges outwards, interfering with the relationship between the ovaries and the tubes. If they are clearly causing uterine distortion, and surgical treatment is indicated at all, surgery (rather than treatment with a laser or blocking the blood vessels by a process called embolization) is still the best way of improving fertility.

Adhesions inside the uterine cavity – most common after a complicated pregnancy and delivery, a miscarriage, an abortion, or certain infections such as tuberculosis – are often missed, and yet they are frequently a cause of infertility. It is a tragedy to see a patient who has gone through three or four arduous failed IVF cycles without having had her uterus checked first, as treatment for adhesions is mostly simple and effective.

Adenomyosis is the condition where little pockets of the endometrium begin to develop outside the cavity and grow into the uterine muscle wall, sometimes causing severe scarring. When there are many pockets, the uterus can become enlarged and its cavity irregular. Each menstrual period may be very painful. Treatment is difficult, but a few cases are amenable to surgery, and sometimes hormone treatment may reduce the size of the areas affected by scarring.

Endometriosis is a related condition. As we have seen, endometrial cysts in the ovaries can impair ovulation. But the endometrial lining can grow in a variety of places, causing scarring to the bladder, the bowel and the tubes. Sometimes endometriosis causes adhesions in the pelvis which can prevent the egg getting into the tubes. Although it sounds alarming, this is never life-threatening, and is commonly found in fertile women as well. Nevertheless, when the scarring is extensive, it may be associated with infertility.

Certain cervical problems can contribute to infertility. A narrow or scarred cervix may impair sperm transport. A woman whose cervix is scarred from treatment to prevent early cancers developing – for example, a *cone biopsy* – may be less fertile, while extensive diathermy or laser treatment to the cervix can increase the chances of an infection that in turn may cause inflammation or scarring within the uterus or fallopian tubes.

Possibly 5 per cent of infertile women have abnormalities of their cervical mucus. Its chemistry and composition change during the menstrual cycle so that, in the days leading up to ovulation, it should become watery, allowing sperm the chance of movement into the uterus. Scarring or other cervical conditions may cause the mucus to be thick and unreceptive, thus contributing to infertility. Sometimes there may be an immunological problem, so that the woman produces antibodies against the sperm, killing them in the mucus.

One of the oldest, and now one of the least popular, ways of diagnosing problems with the cervical mucus is to examine the cervix twelve to twenty-four hours after sex. The so-called post-coital test, or PCT, involves examining a small sample of cervical mucus under a microscope. If the test is positive, sperm will be detected in the mucus, swimming vaguely straight. If the sperm are not seen, are swimming slowly or are not moving at all, this may indicate a problem with cervical mucus. But there are so many reasons why the PCT can be negative that this test has been largely abandoned.

If there is a cervical problem, it can often be dealt with by intra-uterine insemination or IUI (see page 74), or occasionally IVF.

A game of two halves: male infertility

In 2004, an article published in the journal *Human Reproduction* claimed that laptops were doing serious damage to male fertility.[13] Dr Yefim Shenykin, associate professor of urology and director of male infertility and microsurgery at the State University of New York at Stony Brook, examined the testicular temperatures of twenty-nine healthy volunteers as they worked for an hour with or without laptops. He noticed a rise of between 4.6 and 5 degrees Fahrenheit in the group using laptops over the group not using them. Earlier studies suggest that temperature rises of between 1.8 and 5.2 degrees could affect a man's fertility.

Professor Shenykin went so far as to say that boys and young men should limit their use of laptops if they want to stay fertile. Personally, I'd just advise them to place their laptops on a solid surface, rather than balance them on their knees.[14] Typing on a machine balanced on your lap may wreak havoc with your back muscles – which can scupper your sex life, and thus do far much more damage to your chances of siring a child than a pair of over-heated testicles. But what I find interesting is the huge coverage that reports like this get from numerous journalists. It is probably nonsense to suggest that using laptops causes infertility. There is no evidence to suggest this. Dr Shenykin does not actually seem to have tested the sperm counts of his subjects – or even to have asked each of them if he had fathered a child.

So it is with despair that I report that men whose partners are failing to conceive are so often advised to wear loose-fitting underwear, or even to bathe their testicles in cold water. Nature has already provided a cooling system, by placing the testicles away from the body inside the sack of the scrotum. Data about testicular temperature rises have not, to my knowledge, been cross-compared with fertility rates in countries where the climate is hotter, or among people who have moved from cold to hot countries and vice versa. Concerned males might also remember that many species of animals, such as elephants, have their testicles inside their bodies, without their fertility suffering.

However, I do not mean to make light of the problem

of male infertility, which blights the lives of many couples. Many men feel useless and inadequate on learning that there is some problem with their reproductive apparatus. They become depressed, even suffer from impotence, so great is their inherited cultural sense that a 'real man' is necessarily a fertile man.

Male fertility has been of symbolic importance throughout the ages. In ancient Babylon, a yearly ceremony took place whereby the king fertilized the lands of his dominions in an act of intercourse with a goddess. Goddesses being notoriously hard to track down, a specially initiated female temple functionary usually took her place. Although the evidence is hotly disputed, the Old Testament abounds with references to similar practices in the religions of the Israelites' neighbours, the Canaanites. In a manner reminiscent of the antics of the god Set in ancient Egypt, an act of intercourse by a mortal male had power to renew the plants and trees, ensure fruitful live-stock and guarantee ample rain. Mere mortal men must have felt pretty pressurized.

The thread connecting sexual potency, fertility, strength and manhood is deeply woven into every culture, marking a corresponding fault-line of insecurity. In the nineteenth century, this vulnerability was targeted by some doctors, both respectable and otherwise, who promised they could restore men's flagging virility with various dubious extracts. The champion of the virility cure has to be John Romulus Brinkley, a man who referred to himself only as 'Doctor', although the likelihood of this former telegraph operator having any medical qualifications is slim. In the early years of the twentieth century Brinkley teamed up with an armed robber to offer hope to the men of North Carolina. 'The Greenville Electro-medical Doctors', as they called themselves, took out adverts in the local press, asking the poignant question, 'Are You A Manly Man Full of Vigor?' Those who suspected otherwise could have their testicles injected with a coloured fluid at $25 a shot.

After a spell in the clink, Brinkley purchased a $500 licence from the Eclectic Medical University of Kansas City, entitling him to practise as a doctor in Kansas – a state whose *laissez-faire* attitude to medical matters made it a byword for quackery. In an area teeming with fellow snake-oil peddlers Brinkley had a tough time, and was forced to take a job tending to the wounds of employees at a meat-packing firm. Here he spent time observing the highly sexualized behaviour of the penned-up billy goats, which provided him with inspiration for his next move.

An ageing farmer allegedly visited Brinkley, complaining about his lack of 'pep' in the bedroom. Brinkley decided to implant a pair of Toggenhorn goat testicles in the man's neck. Two weeks later, the farmer reported a full return to his former glories, and some years further on he sired a son – inevitably called 'Billy'. By 1923 Brinkley was earning a fortune from the goat 'cure', even performing his operations on such influential figures as Harry Chandler, owner of the *Los Angeles Times*. Treating Chandler, who had just set up LA's first radio station, encouraged Brinkley to become a media mogul himself. Returning home from California, he set up KFKB – 'Kansas First, Kansas Best' – a broadcaster devoted, unsurprisingly, to advertising the wondrous benefits of Dr Brinkley's procedure.

Brinkley never published a study of the gland operation, nor did he invite any properly qualified medical professional to come to see what he was doing. Despite this, many men were so moved by his claims that they were prepared to pay $750 – cash, up-front, no credit – to undergo the procedure. By 1927 Brinkley's clinic was performing around fifty operations per month. Ten years later, he owned a private plane, a dozen Cadillacs, three yachts, a number of oil wells – and a 16-acre estate, roamed by Galapagos tortoises and penguins, on which he reigned supreme, his clothes adorned with diamond-encrusted pins. In 1941 he died, disgraced and bankrupt, after a number of successful malpractice suits and a crippling twenty-year bill from the American taxman.

Brinkley now has a respected place in the quacks' Hall of Fame. But could there have been any real value in his operation? Brinkley was not the only person convinced that gland extracts could restore both male and female health. His million-dollar business had been inspired by the work of Professor Edouard Brown-Sequard, a world-famous neurologist who fed himself a mixture of guinea pig's and dog's testicles. The 72-year-old professor published an account of his research in the *Lancet* in 1889, claming that, within two weeks, he'd noticed an increase in strength, a boost in mental functioning and regular bowel movements. He reported similar effects on three other men of advancing years.

It is highly unlikely that there can be any observable benefits of such treatments, whether you implant testicles inside someone's body or consume prepared extracts. It is similarly doubtful whether injecting people with testosterone – a much-vaunted 'cure' for flagging male libido – does any good. It seems more possible that the success of Brinkley's and Brown-Sequard's treatments was largely attributable to the placebo effect. In 2002 an Australian team replicated Brown-Sequard's experiments, using testicles obtained from dogs that had undergone castration by a nearby vet. They concluded that the concentration of testosterone in Brown-Sequard's extracts was about a quarter the strength of the dose now given to men in testosterone replacement therapy. It was also unlikely that drinking a liquid extract would do any good, since any useful hormone in the testes would not be water-soluble – and though testosterone is made within the testicles it is not stored there. In other words, people probably felt better after these treatments because they believed in them, rather than because they were having any physiological effect.

There are no miracle extracts that boost male fertility. As a general rule of thumb, it is often far easier to intervene, surgically or chemically, where there are problems in the female reproductive system. For most male problems, IVF is usually a better solution.

Journalists have speculated about the supposedly widespread decline in men's sperm counts over the past forty years. It is suggested that we are being emasculated by the amount of toxins in our environment, even that the level of oestrogen in our water supply is turning men into women. There is, however, no really solid evidence that men are now more or less fertile than previously. For one thing, the first sperm counts were taken from men of very different backgrounds from the sort of men attending fertility clinics today. For another, the methods used for measuring sperm counts – although still not terribly efficient – have changed, so there can be no valid comparison of results obtained at different times. And for another, the low sperm counts reported may not always cause infertility.[15]

In rare cases, the testes may not be producing any sperm at all – possibly because they have been damaged, or because the pituitary gland is not stimulating them properly. Alternatively, it may be that the testes are producing sperm, but those sperm are not entering the semen, either because there is a blockage in the network of fine tubes inside the testicle, or because the muscles are directing the sperm into the bladder. This rare condition usually results from nerve damage, although some prescription drugs can cause it to happen.

But mostly, the testes are just producing fewer sperm, and poorer sperm. Low-quality sperm are a major cause of male infertility. Sluggishness is the most common defect; even when the number of sperm is normal, a large percentage will be malformed, or barely mobile. There are a number of reasons why abnormalities may be more pronounced in some men. Drugs such as tobacco, marijuana and alcohol can have a negative effect on sperm counts, as well as some prescription drugs like anti-depressants and anti-malarials. So can environmental factors – which is why men in certain occupations are at a greater risk than others of fertility problems. Men whose jobs expose them to harmful quantities of lead – such as petrol station attendants – have been said in the past to be more likely to have low sperm counts, as may those whose occupations expose them to high levels of vibrations, such as welders, drill operators and boilermakers. It is possible, too, that stress and fatigue may also contribute, though the evidence here is less good. Common infections – such as mumps contracted in adulthood – can also affect the quality of the sperm, either temporarily or permanently.

A doctor's first recourse should be an examination of the testicles, to check for any obvious abnormalities. A common one is called varicocele, essentially an enlarged vein around the testicle, which is thought by some to be linked to sluggish or immobile sperm. The evidence that varicocele causes male infertility is not very good – the idea seems to be related to the unproved notion that heat damages the testes, and that the heat of the blood pumping through this vein has an adverse effect on the sperm.

But around 20 per cent of both fertile and infertile men have a varicocele – so the link is not obvious. Physical examination is also important because infertile men have a very slightly raised risk of testicular cancer.

A first step will be to send several samples of semen for examination under a microscope. This will reveal how many sperm are present in the semen and how mobile they are, and give an idea of the chemical make-up of the seminal fluid. If few sperm are found in the seminal fluid, the temptation may be to decide that the man has a 'low sperm count'. But the number of sperm in each roughly teaspoon-sized ejaculation can vary from one day to the next. An anxious male will probably have to produce repeated samples on different days before he knows for certain whether his sperm count is truly low.

Around 40 per cent of a fertile man's sperm will normally have poor motility – the most readily seen finding is that the sperm are unable to move in a straight line. Two tests have been developed to look at sperm movement, both of which involve placing the sperm in a special fluid and observing them closely. In the 'swim up test', sperms are placed in a tube with the special fluid at the top. The healthy sperm should swim upwards into the fluid, leaving the more challenged ones at the bottom. The 'sperm velocity test' uses computed measurement and assesses the speed of the sperm's movement by time-lapse images.

If the count is low, doctors may also take blood samples to look at the levels of FSH and LH. As with ovarian failure, the levels of FSH may be increased if the testes are failing. Occasionally there may be a rare pituitary problem. For example, around one in every 60,000 men suffers from a genetic condition called Kallmann's syndrome, in which low levels of hormones result in very shrunken testes and poor sperm counts. They also lack a sense of smell – so-called *anosmia*. In general, the symptoms are so severe that they are usually spotted long before a man with this condition starts trying to have a family.

Regrettably, the popular picture of reproductive medicine still means that many people think all 'hormone problems' can be treated with tablets or injections. If a man is found to have low levels of LH or FSH, hormone therapy may hold out hope if there is any abnormality within the pituitary. However, injections of testosterone, in large quantities, actually stop sperm production – although some men experience a sudden surge in their sperm counts after they stop taking the hormone.

Some men produce antibodies to their own sperm – effectively damaging them before they have a chance to fertilize any eggs. In around half of these cases (probably under 2 per cent of infertile men), steroid drugs offer some hope. These have the effect of suppressing the immune system and preventing the production of the killer antibodies. But drugs like this need careful monitoring and certainly should not be taken over the long term.

Many private clinics still make claims for the value of costly courses of injections, or herbal or Chinese medicines. Many retailers sell 'male fertility supplements' at a premium. The use of these 'remedies' is not supported by any serious evidence. With the exception of giving a short course of antibiotics to clear up infections, 'chemical' treatments for male infertility are of little use. Over the years, I have often heard feminists claiming that, because of a preponderance of males in the medical profession, there is a bias towards treating and curing men, and overlooking the problems of women. There is always room for improvement – but there is certainly no bias in reproductive medicine, where, if anything, we are far better equipped to treat the problems facing women than those troubling men.

If sperm are completely absent from the semen, or mostly dead, a doctor may also take X-rays or conduct some exploratory surgery. Microsurgical techniques have improved the chances of success, and sometimes blockages within the tubing can be cleared. Success rates for this type of surgery tend to depend on how long the blockage has been in place. If the problem is of long standing, the testes may have ceased producing sperm and no operation will help. In around 20–30 per cent of men with blockages, an operation can allow normal sperm to pass into the seminal fluid again.

Our ability to increase male fertility, then, is pretty limited. But one technique that should sometimes be considered before IVF treatment is artificial insemination (AI). The most effective method for this is to inject carefully prepared sperm directly into the uterus – so-called intra-uterine insemination (IUI). But to place a man's ejaculate directly into the uterus, bypassing the protective cervical mucus, is dangerous, because it may contain harmful impurities and bacteria that might even make the woman more infertile. So once the sample has been collected from the man, the sperm are 'washed' by being placed in a special fluid, which is then spun and filtered to sift out the debris that is also present in the seminal fluid, before they are introduced into the uterus.

Some clinics obtain quite good success rates using a split ejaculate. This method, based on the fact that the first part of a man's ejaculation contains most of the active sperm, requires the man to ejaculate into two separate pots. However, as most men find some difficulty in getting their semen into just one pot, this act of juggling can be difficult.

IUI may be combined with giving drugs to the woman to stimulate her ovaries. IUI can often be used for the relatively high percentage of couples whose infertility has a dual cause, for instance a low sperm count combined with irregular ovulation. It can also be useful in cases where a woman's cervix produces mucus that is hostile to her partner's sperm. The disadvantages of IUI are that it cannot, on its own, do anything about problems with ovulation or the way an embryo implants in the uterus.

How fragile and fraught the process of making babies really is! It is ironic that the species with the largest brain, the species that has the most control over its destiny, *homo sapiens*, is also a species that is desperately feeble at reproducing itself. That might, in itself, explain why we have developed such an extraordinary cortex. We need to counteract the biological reality that we are not very successful at multiplication.

Expert claims are repeatedly made that infertility is on the increase. In June 2005 Professor William Ledger, of the University of Sheffield, suggested that the number of couples with these problems could double within the next decade. I think this is a gross over-estimate; but I agree that our ageing population, the rising incidence of obesity and possibly an increase in sexually transmitted diseases will remain serious hindrances to reproduction. IVF, costing at least £3,000 a cycle, is going to be increasingly in demand for this sub-fertile population.

Maybe governments should put more resources into tackling obesity and infection among young people, to stave off this theoretical 'infertility time-bomb'. But there is a strong tendency to attribute blame to infertile people for causing their own problems when they have not. I remain deeply unconvinced that government health warnings will make much impact on human behaviour; but even if they did, there is little evidence that this would reduce the burden of infertility. Health warnings are for the most part merely an exculpatory exercise by governments that have failed to treat a major medical problem that is not generally the fault of the sufferer.

The therapeutic problem is that each individual cycle of IVF has only around a 23 per cent chance of success. So how and why *does* it work? When should it be used? Is it completely safe – medically and ethically? I will attempt to answer those questions in the next chapter.

IVF: A 'blind assertion of will against our bodily nature'?

In 1898 the *Journal of the American Medical Association* received a report from a Dr Neesen, of Brooklyn, who was taking a postgraduate course in Vienna. Neesen reported extraordinary scenes at the home of a certain Viennese doctor, one S. L. Schenk:

> When I called at Dr. Schenk's house I found the street blocked with carriages of all descriptions. A group of well-dressed people stood on the stoop of the house, waiting to be admitted. The anterooms were crowded to suffocation with visitors, most of them women, richly attired and genteel looking, all waiting to consult the professor.[1]

The reason behind the tumult was that Dr Schenk claimed he could influence the sex of a baby by regulating the diet of female patients before pregnancy began. His view, in a nutshell, was that the healthier parent determined whether the embryo would develop into a male or a female. Healthier men, he argued, tended to produce females, and, conversely, women to produce males. Therefore, he claimed, by boosting the health of female patients, he could ensure they gave birth to healthy boys.

Like many fertility specialists – and I use the term rather broadly with respect to Dr Schenk's claims – the good doctor was alarmed by public reaction to his findings. As he told his American visitor,

> Many persons have been misled by the incorrect reports that have got abroad about this treatment. The rooms out there are full of such. They come here wholly misinformed and expect me to do the impossible. They are pestering me constantly. I have, so far, endeavored to see every one and treat them courteously, but the task is getting too arduous.

Schenk complained that he was now unable to conduct his regular practice because of all the eager prospective parents thronging outside his house. For fear of being mobbed, he was compelled to enter and leave via the back door. Repeatedly he issued statements to the crowds that he was not taking fees for this treatment, and that he was only experimenting on a handful of carefully selected couples.

Although claims like Dr Schenk's have been made throughout medical history, there is little to substantiate any of them. He could not know, at that stage, that embryonic cells (like virtually all cells) have twenty-three pairs of chromosomes, one copy from the egg and one copy from the sperm. The chromosomes that determine sex are called X and Y and are regarded as a pair. A girl has two X chromosomes, a boy one X and one Y. The Y chromosome carries the genes that will produce male characteristics. Although the egg and sperm cells start life with paired chromosomes, they both have to get rid of one half of their chromosomes, otherwise the embryo would have too many chromosomes (see page 337–8). So by the time it is fertilized, the egg contains a single X chromosome. Whether the fetus turns out to be male or female depends on whether the sperm that fertilizes the egg carries a Y or an X chromosome.

The public response to Schenk's claims set the tone for the way so much later work in reproductive medicine would be received. Even today, when the internet and press provide virtually unlimited opportunities for balanced debate, the press and the public are often misled. People assume, quite routinely, that IVF techniques can 'give anyone a baby', and are quite unaware of the associated failings, risks and ethical problems.

Ancestors of IVF

How would the public have reacted if they had known what Dr Schenk was up to in the laboratory some twenty years earlier? In 1878 this Viennese doctor had conducted experiments attempting to create mammalian embryos outside the body.[2] So he deserves perhaps to be considered as one of the grandfathers of IVF. He mixed eggs and sperm from rabbits and guinea pigs, and it seems that he might just have been able to create two-cell embryos in his laboratory – although, without the right fluids and nutrients to nurture them, these first *in vitro* life-forms never developed any further.[3] It is possible, of course, that the fertilization he thought he had observed was simply activation of the egg in a hostile environment. Since that time, many so-called embryos observed in human IVF are simply the result of parthenogenetic cleavage – a phenomenon that repeatedly misled pioneers right through the twentieth century. But Schenk did hit on a remarkable idea: he included scrapings from the female rabbit's genital tract in the culture system – one of the very first examples of the modern use of 'conditioned media'.

But not even Dr Schenk was the first to attempt to fertilize eggs artificially in glassware. That distinction goes to Lazaro Spallanzani, who mixed frog's eggs with seminal fluid to produce live tadpoles almost 250 years ago. Spallanzani was a fine exponent of experimental method, one of the first scientists to use really carefully designed procedures to answer specific questions about biology. His elegant experiments proved beyond doubt that both egg and sperm were necessary for conception. Above all, he was the first person to show that direct contact between egg and sperm was necessary for the embryo to form. In one early set of experiments, he used a barrier to demonstrate that seminal fluid was necessary to fertilize frog's eggs. He designed taffeta breeches and clothed the males so that, no matter how hard they clasped the females, their ejaculate could not reach the eggs. He wrote:

> The idea of breeches, however whimsical and ridiculous it may appear, did not displease me...males, notwithstanding this encumbrance, seek the females with equal eagerness, and perform as well as they can, the act of generation...but the eggs are never prolific, for want of being bedewed with semen which sometimes can be seen in the breeches in the form of drops. That these drops are real seed appeared clearly from the artificial fecundation that was obtained by means of them.

For of course, the clincher (if I may be allowed to call it that) was that the drops of semen from the breeches fertilized the eggs *in vitro*.

Spallanzani's later experiments in or around 1780, particularly his use of a syringe to inseminate a bitch on heat – making sure that the seminal fluid was kept at the blood temperature of the dog – are, without doubt, the beginnings of modern reproductive technology.[4] But I wonder what he, a Catholic priest, would make of our use of in vitro fertilization to produce human babies?

The artificial fertilization of frog's eggs is relatively easy, because normally they are fertilized outside the body in pond water in all weathers – a pretty simple environment to reproduce. I remember as a schoolboy in 1956 being shown the age-old trick of IVF in sea-urchin eggs at Millport Marine Biology Station in Scotland. We were taught to follow the experiments of Lord Rothschild, the scientist who simply used the medium of unprocessed sea water from the Isle of Cumbrae in the Clyde estuary.

Mammals present much more complex problems. Although there were numerous attempts to fertilize their eggs in the century after Schenk, the obstacles were huge. The first was to obtain mature eggs. Next, scientists had to find a way to 'capacitate' sperm, as immediately after ejaculation sperm cannot seek or penetrate an egg: they need to be in contact with various activating chemicals from the female before they are capable of fertilization (see pages 36–7). Then there was the need to develop technology to maintain the embryo outside the body in precisely the right environment. Another problem was how to transfer the embryo to the uterus, without causing any damage and at precisely the moment when the uterine lining had been prepared to be receptive for implantation. Enter a remarkable polymath and Renaissance man: Walter Heape, at the University of Cambridge.

Heape, being independently wealthy, was educated at home. As a young man, he traded in rice, sugar and textiles in Africa, Australia and New Zealand for a while. But ill-health overtook him and his doctors advised a quieter life. Entering Cambridge in 1878, he started to study plants and animal tissues. As a graduate student he started his groundbreaking work on embryos, and in 1890 published his experiments on an albino female Angora rabbit.[5] He flushed two four-cell embryos from her fallopian tubes and, helped by a surgeon, one Mr Samuel Buckley FRCS,[6] transferred these embryos to the tubes of a Belgian hare that had been mated for the first time three hours earlier. The hare was chosen because it is an animal closely related to, and compatible with, the rabbit. Prior mating was crucial, because it was this that stimulated the endometrial lining of the uterus to be receptive to the embryos. Around thirty days later, the hare gave birth to four Belgian hares and two Angora rabbits. That the latter were from the genetic mother was clear because were both albino and had the long silky hair peculiar to the Angora breed. Heape thus proved that it was possible to harvest early embryos and transfer them to a recipient mother without affecting the genetic characteristics or development of the offspring.[7] He repeated these experiments in various other species, including the guinea pig and the cow.

Heape, an interesting character, went on to write books on sexuality, feminism and preparation for marriage. He was totally opposed to the suffragette movement, taking the view that sexual equality was illogical because male and female bodies were so radically different. In other respects, however, he was uncomfortably modern, recognizing the possibility of making money from the use of biological technology and believing in the commercial exploitation of science: he argued that, by understanding breeding habits better, fishermen, farmers and livestock merchants could boost their profits. He died in 1929, having developed a machine for high-speed photography.

We have already encountered the efforts of Professor Brown-Sequard to improve his own well-being with testicular extracts. He may have been mistaken, but his interest in endocrinology, the study of the glands, contributed to the science needed for IVF. The earliest experimenters observed that certain organs seemed to produce secretions which, in turn, were linked to behaviour. Castrating a rooster, for example, reduced his interest in sex – but re-implanting the testicles, even just inside his abdomen, restored his former vigour. In the early days of research in the nineteenth century and the beginning of the twentieth, this discovery led to an unhealthy degree of commercial activity, with various quacks offering extracts of both male and female glands along with the promise that these could boost both libido and fertility. A sandwich of minced ovaries was among the tasty remedies offered to infertile women, along with an over-the-counter elixir called 'Spermin'. Nowadays, of course, we have 'Semenax', which, according to adverts on the web, has ' been proven over hundreds of years of experience to stimulate sexual activity and increase semen and sperm production'. I suspect 'Semenax' has about as much potency as 'Spermin'.

In 1923 Drs Edgar Allen and Edward Doisy at Washington University in St Louis, Missouri, made a vital advance.[8] Through studying the growth and death of the ovarian follicles in mammals, they hazarded a guess that these might secrete substances that were crucial to the whole cycle of reproduction. Using fluid extracted from the follicles, they realized that they were able to control the onset of a woman's period. They had realized that this hormone – later called oestrogen – was a crucial factor in reproduction.

At first, the potential of this discovery seemed to be boundless. By finding out the chemical 'secrets' of reproduction, doctors were gaining the means to treat infertility just as they did so many other conditions – by prescribing the right medication. Sadly, it wasn't that simple. In the first place, Allen and Doisy didn't understand that a live birth depends not only upon oestrogen but upon a whole range of other, carefully timed hormonal secretions as well. Second, the extracts that they hoped would provide a wonder cure were difficult to obtain. Like Brown-Sequard before them, they looked to the animal kingdom to provide a ready supply. But to obtain one-hundredth of an ounce of oestrogen, you needed the ovaries of around 80,000 heifers. Prices – $200 for a gram of oestrogen in the 1930s – reflected the enormous difficulty in obtaining these extracts, which were simply beyond the reach of all but the very wealthiest. And they would not have done much good.

The race was on to find a way of synthesizing hormones, removing the need for skip-loads of testicles and ovaries. A breakthrough came in 1928, when it was realized that the urine of pregnant women was a rich source of oestrogen and could be used to stimulate the ovaries.[9] And while some scientists sought to synthesize the chemical components of reproduction, others were working on the more solid stuff of eggs and sperm.

Towards a receptive culture

It is worth considering the historical picture. In 1932 Aldous Huxley published his extraordinary novel *Brave New World*, in which he fairly accurately described IVF before it had been achieved. (The essential difference was that Huxley envisaged that the embryo would grow right through development until 'birth' in the glass vessel.) Four years later, the *New England Journal of Medicine* published an editorial that read:

84

Conception in a watch glass
The Brave New World of Aldous Huxley may be nearer realization. Pincus and Enzman have started one step nearer with the rabbit, isolating an ovum, fertilizing it in a watch glass and reimplanting it in a doe other than the one that furnished the oocyte and have thus inaugurated pregnancy in the unmated animal. If such an accomplishment were to be duplicated in the human being, we should in the words of 'flaming youth' be 'going places'.

The editorial was not attributed, but was almost certainly written by one John Rock of Harvard Medical School, one of the great medical pioneers of his age. A devout Catholic, he attended Mass daily and kept a crucifix on his office wall. But throughout his life he believed that birth control, for married couples, was right. It was inevitable that he would eventually meet disapproval from the church and ironic that he was one of the first to attempt human in vitro fertilization. In 1938 John Rock teamed up with Arthur Hertig, pathologist at the Free Hospital for Women in Brookline, Massachusetts. Over some years, Rock recovered some thirty-four fertilized eggs from hysterectomy patients whose operations had been carefully timed. Out of loyalty to his university, he named the first successfully harvested egg the 'Harvard Egg'. Subsequently he and Hertig called another the 'Dominic Egg' after the Boston Red Sox centrefield player Dominic DiMaggio, who hit a crucial long double during the 1946 World Series just at the time Hertig was looking at it down the microscope.

Some people have pointed out that what Hertig and Rock were doing was, to say the least, ethically dubious. There is evidence suggesting that Rock did not get informed consent from these patients, and what is more, it is almost certain that many of them were Catholics.[10]

One of the factors that undoubtedly hindered the development of IVF was the difficulty of observing precisely what happens naturally in living humans. The events surrounding fertilization were a black box; there was serious disquiet about the ethics of collecting very early human embryos at hysterectomy, and most researchers were reluctant to attempt to get access to very early human embryos for study. That is why only Pincus and Rock come readily to mind among the few medical scientists who influenced our knowledge about human reproduction before the 1950s.

In 1948 Rock announced that he had recovered eggs from women undergoing hysterectomy who no longer had use for their ovaries. Using sperm left over from artificial insemination, he made eight hundred attempts to fertilize these eggs, eventually getting what he and Hertig believed were three embryos.[11] It is likely that none of these embryos was normal – they probably just looked as if they had undergone cell division, having degenerated because of the inadequate culture conditions that Rock had at his disposal.

At this stage, the problem was that nobody knew what was needed for embryos to survive in a Petri dish. Rock could not provide the conditions to help sustain embryonic life, and had no mechanism to transfer the eggs safely to the mother's uterus. These considerations did not stop hundreds of infertile women writing to Rock, begging him for help. Once Rock had published his work, fertility clinics in the USA, spurred on by the surrounding publicity, multiplied – there were 52 in 1949, 119 in 1955 – even though the technology to produce live births from IVF did not exist.

John Rock lived to the ripe old age of ninety-four, and I had the privilege to meet him once in the last decade of his life. Curiously, at that time I had no idea about his struggles with IVF and probably would not have been particularly interested even if I had known he had pioneered such work. What interested me about him was that he was a great exponent of surgery of the fallopian tubes, a field in which I had immersed myself in the 1970s, believing that IVF would never work. But at our meeting he showed only minimal interest in my attempts to introduce microsurgery into the field, preferring to enthuse to me about the potential of IVF.

By the late 1940s the most active centre for serious research into in vitro fertilization in mammals was undoubtedly Cambridge University, where numerous distinguished biologists were looking at the technical problems involved, albeit only in animals. A big breakthrough came here when Professor C. R. Austin and Dr M. C. Chang discovered that if sperm were placed in the fallopian tubes of rabbits shortly after ovulation, very few eggs were fertilized. If, however, the sperm were placed in the tubes several hours *before* ovulation most, if not all, the eggs were fertilized. This suggested that a special process was going on to potentiate the sperm – these workers called it 'capacitation' (see pages 502–3).[12] Soon after this discovery, a number of scientists around the world found that they could fertilize eggs artificially, outside the body of the animal. By the early 1970s people were able to reproduce the conditions the sperm needed. One scientist from Cambridge, Dr David Whittingham, who went on to develop embryo freezing in mice – a technique that has since been a boon to thousands of human couples – achieved in vitro fertilization in that species after capacitating the sperm artificially in 1968.[13] This was a highly important experiment which must have influenced Robert Edwards's work.

One great leap forward was made earlier by Dr Chang. A Chinese scholar who later worked in the USA, M. C. Chang was the ultimate pure scientist with no interest in the sensational aspects of IVF – in fact, he was not concerned with the idea of treating a human patient at all. But in 1959 he had published a crucial paper in *Nature*, the world's leading scientific journal.[14] In the experiment he described, he had first stimulated the ovaries of a number of rabbits by giving them an injection of sheep's pituitary extract. He collected mature eggs from these animals and then harvested sperm from the uteri of other rabbits that had previously been mated (so as to be sure to achieve capacitation). He then cultured the unfertilized eggs and spermatozoa together at body heat for four hours. After this, the fertilized eggs were kept in warmed flasks containing serum (derived from rabbit's blood) for a further eighteen hours. The embryos were then transferred to the uteri of six rabbits – and, remarkably, four of them delivered a total of fifteen healthy babies.

Chang was brilliant, modest and hard-working, undoubtedly one of the most deeply impressive men I have ever met – but this gentle individual lived in a world of his own. When I met him he seemed very ancient: a small, wizened Chinese, resembling my mental picture of Confucius. I remember being collected by him in his battered old car when I made a visit to his laboratory in Worcester, Massachusetts. My immediate impression was not reassuring – he seemed too old and preoccupied to be allowed behind the wheel on a public road. He was expecting guests that evening and invited me to join him for the Chinese meal he was preparing. He drove me around furiously on a shopping expedition, mounting the kerb inadvertently two or three times – and twice we stopped because some vital bit had dropped off the back of the vehicle. He visited several butchers en route, and eventually found a shop with which he seemed content. It was selling suitable chickens; but he complained that a whole chicken was far too big to feed the sixteen guests he was expecting. Eventually he settled for a half-chicken. When all his guests arrived that evening there was a sumptuous repast ready. By the end of the meal, all were replete – there was actually a huge amount left over – and I was the only vegetarian present.

The next day, Chang gave a seminar. His lectures were a nightmare: brilliant, humorous but quite unintelligible, he would continually shoot off at tangents down obscure avenues where nobody, I think, could really follow him. But his published achievements were far greater than those of anyone else in the field, so he was never criticized – even if, as I suspect, nobody fully understood what he was saying.

By the 1970s, undoubtedly, the world was ready for some of this work to be translated into clinical use. Throughout the centuries, medical treatments have been developed to meet specific needs and circumstances. Our understanding of neurology, for example, and ability to treat head and spine injuries, progressed dramatically because of the numbers of wounded men returning from the First World War. So, although people had probably been experimenting with artificial insemination for centuries, IVF treatment found its place in the culture of the post-war world. Religious beliefs seemed to be on the wane, and people increasingly felt fewer qualms about abortion and contraception. As the power of organized churches receded in the west, governments stepped up their involvement in people's lives, and one facet of this trend was the introduction of far stricter controls on adoption. So, in the post-war years, it became increasingly difficult for infertile couples to adopt – both because the number of babies available was diminishing and because governments were making it such a difficult process.

My mother, Mrs Ruth Winston-Fox, was a pioneer in this area, at one time placing more babies for adoption than any other social worker in the UK. She was determined that regulation should not be too bureaucratic, but also felt she had a responsibility to ensure that the many babies born in the decades after the Second World War were properly cared for and adequately nurtured. During her professional life, she saw the number of babies put up for adoption gradually dwindle as the stigma associated with having a baby out of wedlock began to fade.

Scientists should be interested in the pursuit of knowledge and understanding at all times. But they do not stand outside the societies in which they work. The time was right for IVF, not just because of the advances that had been made in research, but because of what was going on outside the laboratories.

Rebels with a cause: pioneers of IVF

Things moved along rapidly in 1968 when Patrick Steptoe, an obstetrician working at Oldham Hospital, got in contact with Robert Edwards, suggesting they collaborate. Edwards had been working on animal eggs in the same department as David Whittingham at Cambridge and was desperate to gain access to human eggs. I have the strong impression that most of his contemporaries may have been very dismissive about his ambitions to repeat IVF in humans. Perhaps they felt this was morally outrageous, perhaps they thought he was an upstart, perhaps they just thought that doing anything in the human was getting one's hands scientifically dirty. This last attitude was fairly widespread when I was a young doctor in London – and certainly Cambridge was always viewed as a bit of an ivory tower. In any event, it seems Edwards's medical colleagues at Addenbrooke's Hospital were disinclined to help – so he turned to institutions outside Cambridge. For example, for a time he came down to my institution in London, the Royal Postgraduate Medical School at Hammersmith Hospital, then a leader in much medical research in Britain and a place where people have always been inclined to push the boat out and do 'outrageous' things.

I am told that at one time Edwards used to bring a specially prepared culture medium in a flask and sit quietly in the surgeon's changing room in what was called Lower Theatre (it has since been demolished). He might wait patiently for hours while the consultant surgeon – subsequently my boss: Mr William MacGregor, a grumpy expatriate Australian – would decide whether or not to bother to give him any ovarian tissues or the contents of follicles. I did not arrive at Hammersmith until at least two years later, but I soon encountered Mr MacGregor. He was an excellent surgeon technically, but fairly dismissive about infertility and its treatment. Later on, he called the infertility clinic I tried to set up at Hammersmith the 'Futility Clinic'. This attitude was widespread among gynaecologists in those days: few had much sympathy for what was thought to be an unnecessary branch of medicine. It was an attitude that had profound consequences when, years later, the new reproductive technologies were identified as requiring government regulation.

Inevitably, after a number of wasted journeys, Edwards became disenchanted with endless visits to institutions like Hammersmith and empty-handed returns to Cambridge. One of his problems was that few in the medical establishment took him seriously, and many thought the idea of culturing embryos outside the womb both fanciful and reprehensible. Moreover, in those days – this was forty years ago – medicine was a very authoritarian profession and jealous of medical independence. Scientists wanting to waste a busy and important clinician's time were generally treated with disdain.

It was at this stage that Edwards met Patrick Steptoe. Steptoe was an intelligent, vigorous, determined man and it puzzled many people why he had ended up as a consultant in what was a bit of a backwater, in depressed Oldham. He was also a gifted amateur musician, playing the piano notably well in his student days in Oxford. He was clearly able and more than clever enough to merit appointment as a consultant to one of the London teaching hospitals (then the dominant force in British obstetrics and gynaecology, as well as the places where the rich pickings of private practice were to be had – though some might question whether London consultants are indeed particularly innovative or bright). My impression was that Steptoe's forthright views, and his rather aggressive manner in stating them, had annoyed the dominant medical establishment.

Some years later, but still well before Steptoe's attempts at IVF had become successful, I went to a clinical meeting at the centre of British gynaecology, the Royal College of Obstetricians and Gynaecologists in Regent's Park. In one memorable session, Steptoe presented truly mind-boggling pictures of what he could photograph down the laparoscope. This was revolutionary stuff – a telescope that could actually be inserted through the navel and record, in great detail, beautiful coloured images of internal pathology. For the first time, the surgeon could make an accurate diagnosis of what was going on in the abdomen. Before this, doctors either just did a totally inadequate manual examination, or – in more desperate or puzzling cases – opened the abdomen with a large incision to see what was going on inside. Steptoe's approach, which he had learned from Raoul Palmer in Paris (not only a foreigner, but a Frenchman, and therefore to be doubly distrusted), was truly a revolution.

I sat back in my seat, riveted by what Steptoe was saying. It seemed incredible that pictures of mysterious internal diseases like tubal adhesions, ovarian tumours and endometriosis could be seen with such clarity. At the end of the talk, Steptoe sat down – sweating and justifiably excited by what he had presented. After a pause, the chairman of the session, Professor John Bonnar, refused to take questions from the audience. He simply said, 'I don't need a laparoscope to diagnose endometriosis – I can feel it with the eyes at the tips of my fingers.'

I suppose it is no surprise, then, that these two innovative 'rebels', Patrick Steptoe and Robert Edwards, eventually teamed up. For Edwards, somebody who could gain expert access to the ovary at any time during the menstrual cycle, without doing a major operation, was clearly valuable. For Steptoe, with his genuine concern for infertile women and the idea of working with a really brilliant scientist and doing something truly epoch-making, I imagine the collaboration was irresistible.

In 1969, using the laparoscopic method of egg collection, by which individual follicles could be sucked out, Edwards and Steptoe managed to obtain a number of eggs. The challenge then was to fertilize them. It was clear that one problem lay in keeping the eggs alive outside the human body. In this respect, Robert Edwards had the advantage that scientists at this time were trying to work on these issues in the mouse. Anne McLaren and David Whittingham, as well as other Cambridge colleagues, provided useful information.

Anne McLaren is one of Britain's very best scientists and our leading embryologist. When I went, extremely timorously, to ask her advice in 1979 as to why my very primitive attempts at fertilizing the human eggs I was collecting at Hammersmith were getting nowhere, she looked at me quizzically and said she was sure 'it was something to do with the sunspots'. It was a long moment before it dawned on me that she was trying to be reassuring, and that in vitro fertilization was at that stage open to all sorts of variables – including luck.

Robert Edwards put effort into improving the recipe for the culture medium, the liquid in which the eggs are matured. Meanwhile Patrick Steptoe refined the methods of egg collection and tried to find ways of timing precisely when ovulation might occur – that is, when the eggs were most likely to be ready for fertilization. In 1971 he and Edwards applied to the Medical Research Council (MRC) to get support for their research. The grant was refused. Apparently, the MRC had 'serious doubts about ethical aspects of the proposed investigations in humans, especially those relating to the implantation in women of oocytes fertilized in vitro, which are considered premature in view of the lack of preliminary studies in primates and the present deficiency of detailed knowledge of the possible hazards'. So Edwards and Steptoe courageously persisted without research funding. Some authors have been heavily critical of them for pursuing this work at this stage, but I presume they had full approval from the authorities at Oldham General Hospital. Steptoe was certainly convinced that the patients he wished to treat would not get pregnant by any other means. Eventually their persistence was rewarded, and in 1973 their work led to the first pregnancy from IVF – but unfortunately it was ectopic. They partly succeeded again a year later; but this time the pregnancy spontaneously aborted at an early stage.

While Steptoe and Edwards, ably assisted by Jean Purdey (the third original member of their team), were having tribulations in Manchester, significant progress was being made in Melbourne. Australian universities have outstanding science and their interest in reproductive biology, possibly deriving from the country's economic reliance on sheep breeding, made Australia a formidable force in the field from an early stage. Professor Carl Wood at Monash University in Melbourne was becoming convinced that in vitro fertilization might be better than tubal surgery in dealing with the common problem of blocked fallopian tubes.

At the time I was developing microsurgery in London, Wood attempted to make an artificial tube – a silly idea and a signal failure as it turned out. I say dismissively 'a silly idea' because the fallopian tube is a very complex dynamic structure that actively transports the egg towards the uterus while helping to propel sperm in precisely the opposite direction. Moreover, it helps capacitate the sperm, feeds the egg and embryo, and determines the precise timing of the embryo's entry into the uterus. The idea that this multifunctional organic mechanism could be replaced by an inert piece of plastic seemed ridiculous. More promisingly, Carl Wood also attempted tubal transplantation in an animal model as a way to get round the problem of irreversible tubal damage, publishing his paper in 1980 – about five years after John

McClure Browne and I had published successful tubal transplantation experiments in rabbits. But he had sufficient foresight to become increasingly convinced of the potential of IVF, and this became his main focus. In 1972 he had appointed his own embryologist, Alex Lopata, to a post at Monash, and had been consistently trying to do similar work to Edwards and Steptoe. Over the next few years, the Australian team implanted a remarkable number of embryos – between thirty and fifty a year between 1974 and 1978. But there were serious snags. By 1978 they were witnessing embryo formation in just 10–20 per cent of fertilized eggs; and many of these either failed to grow beyond a certain point or were seriously abnormal.

By this stage, the work had become a collaborative effort between the doctors at Monash University and a team led by Ian Johnston, a highly competent clinician, at the University of Melbourne. Sadly, the competition among the various team members would lead to problems.

The procedure used in Melbourne was different from that applied in Oldham. Wood and his colleagues were convinced that it was best to stimulate the ovary with drugs to mature the eggs. Steptoe preferred natural, unstimulated cycles, timing his laparoscopic egg collection by regularly measuring the changes in the blood levels of oestrogen and LH. Steptoe and Edwards reasoned that allowing the eggs to be matured by a woman's own naturally produced hormones would be better than 'bombing' the ovaries with large amounts of FSH – the approach then being used at Monash.[15]

In the event, Steptoe landed on the moon in November 1978. Louise Brown, the first human to be born as the result of in vitro fertilization techniques, was delivered at 11.47 in the evening: the first live birth from human IVF, to a Bristol mother who had been trying to get pregnant for nine years and failing because of tubal damage. This blonde, blue-eyed baby was a bit small – 5 pounds 12 ounces – but she appeared to be a completely normal little girl.

In one sense, this hugely important success was a bit unfortunate. Collecting eggs after an unstimulated cycle requires much more careful timing than when stimulants are used. Also, stimulating the ovary gives the embryologists several eggs to attempt to fertilize. But because Steptoe and Edwards had been successful with a natural cycle, and not having much information, the Australians seemed to decide to mimic the Oldham approach – possibly against their better judgement. This delayed further successes and also meant that egg collections had to be done at unpredictable times, day or night, around the clock. All this made these rather obsessed and somewhat eccentric doctors doing 'futility' research very unpopular with hospital managers and back-up staff.

At this stage, a most significant figure joined the Australian team. As in so many scientific stories, for key progress to be made, the right person had to wander through the door. In this case, Mr Right was Dr Alan Trounson, a reproductive biologist who had been working in Cambridge. When he joined the Monash team around the time Louise Brown was born, he took a keen look at all the laboratory procedures, went through every one of them very methodically and cut out a great deal of nonsense – what he called 'bullshit'. In my view Alan Trounson is one of the great heroes of the whole IVF story. A formidable intellect and an equally formidable drinking companion, Alan is generous in the extreme with scientific information, an inspiration to older colleagues, and greatly solicitous of younger people and students who wish to learn.

Trounson made critical changes to the way IVF was being carried out. Having improved the composition of the culture fluids, he introduced quality control procedures so that there was absolute consistency of media preparation. He tested every part of the procedure in laboratory mice first. He reintroduced, against considerable opposition, the use of ovarian stimulation. He also had noticed that, rather like sperm undergoing capacitation, harvested eggs continued to mature for some hours after being taken from the body. So he instituted a five- to seven-hour delay between harvesting and fertilization, ensuring that embryo creation was happening at the optimum moment. These improvements gave Australia the honour of producing the world's third IVF baby in October 1979 (the second had been born in the UK, once again assisted by Steptoe and Edwards). Thereafter, the Australians began to outstrip all others in success, and Trounson went on to make a series of signal contributions in the field of IVF that were of key importance.

Scientists are human, and pretty well all early IVF practitioners needed to be dedicated, pushy, determined, self-confident – even a little arrogant – to continue their endless attempts to do something that constantly seemed to have all the odds stacked against it. So, though sad, it is not surprising that there was a degree of rancour after this Melbourne success. The happy mother, Mrs Reed, had been a patient at the Royal Women's Hospital, for the simple reason that it was larger and had better facilities than the Queen Victoria. But after the birth, some members of the Monash–Queen Victoria team seem to have felt that their counterparts had stolen all the publicity. Feelings became worse when one of the team published an account of the pregnancy, crediting no other members of the Monash–Queen Victoria team apart from himself, but mentioning several of their colleagues at the Royal Women's Hospital. In June 1980 the regrettable, but probably sensible, decision was taken that the two groups should split.

There is a happier postscript to this story, however. Having felt rather shoved out of the limelight in the aftermath of the first birth, the Monash team, with Trounson as its scientific director, pursued success vigorously. Accordingly, though starting later, they were responsible for the majority of the ninety-two IVF births in Melbourne over the next three years, making a number of important innovations – and in 1983 became the first to deliver babies from donated eggs and frozen embryos.

I remember visiting Alan Trounson in about 1980, when I was briefly invited to Monash as a visiting professor. My most vivid recollection of him is in the laboratory, which he had completely overhauled and which was completely sterile. No white coat, no ceremony, no arrogance and no false claims. He was boyishly examining some mouse embryos (at least, I think they were mouse) down the microscope, a cigarette hanging out of the corner of his mouth. Periodically, the ash at its end dropped into the Petri dish under the lens.

I remain exceptionally fond of Alan. One of his most endearing features – the mark of a real scientist – is the generosity with which he shares his thoughts in a very uncompetitive manner. Many years ago, he flew to London to attend a meeting in Manchester where much of the science being presented was rather low-level. He never openly showed his frustration, though I think he felt querulous about certain contributions from some of the biggest names in British reproductive medicine. The meeting ended with a gala dinner and an evening where the alcohol flowed more abundantly than the science had during the preceding day. Afterwards the two of us decided on a brief nightcap and eventually found ourselves sitting on the floor of a hotel bedroom, propped up against a wall. The nightcap turned out to be roughly three-quarters of a litre bottle of duty-free whisky, which stood still unfinished at about four in the morning. Alan captivated me with ideas about early human development and innovative notions about the possibility of investigating human stem cells. We continued our seminar on the train to London quite early the following morning, finally draining the bottle queasily some miles from Watford. At Euston Station we reluctantly parted company, hugging each other tightly with tears rolling down our cheeks. I have not the slightest recollection of ideas permanently lost to science on that occasion.

Carl Wood, the head of the Monash team, had similar instincts of generosity. A man in whom compassion outweighed hunger for success, Wood became troubled that, elsewhere in Melbourne, needy women were losing out just because their doctors didn't have access to the ever-improving body of information about IVF techniques. Eventually, he rang up the rival team and told them everything. But I heard it said that he kept this conversation a secret from many of his own team members, being too embarrassed to admit to his indiscretion.

It is a matter of regret that such a spirit was not so forthcoming in the UK. I admit that I had been Luddite about IVF, and was wrongly sceptical about its potential clinical value. I felt that tubal surgery was much more efficient and less stressful to my patients. And, to be fair, the rumours surrounding Steptoe's valiant efforts suggested that they had treated at least thirty or forty desperate patients consecutively without success. I was reluctant to put the stressed patients I met through those dashed hopes. But towards the latter part of 1979 I began to realize that Hammersmith, at that time possibly the biggest infertility clinic in Europe, could not afford to ignore an advancing technology of such potential importance. It was also clear that undertaking attempts at IVF would have legitimate scientific value. So while I continued my interest in tubal surgery, I began to put together a serious team of people at Hammersmith with the idea of eventually offering a free IVF service. I was lucky that the funds for this were available by pooling the proceeds from our international practice in tubal surgery, for by this time we were in the fortunate position that fee-paying patients were coming to us from all over the world.

In 1980 I teamed up with Stephen Hillier, a gifted scientist who had been doing important research on the ovary and the hormones it produced. His experience with hormone testing, his scientific vision and expertise, and his absolutely single-minded approach to running a disciplined laboratory were priceless assets. Stephen now works at Edinburgh University, where he has continued to make the most important contributions to reproductive endocrinology.

It was obvious to Stephen that if we were to be successful we needed information from Patrick Steptoe about what he was doing. He and Edwards were by now establishing their private clinic at Bourn Hall near Cambridge, but we could not get an invitation to see their set-up. Letters to Patrick were not answered. In September 1981 the Bourn Hall group organized an international meeting on IVF, the first of its kind. The key workers from across the world were all invited, but we at Hammersmith were ignored.

I knew that I had seriously irritated Patrick Steptoe. Four or so years earlier, I felt he had meted out to me the same kind of treatment that I had seen him receive at that Royal College meeting when he had demonstrated his excellent work on laparoscopy. I had been invited to an international meeting in Oxford to present my work on tubal surgery in a rabbit model. The audience was highly engaged. Patrick was chairing the session, and when the time for questions came, he rounded on me, saying this 'is a clinical meeting, and work on rabbit tubes is of no interest to this audience. We expect clinical data.' I responded by saying that I had human data, but hadn't shown them because I had been specifically asked to talk about my animal research. The audience laughed and demanded time to ask questions. Patrick went puce in the face but had to back down while for the next ten minutes I showed some slides of my quite promising results in response to questions.

I am sure that Patrick Steptoe resented my criticism of IVF to the press and my emphasis on what seemed to me (wrongly) to be its limited potential. Moreover, I suppose, advances in tubal surgery were a form of competition for him. I am not aggrieved at his combativeness – indeed, I believe that he and Edwards should undoubtedly have been awarded the Nobel Prize for their achievement. I also think that without Steptoe's aggressive single-mindedness, and Edwards's intelligent and equally dogged commitment, their great success would not have been possible. And both of them were continually subjected to the harshest criticism, mutterings and sniping from the medical establishment, even some time after Louise Brown's birth.

After the Bourn Hall meeting, to which neither Stephen Hillier nor I gained an invitation, Stephen insisted I took every opportunity to apologize to Patrick for what Stephen thought was my aberrant behaviour. At one meeting I subsequently tried to do that in a very public way, embarrassing myself in front of several colleagues. But it was hopeless – and thereafter Steptoe and I hardly ever exchanged a word; whenever we did communicate it was by formal letter only, about individual patients. But Patrick did Stephen and myself a favour. Because we could not visit the lab at Bourn Hall, we were forced to work from first principles. We now knew that IVF could work, and that was all the information we really needed. And in having to devise our own system, we avoided the trap of automatically making the same errors as others. Our IVF programme undoubtedly steadily became more robust in consequence, which is why Hammersmith became, and for a long time remained, the most successful British clinic offering IVF.

The story of IVF glitters with brilliant contributions from several countries and various branches of science and medicine. I can give only a partial account. But Jacques Testart, whose work led to the birth of Amandine, France's first IVF baby, in February 1982, should certainly be mentioned. Among his many contributions was developing a more accurate way of testing a woman's LH levels. A thoughtful and humane doctor, he was one of the first to observe and comment upon the emotional journey undergone by men during an IVF procedure. He commented – and I have been unable to ignore his words ever since – that a man is at a distinct disadvantage because he can't feel the baby inside him, as the mother can. Ultrasound, he argued, was not just a diagnostic tool, but a vital way for expectant fathers to bond with their babies.

In December 1987 Testart issued a statement saying that he would perform no more IVF procedures. He claimed that many of his colleagues were taking part in a sustained campaign of disinformation, not telling the public how slim the chances of successful pregnancy and birth really were. A former communist and committed atheist, he had also come to feel, along with many religious people, that there was something deeply wrong with the technology. In his view, it was the beginnings of eugenics – of parents being able to select desirable characteristics in their offspring, with the richest being able to create an *über*-class of strong, beautiful, intelligent, flawless children.

Some political and social issues

In 1716 Londoners were bemused by the circulation of a pamphlet called *Onania*, which devoted itself to listing the dangers of masturbation. It was full of pitiful testimonials from young men who had over-indulged and now lay on their deathbeds, suffering from a range of complaints such as gonorrhoea and epilepsy. Hope was at hand, however, for readers prepared to take heed – a 'Strengthening Tincture' was available from a certain London pharmacist, at ten shillings a bottle.

Although *Onania* was a sophisticated marketing gimmick, it drew on attitudes that continued to influence people until very recently. Kant and Voltaire, free-thinkers in many ways, considered masturbation to be harmful. As late as 1994, the US Surgeon-General faced calls to resign when she tentatively suggested that US schoolchildren should be told masturbation was safe and even healthy.

Onania is thought to be the first recorded use of the term 'onanism' with reference to masturbation. This comes from the Old Testament story of Onan,[16] who was required to sleep with his dead brother's childless wife. Onan withdrew from her vagina before he ejaculated, spilling his seed upon the ground – for which offence he was struck dead. Onan was observing a custom later known as Levirate marriage, practised not just among the ancient Israelites, but by the Mongols and Tibetans too. When an elder brother dies, the younger brother marries his wife so that his genetic line may be continued. In biblical times, any resulting child was regarded as that of the elder brother; in some cultures, it is considered to have been created by the dead father's ghost. So the story of Onan was not really about masturbation at all – even though it gave rise to a now obsolete term for the practice. Nevertheless, the idea that sperm should be expended only in the context of a married couple creating babies certainly influenced centuries of western thinking. Even today, it has an impact on the way some people view IVF treatment. Indeed, some people – because of powerful religious feeling – refuse IVF because collection of sperm generally involves masturbation.

Even before the birth of Louise Brown, there was an element of disquiet about the moral implications of in vitro fertilization. But because most people felt that Edwards and Steptoe were highly unlikely to succeed in their attempts at human IVF, few discussed the ethical implications seriously. It is interesting that one of those few who did was Robert Edwards himself, who wrote an important paper on the subject several years before the first birth. He foresaw very clearly the objections that 'right-minded' people might have if he and Steptoe were ever successful, and wrote to my mind the best defence of what he was attempting well before his work came to wide public attention.

The mutterings by my senior professional colleagues against Edwards and Steptoe were often unpleasant. I remember one incident at a massive congress in Buenos Aires in 1974, where Edwards presented some of his work and showed photographs of early cleaving human embryos he had cultured in his laboratory. As I left the lecture theatre, I encountered one of the most senior figures in reproductive biology in the world, also leaving the presentation. As this distinguished Cambridge professor passed me, he said, 'Well, what do you think of that, then? All those so-called embryos were dead, don't you think?' As a young researcher with no expertise, I was flattered to be asked, but I can't help suspecting that his reluctance to credit what he had just heard and seen was led as much by moral concern as by any scientific objection.

But the real furore began in 1978, as the world woke up to the reality of IVF. A number of physicians both in the UK and in the USA (undoubtedly led by a sense of outraged morals in many cases) cruelly claimed that Edwards and Steptoe had not been successful. In their 'considered' judgement, the pregnancy was spontaneous and occurred because Mrs Brown had had sex during her treatment cycle. But in the main, the public response to Louise Brown's birth took the form of massive congratulatory headlines in the newspapers, mixed with grave comment about the morality of what the doctors had done. At the University of Chicago, biologist Leon Kass declared to the press: 'This blind assertion of will against our bodily nature, in contradiction of the meaning of the human generation it seeks to control – can only lead to self-degradation and dehumanisation.'[17] His words were echoed by the Protestant ethicist Paul Ramsey, who had written, 'Men ought not to play God before they learn to be men, and after they have learned to be men, they will not play God.'[18]

Meanwhile, feminists split into opposing camps around the issues raised by IVF. Shulamith Firestone declared that the technology was a major step in releasing women from the constraints of biology.[19] Her views were opposed by some other feminists, who argued that IVF was the result of collaboration between the male ego and market forces and perpetuated the idea that women could only fulfil their potential as humans through motherhood, and that they were dependent on a largely male elite of scientists and doctors to achieve this.

In Britain, Australia and, subsequently, the United States, governments launched long-running and high-profile inquiries into the ethics of IVF, reaching an array of widely conflicting conclusions. In July 1982 the British government set up the Committee of Inquiry into Human Fertilization and Embryology, chaired by Dame Mary Warnock. Its report, published two years later, in July 1984, favoured IVF, but the committee recommended that, because there were so many unresolved ethical and medical issues surrounding the technology, a statutory licensing authority be established to regulate research and infertility services involving IVF. Some of those serving on the committee were opposed to surrogacy, which had been one of the topics the committee discussed. The real concern, though, was the question of research on the human embryo, and shortly after the report was published a huge campaign was organized to stop it.

Many of those opposing research were also totally opposed to IVF. Members of Parliament were subjected to a barrage of mail and petitions calling on them to demand a ban on all research. The media activity was massive: newspapers excelled themselves in moral outrage, and every television and radio channel had daily debates about the iniquities of IVF. This was an ideal situation for a populist piece of legislation; and sure enough, in November 1984 the Member for South Down,

Enoch Powell, tabled a Private Member's Bill – the Unborn Children (Protection) Bill. Had his measure passed into law, embryo research in Britain would have been halted, and I think IVF would never have continued as a viable treatment.

Powell's Bill was debated in stages in the House, always in a full chamber. But it did not get further than its report stage in the spring of 1985. It eventually ran out of time because of a remarkable filibuster organized by that old lag of a parliamentarian, Dennis Skinner. Enoch Powell was acknowledged as the great master of parliamentary procedure, and must have thought that afternoon that he had an important victory firmly in his grasp. But Skinner outmanoeuvred him by suddenly moving the Brecon and Radnor by-election writ. Moving a by-election writ takes precedence over other parliamentary business, so Powell was left fuming and white-faced in his seat, waiting to start his final debate. Skinner's speech took nearly two hours to deliver, during which time he explained eloquently how the whole country was consumed by one overwhelming ethical issue, a grave matter that was filling the minds of every moral, thinking person in Britain – the fact that the good people of Brecon and Radnor were not represented in Parliament.

This was one of the most remarkable parliamentary performances I have ever heard. Afterwards, I asked Dennis Skinner how he had felt. Dennis is well known as a hardened political bruiser, but this blunt, tough politician turned to me almost with tears in his eyes, saying that he hadn't slept a wink the night before, so important did he think the subject and so anxious had he been that he might let so many deserving people down.

But all this emphasized the fragility of our position. A rigid, anti-libertarian Bill had been defeated; but it was obvious that the measure had had a massive degree of support in Parliament. Around three-quarters of the membership of the Commons had voted for it on its second reading. A great deal of work would be needed if we were to get public opinion on our side. The 'pro-life' groups were deeply opposed to our work and were doing their best to destroy it.

As IVF gradually became established, it became increasingly obvious that it offered profound benefits to many ordinary people. Infertility is a very common problem – and this fact worked in our favour. Around 10 per cent of the UK population experience difficulty in conceiving at some point during their lives. So it turned out that a large number of MPs and members of the Lords either had had problems themselves or had a family member who had suffered from infertility. And by 1990, at least sixty thousand patients across Britain had received this treatment. So it became a little easier to fob off hostile comments about IVF, destructive Early Day Motions and mini-debates in both Houses.

Three years after the Powell Bill was introduced, while there were still regular complaints in Parliament about the failure of the government to bring in legislation to regulate IVF, the profession decided to show good faith by imposing self-regulation. It set up the Voluntary Licensing Authority (VLA), to which all IVF clinics voluntarily submitted themselves for inspection, and published a code of practice, which all clinics undertook to observe. This body was supported by professionals, and funded by the MRC and the Royal College of Obstetricians and Gynaecologists. At the time I was strongly in favour of setting up the VLA, which took wise decisions and ruled sensibly by consensus. A key factor in its success was the excellent chairmanship of Dame Mary Donaldson who, though firm and strict, operated with a light touch and a sense of humour. Another feature of the authority was the outstanding quality of the advice it was given by its eminent scientific members, people like Sir Douglas Black, Geoffrey Dawes FRS and Anne McLaren FRS, who were universally respected by professionals. When one member of the medical profession who ran a private clinic in London challenged the authority over a relatively minor decision, he got short shrift from every other colleague and eventually backed down.

The VLA gave itself six terms of reference:

1 To approve a code of practice for research into, and medical treatments for, infertility.
2 To invite all centres involved with medical practice or research to apply for a formal licence.
3 To undertake to visit every licensed centre annually.
4 To report annually to the Research Council and the Royal College.
5 To publish information on centres, including work approved and not approved.
6 To contribute to public debate and the legislative process.

Note, too, that IVF was not singled out: all assisted reproductive technology was covered. It is not surprising that this excellent body was eventually to provide the model for the government's own Human Fertilization and Embryology Authority.

When the government had still refused to announce legislation in 1988, three years after Powell's Bill had foundered, the subject became an issue in the debate on the Queen's Speech. Several MPs asked the government why no legislation was forthcoming. Ann Widdecombe,

MP for Maidstone, in particular, castigated the government for failing to take action. Peter Thurnham, a strong supporter of IVF (and an MP who had fought highly effectively against Powell's hard-line views), said this:

I am pleased that at each stage at which we have considered such legislation the number of those who have supported the view of Mr. Enoch Powell has declined. The opinion polls that have been published show that a majority of the public support research when it is designed to prevent congenital handicap or bring other such benefits. Marplan and NOP polls were conducted in 1985, and I feel sure that if an opinion poll were carried out today, a greater number of the public would support research, given that so many more children have now been born as a result of IVF.

In its latest report, the Voluntary Licensing Authority shows that 44 centres have been approved. I challenge those who oppose research to say exactly what harm it has done ... It is only because those who oppose research are unable to reconcile conflicting principles in their own minds that they have been unable to accept the benefits that have accrued from research that has been satisfactorily controlled by the Voluntary Licensing Authority.

Many organisations have come out in support of research. Progress – the campaign for research into reproduction – has been supported by 50 different organisations ... My hon. Friend the Member for Maidstone talked about humanity, but I ask her where is the humanity in forbidding the research ...

As the temporary arrangement has worked so well and is the envy of so many other countries throughout the world, I ask the Government to continue to examine the work of the organisation and to remember that it has done well and achieved the results that we wished it to achieve. The Government should not rush into legislation on the basis of a knee-jerk reaction to unfounded fears expressed by the opponents of research.

This speech represented a glimpse of victory. There was no doubt that public opinion had started to turn in our favour. Peter Thurnham was voicing an increasingly widely held view and there was every likelihood that, when legislation was eventually discussed, it would be reasonably liberal.

The government's own proposal for legislation, the Human Fertilization and Embryology Bill, was eventually introduced in the House of Lords by the then Lord Chancellor, Lord Mackay of Clashfern, in 1990. The subsequent parliamentary vote was influenced by a number of advances (see page 349). It was decided that the technology that had been developed – including the freezing and donation of embryos – did offer a permissible treatment for infertility. Moreover, embryo research was permitted up to fourteen days after fertilization of the egg. The resulting Act of Parliament established the Human Fertilization and Embryology Authority (HFEA), which commenced its work to regulate research, grant licences for treatment and publish the results in 1991. The passage of this legislation was by no means a smooth process, and was never a foregone conclusion. It was constantly hampered by the efforts of various Members of Parliament whose strong views led them to oppose in vitro fertilization as strongly as they opposed more radical research at the fringes of the discipline. Many of them were also strongly opposed to abortion but, curiously, in favour of capital punishment.

What about other countries where IVF was becoming a prominent therapy? In Australia, the federal system of government meant that different states arrived at their own decisions. In South Australia, clinics were licensed only to treat people medically proved to be infertile; in the state of Victoria, treatment was limited to infertile heterosexual couples, while New South Wales was more liberal. These discrepancies gave rise to what became known as 'fertility tourism', with couples moving around from state to state to get treatment. In France there was a relatively liberal attitude and no immediate specific legislation; moreover, IVF became a treatment funded by the state. In Germany, meanwhile, there was widespread hostility to the whole technology of IVF. All spare embryos were supposed to be transferred to the mother and pre-implantation diagnosis of genetic abnormalities (see chapter 7) was banned. In Israel there was no specific legislation, and IVF continued to be funded more liberally in that country than in any other jurisdiction.

In the USA, much of the debate was clouded by the historic *Roe* v. *Wade* decision, which led to the legalization of abortion across all states in 1973. Violent opposition to abortion led to similar feelings about embryo research, and caused the federal government to stop funding this work in 1974. A newly formed National Ethics and Advisory Board (NEAB) was presided over by the Health, Education and Welfare Secretary Caspar Weinberger, who made no secret of his own conservative views. A later NEAB president, Joseph Califano, was equally conservative, but recommended to Carter's government in 1979 that the funding resume. Even so, the issue was such a political hot potato that the government ignored Califano's advice.

Why were there such strong feelings against the new reproductive technology in the most scientifically advanced country in the world? I haven't made any measurements, but I would be prepared to wager that, if you made a pile of the literature *about* IVF, and stood it next to a pile of the literature *against* it, the two columns would be of equal height. The business of making babies arouses strong feelings in people – and not just those with strong religious views. But how is it that IVF was (and still is) considered to be so fraught with ethical difficulties, when other treatments, like neonatal paediatrics, obstetrics, skin grafts or chemotherapy, are not subject to any similar special regulatory treatment?

The chief arguments against fertilizing eggs outside the womb are as follows:

- It involves creating embryos – potential human lives – some of which will be discarded.
- It interferes with the natural processes of procreation and fertilization.
- It is a costly, unnecessary treatment for a condition that doesn't endanger life or produce physical pain.

Each of these points is important, and I would like to devote some time to discussing each in turn. But before that, it might be helpful to look at how we understand the embryo – and how that understanding has been influenced by our religious and cultural backgrounds.

Religious viewpoints: Judaism, Islam, Christianity

In Judaism, we find differing views about the status of embryos and fetuses and what may be done with them. There is, however, one fundamental principle on which the rabbis agree: full human status is not acquired until birth. A baby does not have full human rights until the moment its head emerges from the vagina. With an earnest precision that I admire, the rabbis also wondered what happened when a child came out feet first. In that case, they decided, human status occurred when the majority of the child had left the womb. The status – or the lack of it – enjoyed by the embryo is determined primarily on the basis of the biblical laws about murder. Jews consider that we are all made in the image of God; therefore human life is sacred and taking it is a capital offence. Among the laws set out in Exodus, we find the ruling that if a man strikes a woman and causes her to have a miscarriage, then he has to pay a fine. If, on the other hand, he strikes her and kills both her and the fetus, then he is a murderer and should be put to death.[20] The implication is that the unborn child has some status, but not the full status of a human being.

The late Chief Rabbi, Lord Jakobovits, pointed out that the difference in the Jewish and Christian views on this point stems from differing translations of the passage in Exodus concerning an assault on the mother that damages her unborn child. To the Jew, the Hebrew text is translated 'if there be no accident'. The Greek version of the same text yields a mistranslation, 'if it be without form' – that is, if the fetus has not yet reached human shape. Thus, up until that moment, there is no capital liability in Christian tradition, but thereafter, abortion is a capital offence. Essentially this is the line often taken by Roman Catholics today.

Fortunately for generations of rabbinical authorities – who made a living out of arguing – the situation is not always straightforward. One opinion in the Talmud tells us that formation does not occur until the fortieth day after conception; before that an embryo is *meah b'ulmah* – merely fluid. In other parts of the Talmud, we are told that the fetus is *ubar yerech imo* – that is, like the thigh of the mother, an appendage, rather than a separate life-form. This corresponds to the position later taken in

Roman law, namely that the fetus was *pars viscerum matris*.

One of the interesting ideas is when, precisely, the embryo develops a soul. The Talmud tells how children in their mothers' wombs joined with them in the Song of Moses, after the happy delivery of the Hebrews from the chariots of the Egyptians at the crossing of the Red Sea. Unborn children were also included in the agreement to follow the law as handed down from Mount Sinai. And there is a fascinating account in the Talmud of a prolonged conversation between the Roman Emperor Antoninus and Rabbi Judah the Prince.[21] (I presume that the Antoninus referred to is Marcus Aurelius, as Antoninus the Pius his predecessor died in 161, when Rabbi Judah was still a very young man.)

Antoninus asks the rabbi when he thinks the soul enters the embryo. At conception, or later, at the moment of formation? 'From the moment of formation,' replies Rabbi Judah. 'Not so,' says Antoninus, 'it must be from the moment of conception – can a piece of meat be unsalted for three days without becoming putrid? It must be from the moment God decrees its destiny.' At which the rabbi concedes the point, remembering that the Book of Job says, 'and thy decree has preserved my spirit [i.e. my soul]'. Later Antoninus asks Rabbi Judah when he thinks the Evil Tempter may hold sway over man. 'At the time of the embryo's formation,' replies Judah the Prince. 'No,' says the emperor, 'if so it would rebel in its mother's womb and go forth. But it is from when it issues [from the vagina].'

Rather more pertinent are the rabbinical laws which permit a person to break the Sabbath to attend to a woman in pregnancy, no matter how early. But even here the Jewish ruling is not absolutely clear – medieval rabbis seemed to incline towards the idea that intervention of this kind on the Sabbath was to protect the mother rather than the fetus.

Where does this leave IVF? Having children is a very important principle of Jewish law. 'Be fruitful and multiply' is the first commandment, and maintaining close and happy family units is also regarded as an important *mitzvah* – commandment – in Jewish law. While this principle could be taken to mean that destroying embryos prevents 'being fruitful', there are some strong arguments to suggest that Jewish couples have the right, and even the duty, to seek fertility treatment.

In Judaism, fertility treatment is seen as a medical treatment like any other. It is not singled out as it is, somewhat bizarrely, by current British legislation. The duty to preserve life, to treat people who are sick, to maintain health and to promote procedures that have a therapeutic value is of paramount importance. In Genesis (1: 28), man is to hold 'dominion over the fish of the sea, the birds of the heavens, and every living thing crawling on the Earth'. This 'holding dominion' or 'subduing nature' exhorts man to use his God-given intelligence to exploit the natural world, to use animals for service and for health – providing this is done with the utmost humanity at all times. The use of technology – 'subduing the earth' – is seen as a positive action, provided man uses judgement and understanding, and behaves with responsibility.

In strict Jewish law, the strong support for IVF basically depends on its being used for married couples. Judaism also has very strong concerns about genetic origins – that, for example, 'a son has a right to know who his father was' – so we find some quite different attitudes to the use of donated eggs, embryos and sperm. We will discuss these in more depth in chapter 5.

One area of interest concerns spare embryos. The

rabbis concede that a high-order multiple pregnancy is dangerous, recognizing that it is potentially so perilous to have triplets or quadruplets, for example, that an extra embryo for transfer, instead of being desirable, is a threat. Given that transferring three or more embryos may cause the death of the mother, each of these spares is seen as a *rodef*, a pursuer; and in Jewish law, self-defence is regarded as a moral necessity. So, destruction of a 'spare' or *rodef* embryo – or its use for experimentation – is not necessarily seen as reprehensible or morally wrong by the majority of authorities. Chief Rabbi Jakobovits was of the opinion that the motive for conducting experiments was of critical importance, but he was not in favour of creating embryos specifically for the purpose of experimentation.

In chapter 23, verses 12–14, of the Qur'an, the holy book of Islam, we find the following:

> We created man of an extraction of clay, then we sent him, a drop in a safe lodging, then We created of the drop a clot, then We created of the clot a tissue, then We created of the tissue bones, then We covered the bones in flesh; thereafter We produced it as another creature. So blessed be God, the best of creators!

Islamic scholars have argued that this passage indicates that a fetus is a human life, but one that acquires full human status only at a later point in its development ('thereafter'), when it is 'produced . . . as another creature'. For that reason, much scholarly debate has been focused on deciding when a developing life acquires the full status of a human being.

In Islam, the words of the Qur'an are supplemented by the *hadith* or 'sayings' of the Prophet Muhammed. Among these we find the assertion that 'Each of you possesses his own formation within his mother's womb, first as a drop of matter for forty days, then as a blood clot for forty days, then as a blob for forty days, and then the angel is sent to breathe life into him.' For this reason, some Islamic authorities consider that an embryo does not acquire the status of a human before the age of 120 days. Others put the figure at forty days – but in either case, there is no problem with the fertilization or discarding of human embryos. Also, like Judaism, Islam maintains that since all knowledge emanates from God, humans have a duty to seek it out.

Islam also shares with Judaism a strong concern with issues related to paternity, maternity, and inheritance of rights and duties from one's parents. It forbids not only surrogacy but also adoption. By extension, this stance means that there is nothing preventing the use of eggs and embryos for research purposes if they are not implanted into the uterus of their own mother – because Islamic law forbids their being used by other parents.

The Christian position on the embryo is divided, broadly speaking, into the Roman Catholic and the Protestant camps, which put varying interpretations on the same texts, commentaries and uses of law throughout the centuries. The official Roman Catholic position is clear, following the pronouncement in 1956 of Pope Pius XII that all attempts at in vitro fertilization must be rejected as immoral and unlawful. The Vatican argues, furthermore, that sexual intercourse and procreation are irrevocably linked. Sex without procreation is as forbidden as procreation without sex. It also holds that the embryo is a potential life from the moment of fertilization – and must therefore from that moment be accorded the same status and dignity as any human being. In a statement to biologists attending a conference (to which I was invited) sponsored by the Vatican in October 1982, Pope John Paul II said:

> I have no reason to be apprehensive for those experiments in biology which are performed by scientists, who like you, have a profound respect for the human person, since I am sure they will contribute to the integral well-being of man. On the other hand, I condemn, in the most explicit and formal way, experimental manipulation of the human embryo, since the human being, from conception to death, cannot be exploited for any purpose whatsoever.[22]

The general Catholic position, however, has been that experimentation on the human embryo may be permitted provided the embryo itself is not endangered, and provided the experiments are designed for the benefit of that embryo. This is a difficult position, it seems to me, because all experiments by their very nature must carry dangers, and it is obvious that an embryo cannot give informed consent. A meeting of Catholic bishops at the time of the Warnock Committee also argued that even observation of a human embryo that may endanger it is not allowed. Again this poses difficulties, because there may be a risk with prolonged use even of conventional wavelengths of visible tungsten-source light. These bishops were opposed to freezing embryos, too, unless there were 'genuine or definite prospect of subsequent transfer'. They also condemned any form of embryo selection with a view to transferring only 'the fittest and most desirable embryos'.[23] This would of course eliminate any prospect of pre-implantation diagnosis or aneuploidy screening.[24]

Protestants, with the exception of a few sects, argue that there is a solid justification for in vitro fertilization from the earliest writings of Christian commentators, and the earliest uses of Christian law. In 1988 Gordon Dunstan, professor of moral and social theology at the University of London, argued that, while life is sacred, in none of the monotheistic traditions is life given absolute protection at any stage, and that it would be 'morally odd' to claim an absolute right to life for an embryo or fetus. He also argued that the claim for absolute protection, which had been made by some Christians, was a recent novelty even in Roman Catholic moral tradition – a creation of the later nineteenth century.

Nevertheless, Tertullian (c.160–220 CE), writing in vigorous defence of Christianity against accusation of homicide, states:

> For us, homicide having been forbidden once and for all, it is not lawful to destroy what is conceived in the womb even when the blood is being drawn into the human being. To deny birth is to hasten homicide; for it makes no difference whether you snatch away the soul after birth or destroy it while coming to birth. Even the man who is yet to be is a man just as every fruit is already present in the seed.

This view was maintained by Basil, Bishop of Caesarea, who in the fourth century unequivocally condemned any destruction of embryos: 'A woman who deliberately destroys a fetus is answerable for murder. And any fine distinction as to its being formed or unformed is not admissible.' But his contemporary St Gregory of Nyssa disagreed: 'For just as it would not be possible to style the unformed embryo a human being, but only a potential one – assuming that it is completed so as to come forth as a human birth – so long as it is in this unformed state it is something other than a human being.'[25]

Thereafter, there is constant argument about embryonic status in Christian literature, and it would be out of place to go into huge detail here. But there are striking examples of the recognition that the embryo is not fully equivalent to a person. 'Books of Penance' written in the seventh century set out a list of penalties for various sins, and they applied lesser penalties for the abortion of a fetus if it was 'unformed'. This position would seem to be supported by the writings of St Thomas Aquinas (1225–74), one of the most authoritative theologians and philosophers of the Middle Ages. Aquinas argued that the human embryo did not possess a soul, and was therefore not a human being, and did not have the rights and status of a human being, until the moment of 'ensoulment' – believed to be forty days after conception for males and ninety days for females. He seems to have been influenced by the Greek philosopher Aristotle, who said that the fetus underwent three metamorphoses of the soul during its time in the womb, becoming first vegetative, then animal and finally rational. All of this evidence is sometimes used to support the view that IVF treatment is lawful in the eyes of the church.

It is a pity that religious positions have become so polarized. I have just reread Test Tube Babies – a Christian View,[26] published in 1983. In it, the most vigorous, hostile things are said about in vitro fertilization and a number of half-truths are offered about the science behind it. Writings and statements like this have not contributed to sane dialogue and have certainly not served either side in evaluating the complex arguments surrounding these issues.[27] It is sad that so much discussion in this area – particularly when it becomes virulent, as it sometimes does – seems to lose sight of the great liberal traditions in all religions. I remember learning Aramaic and Hebrew with my ageing grandfather, a great scholar and orthodox Jewish rabbi, when I was ten or eleven years old. One afternoon he suddenly broke off from his lesson on the soul to show me a passage from Dante, in his copy of *The Divine Comedy*, which I now own:

> 'Open thy heart now and the truth expect
> And know the fetus, once the brain
> Is shaped there in each last minute respect,
> The Primal mover turns himself, full fain
> Of nature's masterpiece, a work so fair
> And inbreathes a new spirit, which draws amain,
> Replete with power, all it finds active thee,
> Into its substance and becomes but one
> Quick sentient soul of it own self aware.'[28]

This poetic notion of human consciousness is surely important. We cannot be truly human until development has progressed to true viability. Life is truly a continuum which cannot be said to start at any finite point.

Scientific and medical points of view

This idea of a continuum is one key to this matter. I do not consider that we can say with any clarity at all when life begins. The sperm and egg are human and clearly alive, but they are not persons. Even the embryo, from a biological point of view, has nothing more than the *potential* to become a human being.

IVF is surely justified, but advances in technology should be used to minimize the potential wastage of embryos. I will be discussing embryo freezing in chapter 4, but for now I should like to explain why science, rather more than religious tradition, backs up my view.

In the first place, the assertion that the wastage of embryos is 'wrong' takes little account of what happens in nature. Fertility studies suggest that at least half of all the embryos that are conceived do not lead to a pregnancy. Given that so many naturally fertilized eggs fail to survive, we have to question whether an embryo should really be accorded the same status as a human being.

Second, an embryo before implantation in the uterus cannot be said to meet one of the foremost definitions of a human life. Opponents of IVF argue that each embryo is an individual life-form, and that, in discarding it, we are effectively throwing away a potential human. In other words, they argue that an embryo has a unique identity as a human individual. But this is not borne out by the facts. As noted in chapter 1, before implantation in the uterus each cell of an embryo has totipotential – that is, if we were to separate each cell, it could go on to divide by itself and create further embryos. Identical twins are separate individuals, but they are formed by the embryo splitting some days after fertilization has occurred. In other words, an embryo cannot be said to possess a unique identity.

Third, it is only once it has implanted in the womb that an embryo begins to behave as an individual. During implantation, the cells of the embryo begin to fuse with the cells of its mother's uterus. The embryo itself begins to control hormone activity within the mother – secreting chemicals to block her immune response; secreting HCG to keep the corpus luteum functioning; in most cases, triggering its own birth by producing fetal adrenocortical hormone. It seems to me, therefore, that implantation in the uterus might be a far more obvious point at which to start speaking of the individual.

Fourth, there is a clear distinction between being a

person and being a living being. Imagine you were on a heart transplant list, and received the news that you were about to receive a healthy heart. You would be overjoyed. But imagine you had some inoperable brain tumour and instead the doctors told you that, enabled by some giant leap in medical understanding, they were going to give you another person's brain. I imagine you might not be so delighted at the idea – because you would feel that, whatever happened, 'you' were not going to survive. It follows that there is good reason for our believing, like the religious poet Dante Alighieri, that consciousness is the key. The brain is surely our source of identity – and it is that combination, life plus identity, that deserves to be accorded the status of a human being. An embryo does not have this status, as before the primitive streak develops, at approximately fourteen or so days after fertilization, it has no nervous system.

A fascinating area of neurological and neonatal science concerns the development of the brain within the fetus. It is a subject that raises strong feelings. At around sixteen days after conception, a structure called the neural plate begins to develop. Over the next five days, this element widens to resemble a circle with a groove cut into the top. The groove then closes over, creating what we call the neural tube. The front end of this tube goes on to create the brain; the back end becomes the spinal cord. At the front end, the tube splits into three sections – forebrain, midbrain and hindbrain. At seven weeks' gestation, each of these three structures divides again.

So we are faced with the question: at what point does this brain become capable of registering identity? It is a question of more relevance, perhaps, to the debate about abortion; but it touches upon issues raised by IVF, not least because IVF can lead to a number of early pregnancy losses. Are we guilty of cruelty, if our treatment can cause suffering to fetuses?

It is an interesting consideration that consciousness provides a cut-off point in another branch of medicine. We may not take an organ from a living person for transplantation to another individual, even if the donor is already dying and the recipient will die without an immediate transplant. The key point at which this situation changes is when the potential donor is said to be fully brain-dead. Under those circumstances, and provided there is no possibility of recovery, a transplant may go ahead.

It is well known that it is difficult to register the brain activity of fetuses – and even doing so raises further ethical questions. The clearest evidence comes from studies into pain and stress levels in the unborn child, but it is by no means undisputed. We do know, however, that three things at least are necessary for a human organism to feel pain. There must be sensory receptors to register the pain; there must be nerves to carry the messages from these receptors to the spinal cord; and there must be nerve fibres within the spinal cord to carry the pain message up to a structure called the thalamus within the brain. Current research suggests that none of these is present before the seventh week of gestation. Some studies have put the figure much later, pointing out that the brain lacks the necessary connections between its various parts before twenty-six weeks. This assertion was challenged in 1999, by a study which examined levels of the stress hormone cortisol whenever a needle was inserted into the womb to take blood samples. In my own department at Hammersmith, Professors Vivette Glover and Nicholas Fisk found rising levels of cortisol in fetuses by the seventeenth week of development.[29] I am not fully convinced by the conclusions drawn from these findings, because we do not know whether a rise in cortisol is actively experienced by the developing fetus as pain. But they do suggest that brain activity associated with an identity begins only some considerable time after implantation in the womb. Therefore, I do not concur with the idea that an early embryo is a human life which must not, at any cost, be discarded.

A related argument of IVF's opponents is that the technology potentially allows embryos to be created solely for the purposes of research, rather than to be turned into human lives. Once again, given the very obvious differences between an embryo containing a few cells and a developing fetus, I do not believe that embryo research poses a dilemma.

But what about the wider argument that all of this is tampering with nature? A prominent strand of the Roman Catholic argument is that procreation and sex are linked by a law of nature, and that IVF interrupts that link. But the 'justification by nature' argument is flawed for a number of reasons. In the first place, it presupposes that everything 'natural' is better than that which is 'unnatural'. Yet under the heading 'natural' we could place the killing and consumption of abnormal babies by their mothers – something seen in various cuddly animal species such as rabbits and cats. Under 'natural' we might also place several naturally occurring poisons, such as digitalis, arsenic, cyanide, curare – the ingestion of which can cause a terrifying and agonizing death.

Second, the whole argument in favour of what is 'natural' rests on an idea that technology is something alien and hostile. I often wonder where this idea came from. Perhaps it began with the challenge of Darwin, who informed humans, to widespread horror, that we are very close to apes. Undoubtedly it developed into a more full-blown distaste for science and scientists after the Nazi experiments in the Holocaust and the dropping of atomic bombs on Japan. Whatever its roots, there is today a widespread basic belief that 'technology' is some sort of menace. But no machine ever started working without a human inventing it and switching it on. Technology is itself natural, being the product of human ingenuity. I might add that, at different stages in our history, the wheel and the axe and the bucket have all been examples of technology – but it would be hard to find any techno-phobes who see these inventions as a menace.

Those opposing IVF sometimes assert the unnaturalness of the treatment, yet does this not apply to other branches of medicine? Having a filling in your tooth is 'unnatural' in the sense that it is an intervention with the course of nature. Wearing spectacles is a similarly unnatural tinker-ing, as is taking an aspirin for a headache. Even if you chose to treat your infertility by gathering herbs under a full moon and brewing them into a tea, you would still be making an unnatural intervention, by introducing certain chemical compounds into your body that would not normally be there. It is true that some followers of the Christian Science sect reject almost all medical treatment on these grounds, and they are, not surprisingly, opposed to IVF. But most opponents of IVF seem to be quite happy about most, if not all, medical treatments – other than the ones that give people babies.

To me, there is also some faulty logic at the heart of the Roman Catholic view that sex and procreation are in-extricably linked in the way the Church maintains. If there cannot be sex without procreation, it is argued, then there cannot be procreation without sex. In the first place, it is by no means universally agreed that procreation is the only purpose of sex. Intercourse promotes strong bonds between husband and wife. In the second place, looking at it from a purely mathematical point of view, A Æ B may be true, but that doesn't mean B Æ A always is, too. For example, there can be no life without water. But there can be water without life. There can be no St Patrick's Day without Guinness. There can, as you will see if you step into your nearest hostelry on any day other than 17 March, be Guinness without St Patrick's Day . . .

Finally, let us examine a third hostile argument: the expense of IVF. Given the currently low success rates, it is advisable for most couples to try at least four cycles of treatment – a maximum of three of which, under current guidelines, may be funded by the NHS. The drugs, the monitoring equipment, the blood tests, the salaries of the attending gynaecologists, haematologists, ultrasound specialists, embryologists, nurses, etc., make IVF a hugely costly venture. In addition, it is argued, with so many IVF babies born pre-term, valuable neonatal facilities are being swallowed up by increasing numbers of people who have sought this treatment for a condition that is not life-threatening and does not cause them physical pain.

The last argument, I have to say, tends to be advanced only by people who have either never had children, or never experienced any difficulties in having children or met anyone else who has. No physician who has witnessed the anguish of infertile couples can doubt that the condition causes extreme suffering – even if the suffering is not conveniently located in the pancreas or the foot! If you would like a short introduction, I suggest you visit some of the popular websites for people with in-fertility problems. Most of these are completely devoid of baby photographs – or, if any can be found, they are tucked away in a clearly labelled sub-section of the site.

This is for the simple reason that infertile people, men and women alike, can go through mental agony when surrounded by reminders of other people successfully getting pregnant and having babies.

To argue that any mental pain is necessarily of a lesser order than physical pain is an obvious fallacy, since we have long recognized depression as an illness and spent large sums on research into its causes and possible cures. We also recognize that mental stress is a relevant factor in a range of physical conditions, from psoriasis to heart attacks and our immune response to infections. Medicine clearly accepts that 'mental pain' is real and that it should be treated.

I am not personally convinced that there are any solid objections to IVF on ethical grounds. The main reason for caution, in my view, is that the treatment has a low success rate and is not suitable for everyone with fertility problems. In both Judaism and Islam, the view is that treatment, within the context of a married man and woman trying to have children, is permissible and advisable. But times change, as does technology. Some people want the right to have children without a marriage, or even a partner. Others, married or not, demand the right to have children, even though some aspect of their biology means that the straightforward IVF treatments described in this chapter and the next cannot help them. What hope does science offer to such people – and should we be treating them at all?

3

IVF: When, Why and How

IVF is a contemporary tragedy unfolding in our country and else-where in the world. This tragedy could be called 'The Technical Child.' It is produced and directed by medical and research scientists and technicians. The main characters are the infertile couple, embryonic children, a doctor and his team of medical technicians, the scientific research community, and God, the Creator. To uncover the tragedy it is necessary to look at each of these as a personal subject and to examine the role they take and the actions for which they are responsible.[1]

So asserts the Catholic writer Kathleen Curran Sweeney in her vigorous polemic about IVF. But IVF today is a treatment used across the world. There are IVF practitioners not just in the rich countries of the west, but in most of Asia, all countries in South America and increasingly in Africa. It has been suggested that, around the globe, some one million babies have been born from IVF. What kind of 'contemporary tragedy' is it that results in giving life to those children, and enhancing the lives of their parents? Surely the real tragedy is that this treatment is beyond the financial resources of most couples. Even in Britain, with its much-vaunted National Health Service, this year fewer than four thousand couples will receive publicly funded IVF treatment. In the same year, another thirty thousand women will embark on this treatment in the private sector, and at least four times that number of women will be denied access to IVF because they cannot afford treatment.

One of the reasons why IVF is inaccessible to so many is that assertions about the 'evils' and 'tragedy' of IVF have made it seem reprehensible, a fringe treatment in a grey area of morality. Voices raised in moral outrage, like Sweeney's, have contributed to a reluctance on the part of health services in several countries – including the UK – to support this area of reproductive technology.[2]

But these are not 'technical' children. How can any child be *technical*? This canard has done a disservice to those sad couples who have sufficient love in them to stake everything, against the odds, for an ordinary child. I recently asked Amanda, the daughter of Beth Hornett and my first IVF baby, now married with two naturally conceived children of her own, whether she felt different from any other person. The answer was an obvious 'no' – and, more important, she had never felt different at any time of her life. None of her friends at home, school or university had ever seen her as anything other than normal, ordinary; she was simply somebody who was loved and liked, and had the usual human attributes, skills and weaknesses.

The Grand National of reproduction

Of course, IVF itself is a 'technical' process, and it is unpredictable. It is always uncertain how many cycles may be needed to give an individual woman a reasonable chance of successful pregnancy: despite its widespread use, the success rate is still, on average, roughly 23 per cent per cycle – a bit better in very skilled hands, a little worse in others.

But the technical aspect is ultimately not that important. A successful outcome is in huge measure due to chance. Undergoing IVF is mostly like embarking on a gigantic obstacle race, or a prize steeplechase. It is the Grand National of reproduction. Coming to terms with this element of chance is very difficult for many couples. Not surprisingly, when an IVF treatment cycle has failed, couples want to know why. But fertile people never consider for a moment why pregnancy did not happen after a single act of making love.

The first hurdle to cross is that of ovarian stimulation – there is perhaps 10–20 per cent chance of falling here. Then comes the relatively minor obstacle of egg collection. Perhaps 5–10 per cent of women do not yield eggs at egg collection. The open ditch of fertilization is next. On average, only 60 per cent of the eggs collected will fertilize – and occasionally, there will be no fertilized eggs at all. Next, although fertilization may occur, the embryos may be fragmented or dividing so poorly that they are not viable: perhaps 50 per cent of embryos may be visibly defective in this way. And following embryo transfer, we come to a fence as challenging as Becher's Brook and The Chair combined – for implantation after embryo transfer is the most chancy part of all. On average, only 18 per cent of embryos implant and become babies – the rest disintegrate. The successfully pregnant woman who is delivered of her healthy baby is truly a Grand National champion.

The first thing to observe is that failure at any one of these hurdles and fences may simply be due to bad luck on the day. But of course, it may also be due to one or another underlying factor, and the chances of this happening need to be evaluated. The chances of a fall – the odds against success – are increased by various factors. The age of the woman is the most critical. Women much over forty seldom have more than a 5 per cent chance of success, and are more likely to fail at any stage. And there is another chilling statistic: women over forty, if they do get pregnant, are more likely to miscarry. Around 50–60 per cent of women aged over forty-two have miscarriages after IVF. Severe tubal disease, particularly if the ovaries have been inflamed in the past, also lowers the chances of success in IVF. Women with dilated fallopian tubes – a so-called hydrosalpinx ('water on the tube') – seem to have a reduced likelihood of implantation, and some units recommend removal of the tubes before IVF if they are very dilated with fluid. But the indications for surgery must be considered carefully; sometimes tubes that might have been successfully reconstructed by a skilled surgeon have been removed unnecessarily.

Abnormalities of the uterus – such as fibroids, adhesions inside the cavity, congenital malformations and adenomyosis (see page 32) – all reduce the chances of success. But most of these conditions can be treated before IVF is undertaken, and where this is done the success rate can be substantially improved. Various constitutional problems, particularly obesity, can have a profound negative effect, and there is little excuse for a woman going into this treatment more than marginally overweight.

There are some factors that give certain women better than average odds. Women who are married to infertile men, provided the sperm fertilize, generally have a slightly better than average chance of success. And women who have previously had an unaided pregnancy – even if it has been ectopic or ended in miscarriage – are usually more likely to get pregnant with IVF. Menopausal women receiving donated eggs also stand a very good chance, because donated eggs come from young women and eggs from young women have a much better chance of fertilization and implantation.[3]

Although IVF is called a treatment for infertility, strictly speaking it is not. IVF does not treat the underlying cause of a couple's inability to conceive. So, if there is tubal disease or a low sperm count, for example, that remains the case after the treatment cycle – the IVF just bypasses the damaged tube or achieves fertilization by artificial means on that one occasion. IVF is a one-shot method for getting a baby. It is magical when it works; but, like most Grand National winners, that baby is certainly a triumph achieved against the odds.

There are many reasons why IVF has a relatively low success rate. As we have seen, humans are naturally infertile. The human female has only a fairly narrow window of time each month when she can get pregnant. Moreover, men frequently produce many aberrant sperm, many eggs have defects, human embryos often grow abnormally and die in the early stages of development, and many (if not most) embryos do not implant. Approximately one in five pregnancies ends in some form of miscarriage. All in all, then, there is a great deal of luck in getting pregnant. We do not know how many embryos fail to make it after natural conception, but there is some evidence (from routine early pregnancy tests done in a large number of women who were not using contraception) that at least 50 per cent of normally conceived human embryos do not survive more than a few days.[4]

This natural low fecundity means that couples not

using contraception have, on average, about an 18 per cent chance of conceiving normally. That, at least, is the figure for British women: it is a bit higher in Sydney, Australia, where there is about a 22 per cent chance of conception. But then, the Australians have a great deal of sunlight and a great many sandy beaches and they may even have sex more often. The evidence suggests that the more frequently you have sex, the more chance you have of a pregnancy. One study showed that couples having sex once a month have no more than a 5 per cent chance of conception in any one month.[5] Couples having sex more than twenty times a month have around a 35 per cent chance. So I am pleased to report that the figures from France offer the British a frisson of smug satisfaction. Although there is a Gallic tendency to boast about being the sexiest people on the planet, the chance of conception there averages about 16 per cent per month – though it might (I have no statistics) be a bit higher in Nice and Cannes.

Taken together, these figures mean that, roughly speaking, the average length of time it takes an average couple of child-bearing age to achieve an established pregnancy is about four to five months. Seen in this light, a 23 per cent success rate (the average per cycle in Britain) with IVF is actually a good deal better than nature.

When is IVF likely to be the best treatment?

It worries me greatly that too many people are advised to try IVF before first having a thorough investigation of the cause of their infertility. In so many cases of infertility there are various treatments that are more appropriate, less expensive and demanding, and more likely to succeed;

and even where IVF is advisable, the likelihood of its succeeding may depend on having a prior treatment in preparation for an IVF cycle. Also, adequate investigation beforehand may show that no amount of IVF treatment will result in a pregnancy, and it is pointless going through treatment if there is no real prospect of success.

Infertility is a symptom, not a disease. We would not dream of going to our doctor and demanding a coronary bypass because we had suddenly developed acute chest pain. The pain might be due to indigestion, a bruised rib-cage after a fall, gallstones, pneumonia, shingles, or a simple muscle strain after a game of tennis. Every day women are put through IVF without any understanding of what the underlying problem is. Free-standing IVF centres have encouraged this sloppy attitude among professional people who would be struck off the medical register in other disciplines if they did not attempt to establish a diagnosis before embarking on treatment.

This is one of my biggest concerns about IVF. It has been isolated as a special treatment, largely confined to the private sector in dedicated IVF clinics – or, in the case of many NHS hospitals, isolated from other treatments – and regulated separately from all other medicine. Consequently, IVF often tends to be practised in a medical vacuum. Few IVF clinics offer comprehensive treatment, such as expert ovulation induction or skilled tubal surgery. They are discrete units established solely to carry out IVF – so that is, for the most part, all they offer. It may seem an unpleasant analogy, but think of a garage: it is much simpler to have a mechanic replace a damaged starter motor or clutch than to use a skilled engineer to find out what is wrong and repair the fault. There is a huge tendency towards redundancy in our society because it is commercially the most effective way of trading; and, sadly, it is apparent in many IVF clinics. While many clinics pay lip service to the idea of offering comprehensive investigation and treatment, almost invariably patients (or, to use the regrettably fashionable current term, clients[6]) attending them will be offered IVF – irrespective of whether or not it may be the most appropriate treatment for their condition. It is troubling that the regulatory authority, the HFEA, which was established primarily to look after the interests of patients, has apparently ignored this aspect of IVF. Certainly it does little or nothing about it. Perhaps it cannot – one reason among many why this kind of regulation is pointless (see pages 289–295).

Most couples considering IVF are left wondering whether it is the best treatment for their situation. So when is IVF most likely to help? There are a large number of couples for whom IVF is likely to be the most effective treatment. Knowing that there are clear indications for this treatment is an excellent reason for going to a reputable, stand-alone IVF clinic. The indications can be loosely categorized as follows.

Inoperable damage of the fallopian tubes

The first successful IVF treatment took place on a woman whose tubes were inoperably blocked. As IVF bypasses the tubes by placing an embryo directly into the uterus, it is effectively the only treatment in these cases. But it is important to be aware that tubal microsurgery can still be as successful as IVF if the tubes are not inoperable. I have encountered many women who have been told that their tubes are 'no good' by some physician who either has not examined them properly or appears not to know about the success rates for correctly conducted tubal surgery. Many women with minor blockages, as well as many with adhesions, can be treated more effectively by tubal micro-surgery, or properly used laparoscopic surgery, than with IVF.

Ovulation failure

This is more complicated, and requires careful judgement. While most cases of failure to ovulate are best treated with drugs – either tablets of clomiphene or possibly injections of FSH – there is undoubtedly a place for IVF when these treatments are not working well. I feel that, as a rule of thumb – and depending on the age of the woman – if a pregnancy has not been achieved after six or nine months of ovulation stimulation, it is worth at least considering IVF. Not only can it help where the ovaries fail to respond effectively or efficiently to drug stimulation, it can avoid the problem associated with drug treatment alone for some women, namely that it can be difficult to control the number of ovulations in any month of treatment. Over-stimulation can result in too many eggs, and if they are all fertilized there is the real prospect of a dangerous multiple pregnancy – triplets or more. It is a real advantage of IVF that, irrespective of the number of eggs obtained, the risk of multiple birth is limited by the number of embryos the doctor transfers to the uterus. And if there are more than two or three embryos, these can always be frozen for another treatment cycle at a later stage.

Premature menopause

The average age of menopause in the UK is fifty-two. It is rare for the menopause to occur as late as fifty-eight, and it is regarded as premature before the age of forty. Having said this, most women are infertile about ten years before their menopause. Also, some very young women can have a premature menopause – distressingly, even in their twenties. This may occur out of the blue, and in most cases the cause is unknown; there may be an underlying genetic or other cause which cannot be identified.

Other women may become menopausal because they grow multiple ovarian cysts that have to be removed, and the ovarian tissue left after surgery does not contain enough eggs. This is one of the reasons why I am such a strong advocate of microsurgery. Really carefully done surgery by someone with proper expertise can often preserve ovarian function so that ovulation afterwards is reasonably efficient. Surgery for dermoid cysts and endometriosis is particularly likely to result in damage to ovarian tissue, and microsurgery can occasionally limit this. Cancer treatments – especially if they involve radiation to the pelvis or extensive 'heavy' chemotherapy – also damage ovarian function.

In all these cases it may be possible to take an egg from another woman, fertilize it and transfer it to the uterus of the menopausal woman. This requires appropriate hormone treatment to 'prime' the uterus so that it is receptive to the embryo (see pages 41–43); and it may be difficult to achieve after radiotherapy for cancer, which can occasionally affect the uterine lining so that, even if a fertilized egg is obtained, it may not implant.

Cervical damage

The cervix, if scarred really badly or producing totally inadequate mucus, may sometimes prevent the onward passage of sperm. While artificial insemination directly into the uterine cavity is the cheaper treatment, IVF has a place in some difficult cases. Placing the embryo directly into the uterus can have a significantly better success rate than artificial insemination each month. If the cervix is very badly scarred, dilation under anaesthetic several weeks before the embryo transfer may be needed.

'Unexplained' infertility

Some couples face the rather deflating diagnosis that they have 'unexplained infertility'. In other words, after repeated investigations, the cause of failure to conceive cannot be pinpointed. Sometimes, it must be said, the so-called 'diagnosis' of unexplained infertility is in fact 'no diagnosis' – that is, no conclusion has been drawn because whatever medical investigation has been undertaken has been inadequate. But when exhaustive tests have been done first, IVF usually has a good success rate in these cases. The main exception is in older women, where age plays a role in the 'unexplained' causes of the infertility. I have said that women become infertile on average about ten years before menopause. So a woman of forty-two with 'unexplained infertility' is most likely to be pre-menopausal – but there is no way of testing for this. Unfortunately, in spite of claims to the contrary, tests aiming to detect the number of eggs left in the ovary are very imprecise. Indeed, though there are now commercially available kits on the market, there is very little evidence that these are useful in predicting when the menopause is really due. In practice, the best warning bell is observation during treatment. If very few eggs are obtained during ovarian stimulation, this is an ominous sign.

So, one reason for undertaking IVF in cases of unexplained infertility is that it can be helpful in making the missing diagnosis. If, for example, during an IVF attempt few eggs are obtained, or no fertilization occurs, or abnormal embryos are repeatedly produced, this information, though distressing, may lead to the cause of the infertility finally being identified.

Male infertility

IVF has revolutionized treatment of male infertility. Until recently there were some two hundred drugs said to improve male fertility. When there are two hundred different preparations, it is almost certain that none will work. IVF has made a massive difference; simply putting the sperm in really close proximity with the egg in culture increases the chance of fertilization. If the sperm count is really poor then it may be better to use one of the other treatments, such as intracytoplasmic sperm injection (ICSI), that have radically changed infertility treatment (see page 107). But IVF cannot help if the testes are producing no sperm at all, or if the sperm are all severely abnormal. In such circumstances the best chances may come from using donated sperm (see pages 237–45).

When both partners are infertile

Infertility is so common that it is inevitable that, in a large number of cases, both the man and the woman are affected. Results of infertility tests carried out in a thorough manner show that in at least one-third of couples tested, both partners will have a problem. When this is the situation – for example, when the husband has a low sperm count and his wife is suffering with tubal damage – IVF is generally the best option.

Genetic or chromosomal disorders

Some families with a damaged gene may be at high risk of having an abnormal baby that may die of the disease to which the family is prone. Many other patients with chromosomal problems may be quite fertile but tend to miscarry repeatedly, or may suffer repeated failure of fertility treatment. A treatment that may be considered in such cases is IVF associated with screening of the embryos. Only embryos free of the specific defect causing the problem are replaced in the uterus. This procedure is called 'pre-implantation diagnosis' (or sometimes, in the case of chromosome problems, 'aneuploidy screening'). This is a complex area of medicine, and the scientific and ethical problems associated with these techniques are discussed separately in chapter 7.

If a couple falls into any of these categories, or if they have repeatedly failed more conventional treatments, most clinics will recommend they embark on an IVF cycle. The treatment itself can be thought of as a 'ten-step' programme. I believe it is worth sketching out in some detail, because relatively few people not involved with IVF seem to understand what a lengthy, costly and demanding procedure it is. And it properly begins well before the first step of actual treatment, with due consideration to the need for adequate preparation.

Preparing for IVF

It must be emphasized that the chances of success with IVF treatment – like all fertility treatment – can be increased substantially by appropriate attention to health and 'lifestyle'. It must equally be stressed that many things which are claimed to be damaging to the chances of becoming pregnant are almost certainly not. A very large number of 'remedies' purporting to improve the results are based on incomplete or false evidence.

Obesity
Obesity is the modern curse. Around 64 per cent of American women are overweight and 30 per cent are said

to be obese;[7] in Australia around 52 per cent are over-weight. The figures are a bit better in the UK, but the problem is growing. If it is true that infertility is really on the increase,[8] this is probably a major factor. There is no question that being overweight is bad for either sex, and in women it increases the likelihood of being infertile threefold. In many women it can have a disastrous effect on fertility, and in many men is certainly associated with poor sperm counts. In women, excess body fat is associ-ated with early puberty, irregular periods, and long cycles in which ovulation may not occur. Women who have polycystic ovary syndrome (PCO) are very frequently overweight, but even those who do not are more likely to have disordered hormones that lead to poor ovulation. Women who are overweight respond much less well to the drugs used to start ovulation, and are more likely to have an IVF cycle cancelled because of failure of response. Even when eggs are collected they may fertilize abnormally. Miscarriage is also more frequent if a woman is over-weight; pregnancy is more likely to be complicated; and the risk of losing the baby around the time of birth is higher.

Many studies show that women who lose weight by diet and exercise often find that formal medical treatment for infertility becomes unnecessary. Very often, after weight loss, ovulation returns to a more normal state and pregnancy occurs without drugs being given, let alone IVF. Numerous studies also show beyond doubt that weight loss substantially improves the chance of IVF being successful. Why spend three thousand pounds on IVF, when taking exercise and eating properly will give you just as good a chance of achieving pregnancy – with the added bonus of a longer, healthier life?

I recall how a friend of my family came to

Hammersmith for treatment having failed IVF three times in another unit. Because she was a friend, I thought it best that she saw one of my colleagues, Raul Margara, rather than me. She had been told before that her weight was a problem – but probably too kindly; she did not take the advice remotely seriously. Raul, himself an extremely kind man, realized that firmness was really necessary and told her in no uncertain terms to lose weight. She was very affronted at this, having come to us because she was convinced we would do IVF in some expert way that would result in a pregnancy. So she persistently asked for IVF treatment, and this was equally persistently refused. Relationships between my family and hers became very fraught because they, like her, felt we were not taking her plight seriously. Eventually she did lose two stone – and within two months of doing so, conceived spontaneously and needed no further treatment.

Stories like this are so very common. The message should be completely clear: weight reduction in overweight people is the single most important preparation for any infertility treatment – and may make it unnecessary.

Underweight

Just as being overweight can inhibit conception, so can being seriously underweight. Research shows that even women who are 10–15 per cent underweight because of insufficient body fat are very likely to have irregular or no periods and not to ovulate. Athletes and dancers are very likely to be underweight. A survey of 394 women who entered the New York Marathon showed that a quarter of them did not have periods while training. Infertility treatment in such women is not likely to succeed until they gain weight, and their risk of losing any pregnancy that does result is also higher while they are underweight. Most good clinics try to ensure that people are at a normal weight for their height and build before beginning IVF treatment. So clear are the data on this point that I think it is irresponsible not to do everything to encourage a person to get as near to normal body mass as reasonably possible before any form of infertility treatment is attempted.

Smoking, alcohol and tea

Most doctors argue that if you are trying to get pregnant, you should stop smoking. But, for a woman, there is no really clear evidence of much effect unless she is a heavy smoker. A number of studies indicate that smoking – by either partner – is associated with infertility. One of the more comprehensive was by Mohammed Hassan and Stephen Killick from Hull.[9] These authors studied women attending an antenatal clinic and asked them how long it had taken them to conceive after they stopped using any contraception. Older women and women who were overweight took longer to conceive. Women who smoked more than fifteen cigarettes a day took twice as long as average to conceive, but lighter smoking did not appear to have any obvious effect. Heavy drinking by men (more than twenty units a week) made it twice as likely they would be somewhat sub-fertile – few of the women in this study drank a great deal, and no adverse effects were noted. However, while coffee made no measurable difference, women who drank a great deal of tea (more than seven cups a day) took 50 per cent longer to get pregnant on average.

The Hull study also showed that couples from a socially deprived area with a low standard of living were more likely to take longer to get pregnant than couples in more prosperous environments. As there is an association between lower living standards and excessive smoking and drinking, some of the effects these authors observed may have been more to do with their social situation, including poor diet and a degree of deprivation, than smoking or drinking.

One of the problems with this study – as with many others – is, of course, that the men and women being studied were not complaining of infertility and not attending an IVF clinic. Although most IVF clinics give advice about 'lifestyle', in so far as it concerns smoking that advice is probably unnecessary. Some years ago my colleagues and I studied two thousand couples, many of whom smoked and who were undergoing IVF. Admittedly, most of the people we followed were not very heavy smokers. But we found no evidence that smoking was associated with a lowered pregnancy rate when they had IVF. However, it was interesting that giving up smoking immediately before or during a treatment cycle did seem to have a negative effect. It is deeply politically incorrect for a doctor not to pour venom on smoking, but the evidence that light smoking is a serious cause of infertility is simply not established.

Radiation and other 'hostile' environments

A number of studies have attempted to assess the effect of background radiation. People who fly at high altitude are exposed to rather more X-rays than others, as are those who travel on the Underground or live in parts of the country where there is a higher than average background irradiation. Some of the studies have been done on people working or living close to nuclear power stations. No adverse effect on fertility has ever been clearly demonstrated. It is true that fertility problems are common among air crews, but this may have as much to do with irregular times at home as with any specific effect of being at high altitude.

Stress

This is one of the most contentious areas of comment on human fertility. Many people are convinced that stress is associated with infertility, particularly in women; but the evidence overall is rather poor. Unfortunately, there are many charlatans out there, offering dubious stress relief packages with great claims that they can improve human fertility or treatment of infertility. Being relieved of stress may make somebody feel better, but it does not seem to improve the chance of a baby. Although stressed animals seem to have lowered fecundity in some circumstances, there is really no good evidence that this is true for humans.

Nevertheless, people jump on bandwagons, and it is no surprise that a recent article by the psychologist Daniel Campagne from Madrid has attracted attention.[10] He argues strongly that any IVF treatment should be preceded by stress management, asserting that anxiety and stress may reduce sperm counts, change hormone secretion and decrease the chances of an embryo implanting. But though he cites a number of studies in support of these assertions, the data he puts forward are very unconvincing. I think a much better assessment is that by Dr Anderheim and colleagues from the University of Göteborg in Sweden.[11] They followed 166 women, each going through her first IVF treatment, and assessed their psychological state and, in particular, the level of stress they felt. Using a sophisticated questionnaire, they assessed their levels of feelings of guilt, anger, frustration, contentment, anxiety, depression and so on. They then compared the outcomes of treatment for stressed and less stressed patients and found no difference between the two groups. The numbers of eggs obtained, the fertilization rates, the quality of the embryos and implantation rates were all similar, as was the chance of pregnancy. These results are very similar to those that Enda McVeigh (now working in Oxford) and I found some years ago in our own patients at Hammersmith. While many of our patients showed considerable evidence of stress and anxiety, it did not seem to affect the outcome of our IVF treatment.

For many years the HFEA has been very insistent that IVF units in Britain ensure that patients are able to receive counselling – to the extent that clinics that did not offer counselling found it difficult to get a licence to do IVF. Yet while counselling may reduce feelings of anxiety and increase the sense of well-being a couple may have, there is not the slightest evidence that it affects the outcome of IVF treatment. Dr de Klerk from Rotterdam surveyed 265 couples going through IVF in that city, and offered counselling and experiential psychosocial therapy to some of the patients on a randomized basis.[12] There was no evidence that those who underwent counselling experienced more success from IVF, or coped better with the added stress of undergoing treatment. It seems that the well-meaning insistence on counselling by the HFEA has a limited rational basis – at least for the great majority of patients who undergo IVF.

While counselling may not have measurable benefits, two recent studies indicate that stress may prevent human ovulation. Dr Sarah Berga of Emory University in Atlanta studied women whose periods had stopped. Half of these women were given no treatment and half were given cognitive behavioural therapy – a form of psychological treatment. Dr Berga says that 'a staggering 80 per cent of those treated had a return of their periods, while only 25 per cent of those in the untreated group showed similar improvement'. Sounds good, doesn't it? But only sixteen women were studied – eight in each group. Eighty per cent of eight is presumably seven, possibly six. Twenty-five per cent of the remaining eight is two. The differences may be 'staggering', but the numbers are too small to prove anything. Dr Berga has certainly shown that stress adversely affects monkeys' fertility – but it is well documented that most stressed mammals reproduce poorly: this is only one of the reasons why we have to be meticulous about the care we give to animals studied for research. But the human evidence remains elusive.

What about the use of jokes to relieve tension? Dr Shevach Friedler from Tel Aviv in Israel recently announced the results of his research at the European Society of Human Reproduction and Embryology. He arranged for his friend Shlomi Agussi – a restaurant chef and clown – to visit patients after embryo transfer at Assaf Harofeh Hospital. Shlomi decided not to go down the red-nose, trick-cycle and water-bucket route, but he performed magic tricks and told jokes about cooking. The ninety-three patients who were forced to watch Mr Agussi's performance had a 35 per cent pregnancy rate after transfer. The other ninety-three women who sat in a humourless environment had a 19 per cent pregnancy rate. It could all be an example of making the treatment worse than the disease.

Alternative and complementary medicine

There are a large number of practitioners of alternative medicine who make outrageous and dishonest claims about their success in treating infertility or preparing people for IVF. They never hold their data up to statistical evaluation and virtually never publish even the smallest piece of convincing evidence to support their assertions.

There simply is no evidence that, among many others,

reflexology, acupuncture, aromatherapy, crystals, faith-healing sessions, herb treatments or adjusting minerals in the diet make the slightest difference to human fertility or improve the outcomes of fertility treatment. Yes, of course, people get pregnant while doing these things – but all the evidence suggests they were just as likely to have got pregnant anyway. From time to time people have come to my clinic for an initial consultation having been infertile for eight to ten years. Occasionally, when one of these women comes back for her next appoint-ment, she is pregnant. From time to time it is claimed (often by the patients themselves) that this has happened because in some strange way my very presence in the clinic has helped them to think positively or has lowered their stress levels. This is sheer nonsense. It is easy to see how a practitioner can soon convince himself he is doing something special; but statistical evaluation shows that these reports of successful pregnancy are exactly what you would expect with an enigmatic condition like infertility. They occur by pure chance. Total outsiders, after all, sometimes win the Grand National.

IVF treatment: the ten-step programme

1 Final assessment
Once a couple's infertility has been adequately investigated, a first step is often a check on the woman's hormone levels – in particular her FSH. If this is abnormally high, drugs to stimulate ovulation are less likely to work. The higher the FSH levels (a normal upper limit is about 10 units) the more likely it is that there will be a degree of ovarian failure. Levels much above 18–20 units definitely indicate that a woman is already menopausal. Some women are also

unresponsive for other reasons to the drugs that are used to stimulate the ovaries; so a few units may give a small preliminary dose of the drugs, or do various other tests to try to evaluate the likelihood of an adequate response. Such tests are not, unfortunately, clearly reliable, and most pre-menopausal women do not show any abnormality on any reputable tests.

A number of units also conduct a cervical assessment before the treatment cycle starts. This entails passing a fine piece of plastic tubing briefly through the cervix to confirm that there should not be unforeseen problems with the embryo transfer. If this 'dummy run' shows that transfer may be difficult, it may be a good idea for the doctors to dilate the cervix under a brief general anaesthetic before treatment commences. This can be done as a day case.

2 Ovarian suppression

The first real step of treatment is to begin stimulating the ovaries. Most frequently, IVF units make the ovaries more sensitive to the stimulant drugs by first shutting them down. However, not all methods of ovarian stimulation are preceded by suppression, which is an adjunct to treatment that has been in widespread use only since about 1989. The principle behind it is that it tends to make subsequent stimulation more effective because it eliminates background action from the patient's own hormones and makes egg collection easier to time.

Suppression, in effect, produces a very temporary menopause, with the ovaries returning to normal after treatment has been concluded. The most common way of shutting down the ovaries is with a nasal spray – usually called Buserelin – administered every four to six hours. This briefly stimulates the pituitary gland, and then stops it producing FSH. The pituitary may respond not by immediately shutting down, but by producing excessive quantities of FSH for a few days before it stops functioning. One form of IVF treatment capitalizes on this, by giving extra injections of FSH during this high-activity period, which cuts down the amount of drugs needed subsequently. This is the so-called 'short protocol', and it can be especially effective in older women who do not respond so well to external stimulation of the ovaries, and consequently need more drugs.

Ovarian suppression may also be given by injection; there is a new class of drugs called antagonists which are sometimes prescribed. These seem to be slightly better at improving ovarian stimulation in some older women, but there is no clear evidence of benefit.

3 Ovarian stimulation

In most women, the ovaries will have been completely silenced within about two weeks of suppression beginning. This is normally confirmed with an ultrasound scan to check that no follicles are visible, and sometimes a blood test to make sure that no oestrogen is being produced. At this stage, the ovaries are fired up again by giving injections of FSH, once daily. Over the next week or two ultrasound scans, and sometimes blood tests, are done to confirm that the ovaries are responding adequately.

Injections of FSH are one of the most expensive elements of the whole IVF procedure, as the genetically engineered hormones are tricky to make and consequently pricey. Each dose typically costs around £30–50, and between twenty and sixty of these ampoules may be used in an average cycle. This, unfortunately, is one of the reasons why some people – in my experience, seldom any who have experienced fertility problems – consider that the NHS should not carry the burden for IVF treatment. The government body NICE, the National Institute for Clinical Excellence, has recommended that people should be given up to three cycles of IVF treatment on the NHS. But in practice, because commissioning authorities are short of cash, there is hardly a single local funding authority anywhere in the UK prepared to follow the government's own guidelines.

'Natural-cycle' IVF, where no injections of FSH are given, is hardly ever done these days. Although it keeps down the costs of drugs, at best it usually produces just a single egg. The overall chances of natural-cycle IVF producing a pregnancy are usually just 3 per cent, even in the most experienced hands. This low success rate and the difficulty in precisely timing egg collection are good reasons why natural-cycle IVF has virtually been abandoned.

Although FSH injections are now almost always given after suppression of the pituitary gland, there is some evidence that FSH without suppression may be more effective in a few patients. Some patients seem to benefit with their own pituitary hormones running in the background. These patients are often given a combination of clomiphene (the fertility pill) and FSH. The difficulty with this approach is that blood and/or urine tests are needed at regular intervals to ensure that imminent ovulation is detected. Unexpected spontaneous ovulation is always a possibility if the pituitary is not suppressed, and once ovulation has occurred it is usually quite impossible to collect eggs because these microscopic cells may be floating around anywhere inside the abdomen.

4 *Maturing the eggs*
Immature eggs that have not gone through the final stages

of development will not fertilize normally – or, if they do, they will not undergo normal cell division. The final stages of maturation are normally produced by the luteinizing hormone secreted by the pituitary; but because the pituitary has been suppressed, no LH is being made. So, about ten to fourteen days after administration of FSH has begun, and around thirty-six hours before ovulation is forecast to occur,[13] an injection of human chorionic gonadotrophin (HCG), which acts in the same way as LH, is given. This helps any eggs in developing follicles to commence the last stages of maturation necessary to equip them for fertilization. The timing of this dose is tricky, and the regular blood tests and ultrasound scans help the clinicians to decide the best moment.

5 Egg collection

The first successful IVF treatments involved collecting the eggs by laparoscopy. Nowadays, ultrasound-guided collection is less demanding and more efficient. Egg collection is scheduled for about thirty-six hours after the injection of HCG and is generally carried out in an operating theatre; although a general anaesthetic is frequently not necessary, the patient is always sedated. Usually, a local anaesthetic is given to make the vagina numb and the surgeon passes a needle through it and into the ovary. Even if only local anaesthesia is used, the better IVF units ensure an anaesthetist is present for complete safety. On rare occasions no eggs will be obtained, even though a whole cycle of drugs has been given. Occasionally, the doctor may cancel the egg collection procedure if he thinks it is very unlikely any eggs will be collected. However, around 97 per cent of egg collections will yield at least one egg, sometimes more.

Very occasionally, access to the ovaries through the

vagina may be difficult or impossible – for example, if there are severe adhesions around the ovaries displacing them from the pelvis. In rare cases, the bowel may be stuck around the ovaries, making egg collection through the vagina unsafe or impossible. In such circumstances we have exceptionally collected eggs through the abdominal wall – either by laparoscopy, or by a needle puncture under general anaesthetic.

At each egg collection, an embryologist is on hand to examine the fluid that the surgeon has sucked out of each follicle and, using a low-power microscope, to identify any eggs. These are carefully removed and placed in a special culture fluid, formulated to reproduce, as far as possible, the optimal fluids and conditions inside the woman's body. Sometimes the culture will contain some of her own serum, obtained earlier from a sample of her blood. Once in this protective bath, the eggs are placed inside an incubator.

In the 'old' days, when egg collection was a fraught business and done by laparoscopy, we often operated at very unsociable hours to coincide with the predicted time of ovulation. As we were not allowed to use a proper operating theatre for such seemingly 'unimportant' procedures, I used to persuade the midwives to let me use the labour ward, where normally we might do Caesarean sections. Stephen Hillier would come in and arrive at the side of the operating table with his microscope inside a baby's incubator on a travelling trolley – the 'egg-mobile'. There he would stand waiting, ready to take from me the fluid extracted from the follicles and identify any eggs.

Most other units starting out in IVF had great difficulty getting help and co-operation from staff in other disciplines. Stephen and I were incredibly fortunate in having support from so many people. Hammersmith is a wonderful environment for attempting new things. Our anaesthetists saw what we were trying to do and would be prepared to anaesthetize a patient at two or three in the morning, if we felt this was the right time. On a memorable occasion late one night the enthusiastic David Zideman, an expert anaesthetist who was normally involved in the resuscitation of people involved in major disasters, gave a hand. Willing as always, he put Mrs B. to sleep around midnight. The laparoscopy was a difficult one: it was almost thirty minutes before I could identify the ovaries at all among massive adhesions inside her abdomen, and another fifteen minutes were required to gain access to a follicle. Eventually, holding my breath, I carefully slipped the aspirating needle into it. The fluid would not leave the follicle – the needle was blocked. I did what every surgeon does in this situation: I cursed my totally blameless assistant, Masoud Afnan. So he patiently handed me a new needle and I tried again. This time, the intake of fluid stopped after a few seconds and the ovary – held up temporarily for view with forceps – slipped from the hapless Masoud's grasp and sank, like a stone in a pond, without trace. Another twenty minutes passed while I struggled and cursed, Masoud flinching, as I tried to regain access to the ovary. By the time we got successful suction Mrs B. had been asleep for close to two hours; I was increasingly worried at her having such a prolonged anaesthetic and anxious about her general condition. I certainly did not want David involved in a major disaster of a quite different kind. So, as the suction was completed, I took the precious open test-tube containing the blood-stained fluid, passed it to Stephen Hillier's waiting left hand, and nervously asked, 'What's the time?' Stephen involuntarily rotated his wrist to look at his watch, and the entire contents of the tube cascaded onto the floor.

Within a millisecond, David, Stephen, Masoud and three others were on their hands and knees peering intently at the puddle on the floor and dabbing impotently at it with suction pipettes.

Fortunately, Mrs B. had a second follicle ready for egg collection, and eventually we did retrieve one egg, which fertilized. About a year later I met her in a supermarket in Golders Green, wheeling her infant around in a pram. I did not have the heart to tell her that it might have been twins.

6 *Insemination*

Usually, just before the egg collection, a woman's partner will have produced a sample of semen. Most clinics provide a special room on site where the sample can be produced, and many provide glossy magazines from the top shelf of the local newsagent to help men finding it difficult to perform on the spur of the moment. Michael Owens, our senior charge nurse at Hammersmith, used to organize this facility and, in addition to the usual soft porn, helpfully included one or two copies of *Gardener's Weekly*, presumably to encourage any horticulturalists among our clients.

It is an advantage to ensure semen is collected before the eggs, because the partners can then be together during the egg collection – invariably an anxious time. If the male partner is known to be likely to have difficulty in producing semen to order at the time of egg collection, most units will freeze some of his semen in advance.

Like the eggs, the semen needs preparation. It is examined under a microscope, and the sperm are washed to remove debris and impurities. They are then placed in culture. Four to five hours after egg collection, the sperm will be mixed with the eggs and the dish placed back in the incubator. Sometimes there may be problems with the sperm, but usually clinicians will be alert and should have spotted most potential problems well before beginning the IVF cycle. If the sperm need additional assistance to fertilize the eggs, a technique such as ICSI may be used (see page 165).

7 Culture and embryo assessment

After the introduction of sperm to egg – so-called insemination – fertilization can be confirmed under the microscope. After about eighteen hours, the embryologist should be able to detect the presence of two pronuclei in the centre of the egg: the female nucleus and, alongside it, the male nucleus – the head of the sperm. This means that fertilization is likely to have occurred normally. More than two pronuclei means that fertilization has occurred with more than one sperm, and this is not generally compatible with life – so an embryo like this would not be transferred. Alternatively, the egg may have begun to divide into cells without fertilization ever having taken place: this is 'parthenogenesis' (see page 37) and, if it goes unrecognized, could result in some unfertilized eggs being transferred to the uterus, where they will of course not survive.

At this point it is worth mentioning one of the key imperatives for all IVF units: namely, meticulous record-keeping and labelling to avoid the risk of mixing up eggs, sperm or embryos in the laboratory. The thought of transferring the wrong embryo to a couple is the nightmare with which every team lives. All units have very strict ways of reducing the chances of this happening to the remotest possibility. We never allow an embryologist to handle sperm, eggs or embryos on his or her own in the laboratory; two people have to be present to check everything at each stage. The surname on the tube or dish has to agree with the written record and with the unique record number on all paperwork and glassware. Embryos from each couple are kept inside sealed containers in the culture oven. Each patient going through the process has her own colour code, recorded on paper and glassware and in writing. Everything has to tally at every stage – any discrepancy and eggs, sperm or embryo have to be discarded. This system, which we evolved at Hammersmith, has been widely adopted elsewhere.

But no human system is totally foolproof; and no matter whether there is regulation and how carefully management systems are designed and maintained, there will always be a slight risk of error. What I think is important is that there should not be a punitive culture; otherwise there is always the terrible risk that people will try to cover up their mistakes in the laboratory. Far better that an embryologist should openly admit that he or she is not absolutely sure whether particular eggs were fertilized with the appropriate sperm than be so fearful of retribution as to hide the doubt. This is another reason why I am deeply unconvinced that the regulatory role of the HFEA is helpful. Most IVF units – and especially the more junior staff within them who are conducting these procedures – live in terror at the thought of being singled out for criticism, and this creates an unhealthy atmosphere in which any mistakes that are made are far less likely to be recognized in time. After all, if the worst scenario happens and the 'wrong' embryo is transferred, provided the error is known within a day or two and admitted, it is very easy to take quite simple steps to prevent implantation.[14]

8 Embryo transfer

Around forty-eight hours after fertilization, each embryo will normally have divided into a minimum of four cells, and from this point is suitable for transfer to the uterus. Some clinics transfer the embryos at around seventy-two hours after fertilization, but there seems to be little or no difference in success rates associated with earlier or later transfer.

With embryo transfer one of the chief ethical and medical issues of IVF treatment arises. Because of the serious consequences of a multiple pregnancy, most units in Britain now transfer only two embryos. Transferring three is unsafe because of the risk of triplets. The ideal would be single embryo transfer, but this gives a lowered success rate. One of the issues here – one that, as far as I am aware, has never been challenged – is how this is regulated. Of course, practitioners all agree that while single embryo transfer is the ideal, the transfer of two embryos is (with present knowledge) practically inevitable in order to balance the requirements to maximize the chances of success and minimize the chances of a multiple birth. The HFEA is now demanding that embryo transfer be *limited* to two embryos. Of course, this is sensible advice – but its enforcement raises the question of the couple's autonomy. The embryos are, after all, *their* embryos. Supposing a woman insists on having all her embryos transferred; who has the right to refuse this request if it is deemed that the risk of the pregnancy resulting in triplets is very low? I am quite surprised that the HFEA seems not to have been required to face a judicial review on this matter.

Transferring only some of the embryos that are available implies that the 'best' embryos are selected, the ones most likely to develop to the blastocyst stage and beyond;

but that is one thing we are not able to do at present. Embryologists often score the embryos on the basis of what they can see down a microscope – but this can be only an extremely rough assessment. What the embryologist tends to look for are embryos that are dividing at a reasonable rate into cells of equal and regular size. It is said that slow-dividing embryos, and embryos that are fragmented – that is to say, where the cells are broken up into oddly shaped fragments – may be less likely to be viable. Very often patients are told that they have 'very good embryos' or 'grade one embryos'. This means very little. A 'normal' embryo may look aesthetically pleasing but the view is no guarantee of its success. What you can see under a microscope is very limited – it is like judging a horse's ability to win a race by looking only at its teeth, or studying someone's face and saying they look intelligent. The rate of cell division is equally problematic. Generally, the embryos whose cells are dividing the fastest have the best chances of success. But I have also seen very sluggish embryos resulting in healthy babies.

More recently, there has been a vogue of transferring embryos at a still later stage than seventy-two hours or day three, at around five days, when usually they have developed into blastocysts (see page 41). This requires changes in the culture medium to keep up with the more complex environment needed by the developing embryos.

For most women, embryo transfer is a relatively simple and an entirely painless procedure. The embryos are placed in a very fine plastic tube, a catheter, which is gently passed through the vagina and cervix and into the uterus, and are then slowly injected into the uterine cavity. This usually takes just a few minutes, though if the cervix is scarred or has a complex internal shape it can take longer. Occasionally the cervix may need stretching with a dilator first, but if this is necessary it is normally done before the cycle of treatment has begun. At Hammersmith, embryo transfer is done using ultrasound guidance because we consider it gives added confidence that the tip of the catheter is in the right place before the embryos are injected.

One of the most damaging events that can occur at this time is a conflict between the man and the woman over the embryo transfer. I have seen this happen, and it is a horrendous problem. One couple from Norwich that I treated went through IVF apparently quite harmoniously. They attended every part of the treatment together, except that the husband did not come to the egg collection. But when there were five good embryos available for transfer, the whole situation changed. At three o'clock in the afternoon we were ready to do the transfer when the husband – quite out of the blue – stated that IVF was immoral and he did not want the transfer to go ahead. His wife was devastated. The embryos were returned to the culture oven and I pleaded with this man to reconsider his decision. At first I used extreme gentleness. Over the next twenty-four hours I was repeatedly on the phone to him, trying to get him to see a counsellor, a minister of religion, a friend. Eventually I told him I thought what he was doing was very harsh. Essentially, it seemed that he wanted to punish his wife for some unspoken disagreement or misdemeanour. All my cajoling and argument was to no avail and he remained adamant in his refusal.

Were I faced with this situation now, I would be inclined to challenge the law and the HFEA, and just transfer the embryos, but as he would not give permission for freezing either, the embryos eventually perished. This woman has since remarried and has had a boy born after IVF.

9 Awaiting results

In most cases, there is no more stressful period during the IVF process than the two weeks after embryo transfer. Once embryos are placed inside the uterus, most women have the inevitable fantasy that they are pregnant. The slightest disturbance may be quite upsetting, and many women find it very difficult just to get on with normal life at this stage. I have met women who refuse to swim or play tennis, who won't stand on a step-ladder or a chair to change a light-bulb, who will not even carry the shopping in from a car. All this exaggerated fearfulness starts with the hokum talked by some doctors about how the woman should be treated after the transfer process. So, understandably, many feel that if they get up and walk around they may dislodge the new embryo. But, as I have had to remind so many patients, even the most vigorous sporting activity doesn't affect naturally occurring pregnancies. While it is reasonable to have a short rest immediately after transfer, lying flat for thirty minutes or so, there is no evidence that prolonged rest is valuable.

10 Progesterone support and other after-care

For around two weeks after the transfer, progesterone is given to help the uterine lining to provide the ideal environment for adequate implantation. If drugs have been used to 'shut down' the pituitary gland at the start of the treatment cycle, it is generally agreed that it is necessary to give progesterone artificially because the ovary is not being sufficiently stimulated to produce its own hormones. But giving progesterone for longer than fourteen days will suppress the onset of a period in a woman who has failed to become pregnant, and this can sometimes result in a woman believing she has conceived when she has not. As there are no data that clearly show any advantage to giving it for longer, progesterone should be stopped with two weeks of embryo transfer.

The spectre of miscarriage looms over every successful IVF treatment. It is inevitable that every little twinge, especially if it is in the abdomen, inevitably raises serious fears that the worst has happened; and a slight show of blood – which is not that uncommon in any early pregnancy – seems like a disaster. Moreover, as the hormones are quite severely disturbed immediately after an IVF cycle, it is not unusual for women to 'feel' pregnant when they are not.

The level of after-care offered by a clinic may depend on the volume of patients being treated and whether it is a wholly commercial operation. Some women are not offered any sort of pregnancy testing at all. At the Hammersmith Hospital, it is common practice to take a blood sample between twelve and fourteen days after the transfer. This helps determine whether the embryo has successfully implanted, and may warn the doctors if miscarriage is likely.

Special IVF treatments

Intra-cytoplasmic sperm microinjection
The most important of adjunctive treatments is intra-cytoplasmic sperm injection (ICSI). The basic idea of injecting sperm into the egg was first proposed in the late 1980s. At that time, Dr Simon Fishel from Nottingham, who had originally come from the Bourn Hall team in Cambridge, was attempting to inject sperm under the zona pellucida – the outer 'shell' of the egg – in cases of male infertility. He felt it was too dangerous to inject a sperm into the substance of the egg itself as it was thought

that this might cause genetic damage to it. As it was, with even this limited technique, Simon could not get approval to attempt this approach in humans in England. Once the HFEA had been established, Dr Fishel was forced to continue this pioneering work in Italy, to the detriment of the treatment of many English patients. As it happens, within a few years, Fishel's approach was widely used internationally and then abandoned in favour of ICSI, which was even more revolutionary and carried greater theoretical risk. ICSI has turned out to be a major advance and has completely altered male infertility treatment. Even if there are no sperm in the ejaculate because the tubing leading from the testes is blocked or badly damaged, ICSI may be valuable (as we shall see below). A complex microscope with micromanipulators is used to inject a single sperm into the very substance of the egg itself without damaging it. The equipment used is quite sensitive so, to reduce vibration, the microscope and its various attachments are usually placed on a heavy table-top which is carefully positioned and isolated by absorbers to prevent even the most minute unwanted movement being transmitted from the floor of the room. Most units carrying out ICSI also have a closed-circuit television system for keeping records of how injection goes.

In addition to this equipment, very fine glass pipettes are needed. These are specially made, usually by heating the tips of fine glass tubing in a laser beam. One pipette has a slightly wider diameter at its tapered tip – about thirty-thousandths of a millimetre – and gentle suction is applied to its outer end by rubber tubing. The tapered end is carefully applied to the egg to be injected, which is firmly fixed by the suction to the glass tube. A second pipette with a tip tapered down to around seven-thousandths of a millimetre is used to inject the sperm. The tip of this pipette will have been previously sharpened under a microscope so that it will pierce the egg easily when the time comes for the injection. It all sounds absurdly easy, but the ends of these pipettes are so fine that they cannot be seen with the naked eye. Just waving them around in an air-current is often sufficient to snap them.

The first stage of treatment is to immobilize the sperm so that they are not moving around too rapidly. This has been done by various methods such as cooling, but most units nowadays place the sperm in a viscous solution which slows down their movement – rather like wading through treacle. When the sperm are moving slowly, it is easier for the operator to suck a single sperm (usually tail first) into the pipette. With the egg firmly immobilized by suction, a sudden thrust of the injecting pipette will pierce the zona pellucida and, with light pressure from a small syringe attached to the outer end of the pipette, the sperm can now be injected into the substance of the egg. Skill is needed in this procedure, not least to ensure that nothing other than the single sperm is injected and that during injection the egg and its nucleus are not damaged. If all is well and fertilization is successful, the egg should start cell division within the next twenty-four hours. In practice, between 60 per cent and 80 per cent of injected eggs are successfully fertilized in this procedure.

ICSI is not entirely without risk. Even though many groups have published very reassuring figures, there has been a continuous dribble of information suggesting that the procedure is associated with a higher risk of some fetal abnormalities. Because ICSI is a relatively new technique, first performed in 1992, there is no long-term inform-ation; the oldest children conceived in this way are only in their early teens, so reliable data concerning their future health and fertility are not available. However, some important trends are already apparent. First, there seems to be a slight increase in the incidence of hypospadias. This is where the urethra opens under the middle of the penis, rather than at its tip (page 239). In most cases this can be corrected by relatively minor surgery, but in some boys it can cause considerable deformity of the penis. Second, because male infertility is associated with abnormalities of the chromosomes (including translocations: see page 358), forcing fertilization with a sperm carrying a defect of this kind may pass it on when, left to nature, the deformity would have been self-limiting in that family because of the infertility. Third, there may be some genes on the Y-chromosome, similarly associated with infertility and abnormalities, that may cause future problems for the child. Lastly, by forcing fertilization in this way, we may be encouraging infertility (particularly male infertility) in the next generation.

TESA, TESE, MESA and PESA

A number of men have such severely lowered sperm counts, or such severe blockage to the tubing from the testes, that there are very few, or no, sperm in the ejaculate. Various techniques can be used in such cases in the hope of getting sperm for IVF and ICSI. Several rather confusing acronyms for these procedures are in wide usage.

One is TESA (testicular sperm aspiration), which involves sucking sperm-containing material directly from the testes after inserting a needle. It is technically easy to do and requires less surgical training than other methods. The procedure is quick but, because it can be painful, local anaesthesia or some form of sedation, possibly even general anaesthesia, is always required. The healing process is rapid because no skin stitching is needed. TESE (testicular sperm extraction) is more complicated, calling for a skilful surgeon, general anaesthesia and a fully equipped operating theatre. Usually, the surgeon takes several small pieces of testicular tissue for analysis. It is most usually done on a day-care basis, though an overnight stay is occasionally needed. With both procedures, spare material containing sperm may be frozen for future use.

Complications arising from either procedure are relatively rare. Infection or bleeding in the testis can occur, sometimes causing extreme discomfort, and both procedures can cause some scarring in the fine tubes in the testis and changes in the blood supply. So ideally, these procedures should not be repeated too frequently. TESE is the more useful because examination of the extracted tissue may give a clue as to the cause of the low sperm count. Also, because infertile men are prone to malignant tumours of the testicle, TESE offers the surgeon a view of the testis and a chance to make earlier diagnosis of any cancerous condition. TESE is also the more expensive, but a recent Israeli study comparing it with TESA shows that it is up to twice as effective in getting enough good sperm for ICSI to be performed.[15] At Hammersmith, this procedure has given our best results for very severe male infertility.

MESA stands for microscopic epididymal sperm aspiration. Occasionally, if the epididymis has become blocked – for example, after vasectomy or by infection – good sperm can be sucked out. This requires surgery to expose the fine tubing from the testis. The procedure is similar to TESE in many ways; again, it cannot be endlessly repeated because sticking a needle into this fine tubing can cause further blocks. PESA (percutaneous epididymal sperm aspiration) is similar but is done by merely sticking a needle into the tubing, blindly through the skin. Although few men are comfortable at the thought of needles entering the scrotum, in practice this is a straightforward procedure that can be performed on an outpatient basis. However, we have found that the best results are obtained by giving a general anaesthetic, opening the scrotum through a small incision and gaining direct access to the testis and epididymis to enable us to isolate sperm in more controlled conditions.

Zona drilling

Mention should be made here of a technique of more dubious value. In zona drilling (assisted hatching) a minute hole is made in the outer surface of the egg. This procedure, carried out using the same equipment and skills as ICSI, was developed by our team at the Hammersmith Hospital in 1989, originally as part of the technique used to determine whether embryos carried certain genetic abnormalities. Then, a hole was drilled in the surface of the egg, and a single cell extracted through it for further analysis. But just drilling a hole in the egg seemed occasionally to improve the implantation rate. It was assumed that some eggs have a thicker outer coat, and that the drilling gave the embryo a better chance of hatching from this 'shell' before implanting in the uterus. Although some groups still use assisted hatching with this in mind, it seems of only modest value and then only when applied to the eggs of older women. This very slender evidence of efficacy has not prevented this technique being sold to large numbers of patients.

Blastocyst transfer

The advantages of culturing the embryo for five days are disputed. While a blastocyst undoubtedly has a better chance of implantation and becoming a baby (around 50 per cent do in some units), there is a theoretical possibility that culturing an embryo to such an advanced stage outside its natural environment may cause damage that could affect health in later life. Also, only a few fertilized eggs develop as far as the blastocyst. Many do not divide at all, others divide sluggishly, and many others stop developing after about three days. Thus it may be that blastocyst culture is more successful simply because these are the very embryos that would have gone on to further development anyway; in other words, that by transferring a blastocyst, all the doctors are doing is not transferring the failing embryos.

In the view of many people, myself included, it is not justifiable to transfer more than one blastocyst at a time. The risk of implantation is sufficiently high that a twin pregnancy is very likely. Although infertile couples may be led to believe that having twins is the answer to their problems, all multiple pregnancies are fraught with increased risk – and the risk is sufficiently serious not to warrant the gamble.

GIFT (gamete intra-fallopian transfer)

GIFT was first described in 1985 by Professor Ricardo Asch,[16] working at the University of Texas at San Antonio, where years earlier we had researched together. Ricardo, an exceptionally talented, medically qualified Argentinian scientist, had settled in Texas, where he proceeded to publish a large number of highly original and important scientific papers. I came to enjoy the humorous way he would put his head on one side and look querulously at experiments I was planning. I always had the strong feeling he would do my experiments so much better than I could if he got his hands wet, but he most enjoyed looking through the open laboratory door and commentating, directing operations by remote wisdom. Ricardo was greatly acclaimed in the USA and soon won accolades from researchers and clinicians all around the world. By the mid-1980s he was regarded rightly as one of the leading researchers in human reproduction. Sadly, Ricardo was tempted to the fleshpots of California, where eventually, after running a large department very successfully, he seems to have got involved in selling eggs and embryos as an IVF doctor. He was pursued through the courts and it is said that he now lives in Mexico. To this day, I do not know what went on in his clinic, and I remain very surprised that this brilliant and engaging man was accused of these alleged crimes. Many will see this as an object lesson in the temptations and risks in having a hugely profitable market attached to this kind of medicine. Whatever actually happened, I do not believe a better regulatory framework could have prevented this happening in California; the existence of the HFEA in Britain certainly did not prevent the scandal of very similar events at a Hampshire clinic in 2000.

The clinical technique that initially made Asch's name famous was gamete intra-fallopian transfer or GIFT. Ironically, nearly all of the serious science he did was much more important than this, and yet so much of it has been forgotten. For a while GIFT – because it does not involve handling embryos – was regarded as an ideal technique to avoid the ethical problems associated with selecting and manipulating embryos and a perfect alternative to IVF. The main problem was that it turned out not to be as effective. GIFT involves all the initial stages of routine IVF, such as stimulating the ovaries and collecting sperm; but, crucially, when the eggs and sperm are mixed, the resulting cocktail is simply placed directly into the fallopian tubes using a laparoscope. Alternatively, the egg and sperm may be inserted in a thin tube through the cervix, then through the uterus and into one tube from below. Ricardo Asch's basic idea was that the environment of the fallopian tube is more natural than and therefore superior to the culture mediums we can provide artificially, and hence more likely to allow normal fertilization and embryo growth. He also saw this as a way of treating unexplained infertility when the tubes are normal.

One of the many anomalies about the treatment is that, because GIFT does not involve embryos, in Britain it is not regulated by the HFEA. Therefore, practitioners of GIFT are not restricted in any way. This means that it has not been uncommon for some clinics to return more than three or even four eggs to the tubes simultaneously – risking high-order multiple births.

Unfortunately, the success rates of GIFT do not suggest that the natural environment of the fallopian tube offers any real advantages over IVF. GIFT does have some advantages for a religious person who is opposed to creating embryos in laboratories (see below). But for the rest, GIFT is subject to most of the difficulties and costs of IVF without any particular benefits.

I made my own first very primitive attempts at something very similar to GIFT well before Ricardo Asch published his technique – even before IVF was a serious possibility. In my training at Hammersmith Hospital, I was incredibly fortunate to work with the obstetrician and gynaecologist Professor John McClure Browne. At that time, he was one of the most senior academics in

Britain and head of the most academically active department in Europe. Professor Browne was a man of high moral standards, always deeply serious and of forbidding aspect, and in consequence feared and misunderstood by many colleagues – but underneath his somewhat alarming, distant and academic manner there was a really kind man. He was always very supportive of my work, going out of his way to promote it whenever he could. Just occasionally he showed flashes of extraordinary humour – once he muttered about a patient we had just seen together, saying that he thought she 'had just possibly been using her vagina for commercial purposes'.

Browne was very unusual in being seriously interested in infertility and at Hammersmith ran one of the oldest specialized fertility clinics anywhere, to which people came from all over the world to see him for advice. Although many people thought of him as deeply conservative, Browne was always trying new ideas, and it was he who suggested we tried what now would be described as GIFT. We timed open abdominal operations on women wishing to conceive to coincide with when we expected ovulation to take place spontaneously – though sometimes we gave some clomiphene (then a fairly experimental drug) to stimulate it artificially. During the surgery we would suck fluid from the follicle with a crude needle, mix it with sperm (completely unprocessed and certainly not washed and prepared in the way it is now) and inject it into the end of a fallopian tube. I do not remember getting any formal consent from the patients, though they certainly knew roughly what might be happening because of course we had to request sperm samples from their husbands. On two occasions we achieved a pregnancy, and one of these went on to term. But McClure Browne was too conservative and insufficiently self-promoting to consider this approach as remotely worthy of scientific publication.

ZIFT (zygote intra-fallopian transfer)

This involves intervention at a slightly later stage than with GIFT. The embryos are fertilized in the laboratory, but then placed inside the fallopian tubes immediately afterwards – without embarking on any long culture process. The thinking behind this is that by placing an egg we know to be fertilized into its most natural environment, we bypass the uncertainties of GIFT but preserve its benefits. Like GIFT, it is as complicated and costly as routine IVF, and statistics do not suggest it has any greater chances of success.

Risks to the patient

Whatever technique is employed, even a cursory glance at the above material shows what an involved and complicated procedure IVF can be – one that, at so many points along the way, can lead to disappointment. It is a treatment characterized by waiting and uncertainty, and this can place great emotional and physical strain upon a woman and her partner. I believe this is often not fully understood by people for whom IVF has been – rightly or wrongly – recommended. Of course, many individuals will take whatever chance is offered, no matter how slim. Perhaps the greatest problem with IVF, then, is that the demands and stress associated with it can make many couples deeply unhappy and depressed afterwards if it does not succeed.

Another common problem is associated with the drugs administered. These hormone preparations can make many people feel rather bloated, unwell or just a bit below par. Also, it is very common for women who do not get pregnant with the treatment to have irregular periods for a few months afterwards, even to find that periods can stop completely for a while. Nevertheless, it is very unusual for them not to return within around four months.

Giving hormone preparations to a woman to stimulate the ovaries into producing eggs carries a small risk of ovarian hyperstimulation syndrome (OHSS). This tends to be more common if IVF has resulted in a pregnancy, and it is more likely to occur in women with polycystic ovaries. At least 5 per cent of IVF patients experience a mild form of OHSS. The usual symptoms are feeling bloated, some abdominal discomfort and fatigue. It usually passes after a few days and without the need for any intervention. Pregnancy tends to make the symptoms rather worse.

More rarely, in around 0.5–1 per cent of cases, a more severe form of OHSS may cause tiredness, abdominal pain and swelling, and breathlessness. This requires hospitalization and bed rest, with close observation. The ovaries become very swollen and tender, and tend to exude fluid which collects in the abdomen. The fluid build-up may also occur in the chest cavity. If breathing is severely impaired the chest may need to be drained by the insertion of a tube. Similarly, build-up of fluid in the abdomen may require abdominal drainage. Because the fluid balance of the body is disturbed an intravenous drip may be required. Severe OHSS usually gets better after a few days, but it can be a dangerous condition. A very few fatalities have occurred, but prompt recognition of the condition and early treatment avoid complications in nearly all cases. OHSS can be avoided by identification of those at greatest risk, careful monitoring during the IVF cycle and avoidance of replacing too many embryos. If the onset of OHSS looks likely, the dosage of stimulatory drugs can be reduced, or the cycle can be abandoned. As pregnancy can make OHSS worse, an egg collection may be done, the eggs fertilized, and the embryos frozen for later transfer when things have returned to normal. With these precautions, the risks of problems from OHSS are minimal.

Some fourteen years ago several studies from the USA suggested a link between ovarian cancer and the use of FSH to stimulate the ovaries. Further studies have not confirmed this link. Women who have never been pregnant have an increased risk of ovarian cancer, whether they undergo IVF treatment or not. It is also true that there may be a very slight increased risk of cancer if clomiphene is taken to induce ovulation for more than twelve months. For this reason it is not advisable for any patient to continue it for more than a year.

There are a few risks associated with egg collection. As in any other procedure, there are tiny risks associated with anaesthetics. Occasionally the egg-collecting needle may puncture the bowel or bladder. From time to time a blood vessel may be punctured and serious bleeding can occur. This usually stops after a few hours without any treatment, but on rare occasions a blood transfusion may be needed. Exceptionally, it may be necessary to operate if a vessel continues to bleed. There is also a very small risk of causing infection in the pelvis when the eggs are collected, or when the embryos are deposited. Women at risk of infections may be given antibiotics before any treatment takes place.

Ectopic pregnancy is more likely after IVF if the cause of the infertility is tubal damage. Even if both tubes have been removed, this complication is possible because the developing embryos can implant in the part of the tube that runs in the uterine wall, which cannot be completely surgically excised if the tubes are removed.

The greatest risk to maternal health is the possibility of multiple pregnancy. It is very tempting to transfer more than one or two embryos to increase the likelihood of one successful pregnancy, but the risks to both mother and offspring are considerable. Having twins or triplets may sound delightful, but multiple pregnancies can be disastrous for a range of reasons. Since 1990, regulation in Britain has limited the number of implanted embryos to three – in my view almost the only serious justification for the present regulatory framework. In practice, most British clinics consider it responsible to transfer only two; this may reduce the chance of pregnancy to some extent, but does not much reduce the chance of having a normal baby. Elsewhere in the world, for example in the United States, there is no limit on the number of embryos that may be transferred. But multiple pregnancy brings with it a much higher risk of miscarriage, greater risk of premature delivery, smaller babies, a higher risk of brain damage to the babies, a greater stillbirth rate and huge cost for the care of any premature infants. One in twenty-three twins will die at birth and one in twelve will have a serious abnormality. The risks for triplets are very much greater indeed.

The risk to the mother's health is also considerable. Toxaemia of pregnancy with high blood pressure may mean extensive bed rest in hospital. Virtually all the common complications of pregnancy are much more likely with two or more fetuses. Delivery is much more likely to be complicated and Caesarean section is often needed. Haemorrhage and maternal death are also both more likely.

In addition, the presence of twins, and especially triplets or quads, places extraordinary emotional, social and economic stress upon the family.

Risks to the child

Louise Brown, at the time of writing a 28-year-old postal worker living in the Bristol area, tends to avoid the media spotlight wherever possible. Needless to say, her job, like her previous role as a nursery nurse, provides ready material for lots of corny jokes about babies and deliveries. Messrs Steptoe and Brown kept in close contact, and when Steptoe died in 1988 Louise commented that it was 'like losing a member of the family'. Of course, she prefers to be seen as an ordinary individual, rather than a walking demonstration of the potential of science. But almost thirty years after the technology was developed, what do we really know about its effects on the people it brings into being?[17]

Ever since IVF was mooted, people have asked whether it could harm the developing embryo. Could the culture medium used in the laboratory cause abnormalities? Could procedures like ICSI and zona drilling disturb the fragile DNA-bearing material inside the egg? What were the effects of artificial stimulation upon the egg? By assisting reproduction, could we be creating life from embryos and sperms that contained serious defects?

A number of wide-ranging studies have led to the conclusion that there may be slightly elevated risks. IVF does lead to a higher chance of premature birth – according to one American study, as much as 2.6 times as great as the risk in unassisted pregnancies. This may just possibly be because infertile women tend to be older than average when they finally get pregnant. My own work indicated that people who have simply taken gonadotrophin to stimulate ovulation, but have not had IVF, also experience increased risk of premature delivery.

Premature delivery is associated with – but does not always lead to – a range of developmental problems. Premature babies are born with an immature brain and this can be easily damaged. For example, they may experience learning difficulties, severe difficulties with normal movements, or problems with hearing and eyesight in later life. Because their lungs are not properly formed they may have breathing difficulties which can be fatal. Their immune system is also less able to cope with the outside world, so they are less able to ward off infections. Virtually every organ is more likely to fail because of the fragility of their situation after birth. And we now know that people who were very small as babies are more likely to suffer heart disease, stroke, high blood pressure and probably osteoporosis in middle age. Another key consideration is that the care of premature deliveries places huge strains on the resources of any health-care system. Neonatal facilities are in short supply because it costs, on average, well over a thousand pounds a day to keep one baby in an incubator – and each of three triplets frequently requires up to a month or more in an incubator. Clearly, this is another reason why it can hardly ever be advisable to transfer more than two embryos.

Some further studies also suggest that even 'normal' IVF pregnancies may carry an increased risk of fetal defects. The validity of some of these studies is somewhat controversial, but in my view they should be taken more seriously than they tend to be. It seems to me that the IVF community – the 'industry', if I may call it that – is too ready to dismiss reports of complications and abnormalities. Teams carrying out IVF do, after all, have something of a conflict of interests.

One of the best studies is by an Australian group, led by Dr Michele Hansen at the Telethon Institute for Child Health Research in Perth, Australia.[18] She teamed up with statisticians in Oxford to conduct a careful review of all the publications up to 2003 in the medical literature where birth defects after IVF were reported – fifty-one papers in all. Half of these publications were not fit for statistical analysis, for various reasons. The remainder came from Europe, Australia, the Middle East and the United States. This wide international distribution is important, because it has sometimes been suggested by clinicians who feel these kinds of study threaten them that the high incidence of abnormalities could just be due to 'a little local difficulty'. Most of the papers assessed matched the data obtained from IVF pregnancies with data from control pregnancies within a normal fertile population. Seventeen of the publications were of large studies – that is to say, at least five hundred infants had been followed up. The biggest followed 9,111 infants. To make things as unbiased as possible, the statistical assessors of these combined studies were 'blinded' – that is to say, they did not know as they assessed the paper that had been allotted to them what results were being found in the other papers reviewed.

What Hansen and her colleagues found was that IVF babies had a 30–40 per cent greater chance of abnormality than the normal population. This is not a huge difference – it means in practice that an IVF baby has something like a 4.5 per cent chance of being abnormal in some way, compared with a chance of about 3 per cent for a baby conceived naturally. But while the difference is not great it is extremely important, because it is unclear what causes this effect – if it is indeed a real consequence of IVF.

Certain things are fairly clear. IVF mothers tend to be older than average. More IVF babies are from multiple pregnancies. More IVF babies are born prematurely. All these factors are important, as all are associated with increased risk of birth defects. But it does not seem to be the whole story, for, when all these factors are taken into account, more IVF babies than naturally conceived babies are born with an abnormality.

Abnormalities seem to be more common after sperm injection (ICSI) for male infertility. This may be because of the technique itself. Injecting sperm may cause subtle changes in the egg, or the way the genes in the egg and sperm work afterwards. But it may be because infertile males carry genes which increase the risk of abnormality. Alternatively, because ICSI avoids the natural filtration process which removes abnormal sperm in the uterus and tubes, it is possible that more abnormal sperm are injected and this results in abnormalities in the embryo.

ICSI is not the only process which gives rise to concern. A study undertaken some years ago by one of my PhD students, Maria Tachataki, suggests that freezing embryos may cause surprising changes. Maria examined embryos that had been frozen and then thawed (using a method widely applied in various IVF clinics), and compared them with embryos that had not been manipulated in this way. After the thawing had been done, we compared how the working of one particular gene, TSC-2, was affected. TSC-2 is a curious gene, a tumour suppressor. When it works normally it may protect against some cancers; when it is damaged, affected people may be more prone to get these forms of the disease. What Maria found was that the particular method of freezing/thawing used changed the activity of this gene.[19] In some embryos that were thawed, having been frozen two days after fertilization, the TSC-2 gene showed reduced expression – that is, it was not working as hard. But twenty-four hours after thawing had been completed, expression was back to normal. So it is only a transient effect; but there must be some concern if a gene that protects health like this may be altered or its action suppressed – even if only temporarily. Who knows what long-term effect such a phenomenon could have?

There are some other compelling data that may have very important implications. Seemingly innocuous events happening at the very beginning of life may affect a person's health as an adult. David Barker, at the University of Southampton, meticulously trawled through a huge number of records of the births of children delivered in Hertfordshire just before, during and after the First World War.[20] Most of the mothers of these children had normal pregnancies. But inevitably, there were some hospital records of women with complicated pregnancies. Some of their babies experienced a deficient environment while in the uterus and were not well nourished, being born well below normal birth-weight in consequence. In general, the babies who were born much smaller than average, particularly those who were really small, around 2 kilograms in weight, turned out to be at increased risk of ill-health in later life. A far greater proportion of the adults who had been very small babies died from coronary heart disease before the age of sixty-five. They had all begun life in sub-optimal conditions, and the struggle to compensate for this left them with constitutional scars that had grave effects decades later. These babies had tried to adapt to survive; but this adaptation seems to have had serious implications for their future health. Later research by David Barker and his colleagues showed that heart disease was not the only risk for these individuals. They were also more likely to suffer the related diseases of stroke, high blood pressure and diabetes by the time they reached middle age.

David Barker's observations are clearly relevant to IVF treatments for two reasons. First, while there is no obvious lasting change to an embryo that has been frozen (as far as we know), it is now well documented that, overall, IVF babies are much more likely to be smaller than average at birth. It therefore seems probable that they will be more susceptible to these diseases in middle age. But there is another problem as well, which in a way is more sinister. Although Dr Barker's work has been concerned only with the fetus, it is quite likely that an adverse environment much earlier – right at the beginnings of life – could have profound effects as well. After all, this is when growth and change are at their most rapid. And it could well be that an adverse environment during embryonic development might be even more serious, because many of our genes are really only just starting to function at that time. To my mind, this is one of the most important aspects of the whole area of fertility management; and yet it is widely ignored. It seems to me that really long-term studies will be the only way of trying to sort out this potentially serious issue.

Let me take just one hypothetical example, which does not involve embryo freezing. We know from various studies in embryology that human embryos do not flourish very well at the earliest stages of development when there is glucose in the culture medium. In fact, the precise level of sugar needed for optimal embryonic growth has, in the past, been a matter of controversy. When IVF was started (and to a large extent this is still true), the constituents of culture media were arrived at on a purely empirical basis. Because there were no good data on what would be an ideal culture medium for an embryo, much early work on IVF media was based on intelligent guesswork. To this day, the concentrations of the different sugars in embryonic culture are used simply because they seem to work quite well. It is a salutary thought to consider that, if we are challenging the embryo with an inappropriate concentration of sugars early on in life, the handling of sugar metabolism later on, well after birth, might be a problem. The concern is obvious. It is certainly a possibility that babies born after IVF might be more likely to have problems with sugar metabolism, and in particular, it is theoretically possible that they are more susceptible than naturally conceived children to diabetes.

Even more remarkable than Dr Barker's original work are the observations that the environment during early development can produce striking effects two generations later. In 2001 Lars Bygren and his colleagues in Sweden published studies showing how some boys born in 1905 were adversely affected by their grandparents' diet. In particular, their grandfathers' access, or lack of access, to copious supplies of food affected the longevity of these boys. In an isolated, remote part of northern Sweden there had been regular crop failures and bumper harvests in different years between 1799 and 1880, and these had been very well documented in public records. If the paternal grandfather had had access to ample food from a bumper crop during the time he went through puberty, his grandchild – if male – was more likely to die at a younger than average age. No relationship of cause and effect has been firmly established. Yet the higher incidence of diabetes in these grandsons argues that a plentiful diet produced some chemical changes in the genes on the Y-chromosome of their grandfathers' germ cells during a critical stage of their early development. These changes could have affected how the genes in their sperm expressed when they produced their children. So the environment of the grandfather had a deleterious effect on the boys (who carried that Y-chromosome) two generations later. If it turns out that the way genes are expressed is really being affected by such environmental influences, events in early life could change the pattern of the inheritance of disease in future offspring. This should be a cause for worry and is another reason why I feel the activities of the HFEA are so disappointing. This is the body that should be recording and storing data that would elucidate this serious question.

Given that these Swedish boys showed a strange pattern of inheritance after marked changes in their grandfathers' diets, it is not surprising that there was speculation about a recent study suggesting that a rare childhood illness was four times as common in IVF babies.[21] Beckwith-Wiedemann Syndrome (BWS) is related to a process called imprinting, which affects the way in which genes act (or express), depending on whether they come from the mother or the father. Most genes are not subject to imprinting, but many of those that are tend to be associated with the development of the baby in the womb. Those inherited from the father tend to encourage the baby to grow bigger, while those from the mother tend to act to keep the growth process in check. BWS babies have an increased risk of developing certain childhood cancers, in particular tumours of the kidney. Only around one in every 15,000–35,000 children naturally conceived is born with BWS, but the American study, by researchers at Johns Hopkins University and the University of Washington, suggested that 1 in 2,500 IVF children were at risk.

Subsequent work reviewed possible connections between IVF and other imprinting disorders, such as Prader-Willi Syndrome (PWS) and Angelman Syndrome (AS). Prader-Willi children experience uncontrollable appetites and obesity, as well as muscle problems and learning difficulties. In Angelman Syndrome, children may experience epilepsy, curvature of the spine and severe learning difficulties. A 2005 study by Dr Alastair Sutcliffe, at the Department of Child Health at University College, London, looked at the links between all of these conditions and IVF. He concluded there was a probable link between IVF and BWS – perhaps because the genetic problems in the parents were also giving them infertility problems. With the other two disorders, there was no provable link.

The technique of ICSI has raised additional concerns for a number of reasons. As noted above, it has been suggested that, in the natural way of things, abnormal sperm do not fertilize eggs, and that by intervening in the process we risk fertilizing eggs with poor-quality sperm. There is also the risk that minute quantities of the culture medium are passed into the egg when the sperm is injected, or that the business of injection itself may disturb the fragile material inside the egg. Another of Dr Sutcliffe's studies compared ICSI and IVF children with those naturally conceived, and followed them up to the age of five. He detected around a 6 per cent rate of malformations among ICSI-produced children, compared with 4.5 per cent for the IVF group and 2.5 per cent for the naturally conceived babies.

This would indicate that, although the numbers are low, there is a slightly increased risk of malformations for all children conceived artificially, and that the risk is highest in those whose parents used ICSI. However, most of these studies have a number of flaws. The age of the mothers in the IVF and ICSI groups is sometimes higher – and we know that ageing in itself elevates the risk of giving birth to children with abnormalities. Also, the control group of naturally conceived children is often not properly matched. It may be drawn mainly from local primary schools; children with serious handicaps do not attend mainstream schools, so the comparative good health of the non-IVF and non-ICSI groups may be slightly over-emphasized.

The slightly elevated risks associated with ICSI lead me to advise that it should only ever be applied where no other form of fertilization is possible. No doctor should ever recommend a complex and costly procedure when a simpler – and arguably safer – one is viable. But IVF is done in a highly commercial world and in a competitive atmosphere. I regret that some IVF units will tend to offer ICSI when it is not absolutely necessary. If the man's sperm count is low, the expedient action may be to use ICSI because it more or less guarantees fertilization. Also, if (as is true in most cases) the couple are being treated privately, failure of the IVF cycle as a result of fertilization failure is costly. So ICSI may be offered to reduce the risk of a couple having to repeat their treatment. In the next chapter I discuss whether another IVF technique we use may have hidden dangers.

4

Freezing: The Refrigeration Game

Spallanzani, the priestly scientist, among all his other remarkable observations, noticed that sperm slowed down when subjected to low temperatures. As seminal fluid approached freezing point they became motionless, and then were reanimated once the fluid was thawed. His observations forecast a distant but important development in the manipulation of animal and human reproduction – the use of freezing, or cryopreservation, to store sperms and embryos, and more recently eggs and pieces of ovary or testis. Although it was a very long time before it became a reality, it is said that in 1866 another Italian scientist, Montegazza, theorized that it would eventually be possible to use this technology in human conflicts. He suggested that one day men, heading off to war, might first freeze samples of their sperm so that they could go on creating heirs in the event of their demise. Over a century later, this is exactly what some soldiers did before embarking for the Gulf War. But reproductive cryopreservation has become a focus of conflict for very different reasons.

Banking manhood

Between the 1930s and 1950s, a number of scientists experimented with freezing sperm and realized that seminal fluid was able to withstand extremely chilly temperatures. A key advance was made by Chris Polge and his colleagues at Mill Hill in London in 1949 – with, as in so many scientific discoveries, the help of a stroke of luck. At that time, after the deprivations of the Second World War, the breeding of farm animals was a high priority and insemination of cattle was an important area to be explored. Dr Polge was attempting to store bull's semen, and used various sugars in solution as an energy source to improve sperm quality. One of the sugars he and his team added was from a bottle labelled 'fructose'. When this was mixed with bull's semen, frozen, and then thawed, the researchers found they had many sperms of good fertilizing quality. It turned out that the solution Polge had used wasn't fructose at all. The bottle had been mislabelled, and really contained a mixture of albumen and glycerol.

Glycerol is, of course, an anti-freeze. That is to say, it inhibits the formation of ice crystals – and this reduces the risk of damage to any living cells. After Polge's important work, it wasn't long before the same principle was used in human sperm storage. The first successful human pregnancy from frozen sperm was reported in 1953 by two Americans, Dr Sherman and Dr Bunge,[1] but at this time donor insemination was considered to be very close to adultery, and it wasn't until 1970 that the first commercial sperm bank appeared in the USA, in Minnesota.

Initially offering married men a means to store their own gametes, sperm banks developed into increasingly sophisticated commercial undertakings. Above all, they allowed people to make complex choices about the 'quality' of the sperm with which they sought to create their babies. The Repository for Germinal Choice, for example, founded in 1980 and based in California, promised to offer only the sperm of exceptional men, such as Nobel Prize winners and Olympic athletes. Its founder was a self-made man, the millionaire Robert K. Graham, who had made a fortune by his invention of shatterproof plastic spectacles. Like many other Americans of the time, Graham believed that humans were an endangered species. His view, which was widely shared by numerous scientific intellectuals, was that, because our species could now control its environment so effectively, natural selection – and therefore evolution – had stopped working. In his book *The Future of Man*, first published in 1970, he explained how improved medicine and a more protective welfare society was allowing more weaklings to flourish – weaklings who would not have survived when men were still roaming the savannah with club or spear. These 'retrograde humans', mere incompetents and imbeciles, were increasingly swamping the more productive and intelligent members of society. In order to save the species, he felt we needed to breed better babies and to introduce 'intelligent selection'. Superior human specimens were required, above all, to procreate more gifted children. Graham intended his Repository for Germinal Choice to be a prototype for a network of genius sperm banks all over the USA. The imperative, as he saw it, was to generate 'creative, intelligent people who otherwise might not be born': the future opinion-formers, politicians, intellectuals and leaders in society.

So committed was Graham to his eugenicist principles that he was prepared to offer his next service to society without any charge to his customers, announcing with a flourish that he had persuaded some Nobel Prize winners to give him sperm for cryopreservation. He certainly recruited at least one Nobel Prize winner: William Shockley. Shockley was the inventor of the transistor and a founder of Silicon Valley. He was also a confirmed racist – he concluded that African Americans were more stupid than white Americans because of their poorer perform-ance in the US Army's crude entrance tests that measured IQ. In a number of speeches he proposed that individuals with an IQ below 100 should be paid to undergo voluntary sterilization. Running as a potential senator, he stood on a eugenics platform, achieving eighth place in the election. Graham was also successful in recruiting a number of younger scientists (our profession is not particularly known for its reticence) to donate sperm for his bank. Graham was obsessed by intelligence; he advertised for mothers-to-be in a magazine run by Mensa, the organization that caters for people with high IQ (but not necessarily common sense), hoping to attract women whose husbands were infertile to his cause of improving the nation. The campaign included placing advertisements about his donors, such as the one promoting 'Mr Fuchsia', who was, it was claimed, an Olympic gold medallist: 'Tall, dark, handsome, bright, a successful businessman and author'. Another praised the prowess of a 'Mr Grey-White' – who knows if this was his real name, or even his true colour? – who was 'ruggedly handsome, outgoing, and positive, a university professor, expert marksman who enjoys the classics'. No doubt marksmanship is a use-ful attribute for a sperm donor.

By 1980 there were seventeen banks storing frozen sperm across the USA – providing the raw material for some 20,000 babies. Some copied Mr Graham's initiative by priding themselves on providing particular types of sperm. But it seems that most Nobel laureates scorned the idea; as far as I am aware, no babies were ever created with such 'prize-winning sperm'. One of the more interesting aspects of these sperm banks was that they gave a certain amount of reproductive choice to those who used them. Prospective mothers could take control of their own fertility, as these banks capitalized on the fact that sperm donation made a male partner unnecessary. They also offered a service tailored to lesbians, or women who wished to raise a child single-handed.

Although sperm freezing is now a highly efficient procedure and widely used, it is undoubtedly true that frozen sperm does not have quite the same fertilizing capacity as fresh semen, straight from the vesicle. This is one of the reasons why, when artificial insemination or IVF is done in today's clinics, the preference is for the male partner to provide a sample on-site, so that it can be used without additives and with minimum delay. But where donor insemination is concerned, freezing has a very significant advantage – which is why nowadays it is almost universally used: that is, it allows health checks to be carried out on the donor before the semen is used. This has been particularly valuable in the light of recent widespread concern about the human immunodeficiency virus (and indeed other infections as well). HIV has a long incubation period, so simply testing someone before they provide a sample is not an effective means of screening out the possibility of infection, because they could have acquired it just a few days before the blood test. Sperm freezing allows us to take a sample and test the donor afterwards, not just for HIV, but also for hepatitis and other sexually transmitted diseases. The current practice in the UK is for a blood test to be taken at the same time as the donation, and then another one not less than 180 days afterwards; if this too is clear, the sperm is safe to use.

Whose life is it anyway?

While sperm freezing offers certain advantages, the fact that regulations allow us to store it for up to ten years also provides rich dilemmas. Human relationships, like eggs and sperm themselves, are subject to change and decay. If a couple were to store sperm as an insurance when the man undergoes cancer treatment, they might be able to conceive several children successfully afterwards. But what if the couple should split up? Should the woman still have the right to create her former partner's heirs, without his approval? The sensible answer is no – and that a man should give written consent before his sperm can be used. In Britain, this is the line enforced by the HFEA under the terms of the 1990 Act; but in 1995 a case showed that sometimes the law may be applied too rigidly.

Stephen Blood, a religious Christian, developed acute meningitis in February 1995, at a time when his devoted wife Diane thought that she might be pregnant. She had been trying to have a baby for about three months. Although the welfare of her husband was her paramount concern, Mrs Blood was aware that his illness, with its high fever, could render him sterile. Moreover, she was also aware that by the time he was admitted to hospital he was so ill that he might not even live. Some years earlier, Diane and her husband had discussed this awful situation as a purely theoretical possibility and Stephen had told her that, if anything should happen to him, he would like her to have his child after his death. Within just a few hours of hospital admission, Stephen was critically ill and soon he was in a coma. In her very severe distress, Mrs Blood asked the doctors at the Jessops Hospital in Sheffield whether there was any possibility of getting some sperm from Stephen. The consultant in charge explained that they could obtain a sperm sample by electro-ejaculation, and then have it frozen. Mrs Blood, convinced this would be what Stephen would want if he could consent, signed on his behalf and two semen samples were duly taken and stored. During this episode the hospital contacted the regulatory body, the HFEA, by telephone to ask for approval to maintain the semen in its frozen state in storage, and the then chief executive felt that it would inhumane not to agree. Very shortly afterwards, Mrs Blood had her period – she was not pregnant after all. And within a few days of contracting the infection, Stephen died, still unconscious.

Months later, when Mrs Blood decided she wanted to be inseminated with her late husband's sperm, the HFEA backtracked and said that its storage and usage were illegal. The key point was that while Stephen might have given spoken consent in private conversations, he had not signed a piece of paper. So the HFEA also refused Diane permission to export the sperm to another country where the procedure could take place.

It turned out that our worthy regulators found they had a formidable foe in Mrs Blood. She was a committed Christian, with strong views about the importance of raising a family. In her view, rooted in her conviction that this would be what Stephen definitely wanted and in their shared religious beliefs, she had to fulfil her duty of marriage. She was able to produce evidence that this was something the couple had discussed, and that Mr Blood believed having children to be an essential feature of their

Christian marriage. Mrs Blood repeatedly tried to persuade the HFEA of the rightness of her cause, but the authority was implacable and declared its intention to follow the letter of the law. Once Diane Blood, with the full support of Stephen's family as well as her own, had called for a judicial review of her case, there was widespread media coverage. At this stage Lord Lester acted as her barrister and he called me to give evidence as an expert witness.

Public opinion, stirred to high levels of excitement by coverage in the media, was firmly on Mrs Blood's side, but the High Court rejected her case. Iniquitously, the HFEA thrust the knife even further into Diane's body by calling for their legal costs. They were, of course, within their rights; but they had formed their own opinion of Mrs Blood's character, and their request seems vindictive even now. I remember very clearly a private conversation with one of the senior members of the authority who was convinced that Mrs Blood was simply in this for the publicity. In fact, I am certain that nothing could have been further from the truth; but as the HFEA had refused to meet Diane and discuss matters with her, it was perhaps unsurprising that they had a false view of her motives.

Diane, still with the strong support of all her family, decided to carry on fighting. After an agonizing wait, and much unwelcome time in the media spotlight, she was granted a hearing by the Court of Appeal. Chaired by Lord Justice Woolf, the court upheld the rule of written consent; however, it considered that the public interest was not being served by preventing Mrs Blood from exporting her late husband's sperm. In 1998, she was referred to my colleagues Paul Devroey and André van Steirteghem in Brussels, who run one of the best European clinics. They used the frozen semen to perform ICSI and she subsequently fulfilled Stephen's wishes by giving birth to her first baby boy, Liam. And in the summer of 2002 Joel, his younger brother, was born.

In retrospect, one cannot avoid asking what damage was done in allowing this clearly happy family to exist. What aspect of public policy was truly served by the HFEA's rigid stance? It seems a somewhat sad reflection on our regulatory system that, over four years, Mrs Blood went through such a harrowing experience before so simple, reasonable a request could be fulfilled.

If Diane Blood's story shows some of the social problems raised by the freezing of sperm, the experience of Mr and Mrs Davis in Tennessee showed that the issues concerned with freezing embryos can be even more fraught. In the spring of 1979, Mary Sue and Junior Lewis Davis met while doing military service in Germany. After a brief courtship, they returned to the United States for their wedding and then resumed their army posts. Like many couples, they intended to have children fairly quickly – but the initial joy of a positive pregnancy test was soured by the news that the embryo had settled in Mary Sue's fallopian tubes. She underwent surgery, during which she had her right fallopian tube removed.

In those years, while doctors in England and Australia had had some success with IVF, attempts in the USA lagged behind. The doctors informed Mary Sue that she still had a reasonable chance of motherhood by natural conception – after all, she still had one fallopian tube. But here was a cruel turn of events. Usually, when one tube is damaged enough to lead to an ectopic pregnancy, there tends to be similar damage on the other side, even if it is not obviously visible to the surgeon. And so, while Mary Sue did indeed conceive again – five times in all – each resulting embryo implanted in her remaining tube. This was extremely depressing, incredibly debilitating and a constant cause of anxiety. So, for the sake of her health, Mary Sue eventually decided to have her remaining fallopian tube tied off.

Still determined to become parents, the Davises made moves to adopt a child. All the initial stages of adoption were completed when, in another unkind blow, the child's birth mother backed out at the last minute. They attempted to overcome their grief by approaching various private adoption agencies, but soon discovered that, with their basic soldiers' pay, this was an option quite beyond their reach. However, by this time, the American pioneers Dr Howard Jones and his wife, Dr Georgeanna Seegar Jones, had had success with IVF (their first IVF baby, Elizabeth Carr, had been born in December 1981) in Norfolk, Virginia, and IVF was really starting to take off in America. It seemed to offer the Davises their only chance of attaining what they longed for.

But, like many couples, the Davises soon discovered that IVF is no panacea. By now they could just afford to spend some $35,000 dollars (it would have been very much more a few years later), and between 1985 and 1988 they underwent six agonizing cycles of anticipation, hope and disappointment – made considerably worse by Mary Sue's deep-seated phobia of needles and hospitals. All were unsuccessful.

There remained one of the great hopes for people undergoing repeated IVF: a possibility offered by technology recently developed in Australia. The affable Alan Trounson and his colleague Linda Mohr had made yet another breakthrough in Melbourne.[2] Amid convivial celebrations at a large international meeting, Linda (whom I subsequently thought of as the Ice Queen) presented their work. They had achieved the world's first human pregnancy after embryo freezing and thawing. Essentially, what Alan Trounson had done was to use information derived from work he had already done in cattle, and work that had been conducted in Cambridge in mice embryos by Dr David Whittingham. A key to their success was the use of more sophisticated cryoprotectants, and very carefully timed slow freezing, so that ice crystals did not form inside the cells of the embryo and destroy it. In those days, freezing the embryo was done by lowering it into a solution of liquid nitrogen very slowly (over perhaps half an hour). Nowadays, this method has been replaced by the use of computer-controlled, programmed refrigerators. Initially, most clinics found it difficult to achieve what Trounson had done, and it took a few years for the method to become widely established.

It was in December 1988 that the Davises received some very welcome news. Staff at the clinic where they were being treated had now been successful in using the Australian technology to freeze and store any spare embryos created by the IVF treatment. In Mary Sue's final IVF treatment cycle, nine embryos were removed, and two were transferred to her uterus. The rest were stored in liquid nitrogen, where – as Whittingham's mice experiments in England indicated – they could almost certainly remain indefinitely without any obvious damage.

At that stage, no-one at the clinic explained the full implications of the freezing procedure, or what might happen to the embryos in the long term. In fact, in those days very few doctors had considered that frozen embryos might become hostages to fortune. So, in the minds of the Davises and, it seems, of the people treating them, there was no 'long term' beyond the few months necessary to obtain a successful implantation.

Sadly, the embryos transferred on 10 December 1988 did not result in a pregnancy. That Christmas, when much of the rest of the world seemed to be preparing to celebrate the joys of family, Mary Sue had to prepare herself, physically and emotionally, to accept another menstrual period. At this stage she decided that she was going to continue to try for a pregnancy until the remaining stock of embryos had run out. But then Junior dropped a bombshell. He filed for divorce. Later, he told the courts that he had known their marriage was unstable for at least a year – a situation not unrelated to the emotional, physical and financial toll exacted by continuous unsuccessful IVF treatment. He had kept going, as do so many people misguidedly, in the hope that a baby might cement the cracks in their once-happy marriage.

The divorce was a straightforward affair – or rather, it would have been, but for the existence of several four- to eight-cell aggregates of human tissue, invisible to the eye, stored in liquid nitrogen in the freezers of an IVF clinic.

Junior Davis had had a difficult childhood, as the youngest but one of six children whose parents had divorced. His mother had endured a nervous breakdown, with the result that Junior and three of his brothers had been raised in a children's home. They saw their mother every month, but their father became a shadowy presence in their lives, visiting only three times before he died in 1976. In addition to the painful separation from his mother, Junior suffered considerable anxiety as a result of the lack of a relationship with his father. He grew up determined that any child of his would be part of a stable, two-parent family.

While Mary Sue accepted Junior's concerns about her raising any potential children in a one-parent family, she viewed as abhorrent the prospect of discarding the frozen embryos and demanded the right to donate them to another childless couple. But her ex-husband was equally opposed to this. Should the potential adoptive parents divorce, he argued, he would once again be condemning a child or children of his own, as he saw it, to the miseries of his own upbringing. In addition, he would spend the remainder of his life wondering what had become of his potential offspring. Should any of these embryos become people, and discover the facts of their birth, they too might be convulsed by doubt and suspicion about their genetic parentage.

Thus a previously loving couple were entrenched in opposing sets of opinions, either of which may sound quite reasonable from the outside. And therein began a legal battle, encompassing not just the status of a frozen embryo but the rights and freedoms at the very heart of the American Constitution.

The debate was coloured by some strongly emotive language, especially in the case of an expert opinion obtained by Mary Sue's lawyers from the French geneticist Dr Jerome Lejeune. Lejeune, consumed with religious zeal as ever, referred to the minuscule embryos as 'tiny persons' and 'early human beings'. He said he was 'deeply moved' that 'Madame [Mary Sue] wants to rescue babies from this concentration can'. But of course, though he was called as an expert witness, he was not speaking as a scientist; he was expressing an opinion that had nothing whatever to do with his undoubted expertise in genetics. He ought to have understood that, biologically speaking, it is highly doubtful whether one can speak of an embryo at the eight-cell stage as being a person. Apart from any other consideration, each component cell of such an embryo has the potential to divide and multiply into further life-forms. And some people will find it offensive that he could speak of embryos that have no human consciousness, and therefore could experience no suffering, as being in a 'concentration can'. But in Tennessee these opinions carry much weight and, despite contrary arguments from Dr Irving King, the gynae-cologist who had performed the IVF, the court decided that the embryos had the status of in vitro children. So, accord-ing to the laws governing the best interests of children, it was decided that Mary Sue should have 'custody' of them.

When this decision was announced, her former husband Junior lodged an appeal. He was supported in this by a number of concerned legal experts, who realized that the decision could have paved the way for making IVF incapable of being practised in Tennessee. This time Junior won his battle. The Court of Appeal reversed the initial decision. In giving judgment, the appeal judge said, 'It would be repugnant and offensive to constitutional principles to order Mary Sue to implant these fertilized ova against her will,' and 'It would be equally repugnant to order Junior to bear the psychological, if not the legal, consequences of paternity against his will.' Giving the court's ruling, the senior judge declared that the embryos were not children, but the 'joint property' of both Mary Sue and Junior. After this judgment was delivered, the emotional Charles Clifford, attorney for the ex-husband, apparently fell to his knees outside the courthouse and is reported to have said, 'All right, thank you. Justice is done.' One rather mischievous journalist, writing about the case later, said that in his opinion Junior and Mary Sue should each take responsibility for three embryos, and toss a coin to decide what to do with the seventh.

In making its decision, of course, the court in Nashville, Tennessee opened a legal, ethical and practical can of worms. The whole reason why the case had come under the scrutiny of the legal system was that the Davises had made no prior agreement about the fate of the stored embryos; but the court's final decision did not provide one either. The embryos remained in storage at the IVF clinic. Effectively, this was a decision in Junior's favour, since presumably an embryo cannot be stored for ever. Unless there are some radical advances in technology, he has the power, by doing nothing, to keep the embryos in their cryo-limbo until such a point as there is a major power cut, the liquid nitrogen runs out or the embryos become biologically unusable. Admittedly, biological degradation might take a very long time indeed – Whittingham's mice experiments, if extrapolated, suggest that a mammalian embryo might be perfectly viable after three hundred or even four hundred years in liquid nitrogen. What a time capsule that would be! But in the Davis conflict, legal experts on both sides argued that the situation raised by these embryos calls into question some of the most fundamental rights guaranteed to the individual by the US Constitution. The Fourteenth Amendment declares that 'no state shall deprive any person of life, liberty, property without due process of law'. Over the years, court rulings have interpreted that right to liberty in a variety of ways – including among them both the right to procreate and the right *not* to procreate. If Mary Sue is entitled to give life to those embryos, is Junior equally entitled not to give them life? Should any person be forced, by a ruling of the government, into any form of parenthood?

Heirs and spares

These are perplexing issues, and ones that might never have arisen had Alan Trounson not frozen and thawed a human embryo. The nature of IVF treatment and its variable success rates give rise to the conditions under which the frozen embryo can become the centre of a tug-of-love. As we saw in previous chapters, IVF treatment generally involves stimulation of the woman's ovaries with hormones to produce a crop of eggs. In practice, only some of these will fertilize successfully – depending on their own quality and the quality of the sperm brought into contact with them. This raises the potential, in at least 50 per cent of cases, for there to be more fertilized embryos than we wish to implant back inside the woman's uterus. Thus arises a set of complex dilemmas.

But embryo cryopreservation is a very important advance. It offers instant recourse to a stock of almost ready-to-go embryos in the highly likely event that implantation fails during the first treatment cycle. This relieves the woman of the need to repeat the time-consuming and potentially risky cycle of ovarian stimulation and egg collection. It allows people to delay and time conception – which may be particularly useful in cases where one or the other partner might be about to undergo treatment that could render them infertile, such as radiotherapy. It also opens up the opportunity for infertile couples, or those with a high risk of passing on a genetic disorder to their child, to receive a donated embryo from someone else. Because the embryo is frozen, it has an advantage over egg-sharing programmes, where the IVF clinic has to synchronize precisely the ovulatory cycles of two women.[3] It also bypasses some of the ethical objections to the whole business of IVF. Some Christians feel that the treatment is wrong because it results in the creation of so many surplus embryos. In their view, a person is created at the point of conception. Freezing allows that 'life', if that is how you choose to define it, to be preserved and potentially passed on – either to the original couple or to others.

In contrast to the sorts of personal and social dilemmas it raises, embryo cryopreservation is technically relatively straightforward, since the trickiest elements of IVF occur when the eggs are harvested and fertilized. The embryos are mixed with a special fluid to protect them from the potentially harmful effects of the freezing procedure, then stored in a plastic straw or a glass ampoule. This is placed into liquid nitrogen, which is very slowly cooled to –196 degrees Celsius using a computerized programmable refrigerator. When the time comes for thawing, they are simply taken out of the liquid nitrogen and left at room temperature. Once they are thawed, the cryoprotectant fluid is removed and culture medium is added. If the embryos were frozen at the more developed blastocyst stage – which is increasingly preferred, as it gives a better idea of how well they are developing and how likely they are to implant – they can be thawed and transferred on the same day. If they are younger and less complex, they are usually placed in culture overnight until they have reached the right level of cell division for transfer to the uterus.

It would be wrong to describe this procedure as being 'waste-free'. In practice, in most units some 75–80 per cent of embryos survive the freezing and thawing process. Damage does occur, not during the storage period but during the changes in temperature related to freezing and thawing. Therefore, it is usually necessary to thaw out several embryos in order to obtain a few that have the potential for implantation. For some people, this still amounts to a wastage of life – and so negates the value of embryo freezing.

To maximize the chance of implantation, transfer of the embryo to the uterus is carefully timed to fit in with the woman's menstrual cycle, the idea being that the uterus should be ready to receive and nurture its precious cargo. In younger women with regular menstrual and ovulatory cycles, this may be done in harmony with the natural rhythms of the body, but it is common to give at least one dose of progesterone on the day of transfer to help ensure that the uterus is as receptive as possible. It may be preferable to give a mixture of oestrogen and progesterone to take over the role of the natural hormones, and in older women, or those with no ovaries or irregular cycles, this is routine. Sometimes these drugs may be given for up to ten weeks after a positive pregnancy test result. Some women have concerns about the embryo transfer itself, but this generally causes minimal discomfort and is similar to the transfer procedure with fresh embryos.

Success rates for IVF using frozen embryos vary considerably, and are not particularly high in all clinics. Outcomes do depend partly on the quality of the embryos that are frozen. It is difficult to get recent statistics – the figures collected by the HFEA and published on the internet are up to five years out of date, which seems hardly acceptable in a rapidly changing field like IVF; indeed, given the facility of modern electronic data collection, this seems just another example of the incompetence of the regulatory authority. The latest available HFEA figures at the time of writing indicate that the live birth rate varied hugely between clinics, some achieving 29.2 per cent, others only 3.8 per cent. The lowest figure compares unfavourably with the success rates for transferring unfrozen embryos – around 16 per cent. At Hammersmith, where the policy has been to freeze only 'good' embryos,[4] the implantation rate is currently around 17 per cent per embryo transferred; and if two are put in the uterus together the pregnancy rate is around 26 per cent, very similar to routine IVF.

Potential risks of freezing: challenges for research

Freezing itself does not seem to damage the embryos – at least, there is no strong evidence that any of the thousands of babies born from this technique have any abnormalities. But, because the first procedure was undertaken in 1984, there haven't been any long-term studies to indicate problems that may occur only in adulthood or old age. And we now know that environmental changes while embryos and fetuses are growing can possibly affect health in middle age (see page 183–84).

I do have some concerns. The deep-freeze procedure entails the use of potentially harmful chemicals. We humans are, and always will be, mostly composed of water, and the embryo is similar. When we apply an anti-freeze compound which rids the embryos of excess water we might, theoretically, cause problems. Some of these compounds are very powerful – they have the ability to pass through cell walls and, in very high doses, even to cause genetic mutations. The very nature of these changes in the DNA could just possibly mean that they might not become visible in any resulting children for a number of years. Consequently, we still don't know, for certain, whether freezing is totally risk-free. It is just worth remembering that it was some fifty years before we understood that X-rays administered to pregnant women gave their babies an increased risk of developing leukaemia. It is to be hoped that there will be no similar revelations about cryopreservation in 2034.

One French study involving mice embryos did examine the long-term effects of embryo freezing in mice. The researchers noted weight gain, changes in jaw structure and unusual behavioural patterns in the 'frozen' mice.

This study has been hotly debated – and to date never repeated. But while mice may not be similar to humans, this study gives us good reason to continue monitoring the health of people born from embryo freezing for a long time to come.

I described in chapter 3 how one of my PhD students, Maria Tachataki, who joined my laboratory from Crete, examined human embryos that had been frozen and subsequently thawed and found temporary changes in the activity of a particular gene that suppresses the formation of certain cancers (page 182). I am not claiming that freezing embryos would necessarily cause cancer. Indeed, I think that this is very unlikely. None the less, it is alarming that a cancer-suppressing gene was inhibited in this way, albeit only for, apparently, a few hours. And it is worrying that studies of this sort are uncommon and seemingly not welcomed. I would like to think that research workers would now study the action of other genes in such circumstances; but the difficulty in getting approval for even properly conducted research on embryos will put off many young scientists who cannot wait a year before they can do an experiment like this.[5] Surely, studies like this are a good reason for undertaking embryo research. Unfortunately, the HFEA has not shown much enthusiasm for promoting such studies, and many colleagues have taken an extremely hostile stance towards those that have been conducted, which prompts me to reflect on a possible conflict of interest. When my Hammersmith colleagues and I first presented our findings on freezing in front of clinical colleagues at various national and international meetings, they were sharply critical. Perhaps the idea that freezing could cause embryo damage so alarmed them that they reacted defensively. It is interesting that basic scientists never showed the same reservations and showed great interest in the results.

Looking back, I find it interesting that when I first raised my concerns that embryo freezing might possibly lead to unexpected results in children, my colleagues were far from impressed. At one public meeting organized by the British Fertility Society in the early 1990s, I pointed out that, while there was no clear evidence that freezing caused embryonic abnormalities, it was surprising how many abnormal babies appeared to be born after IVF, and suggested that we should look at the data on results achieved after embryo freezing. This was over ten years ago. Because the press were present at the meeting, these comments were widely reported in the newspapers. That may have been unfortunate; but what is surely unaccept-able is that some clinicians made these views themselves seem highly reprehensible, arguing that all the fuss was simply publicity-seeking. Subsequently I received a letter from the then chairman of the HFEA suggesting that I should restrain myself in presenting such views publicly as the controversy was frightening patients.

This was not the first time that Hammersmith had crossed swords with the HFEA over this matter. A year or so earlier, the information pamphlet we gave out to all IVF patients had stated that while we did embryo freezing when we felt it was really necessary, we could not recom-mend it wholeheartedly as a routine procedure. We felt that there might be risks to the embryo, at present unclear, that might just cause problems later on in life. One of the concerns we had was that freezing might render a child more susceptible to late-onset disorders. In particular, I privately felt that there was a theoretical risk, at least, of leukaemia, although our hospital pamphlet did not say this. At a subsequent inspection of our unit by the HFEA, I was informed by the inspectors that, while our clinic was functioning well, they were concerned about the literature we were handing out to patients – specifically, about our health warning regarding embryo freezing. The implication was all too clear. There was a possibility that we might not have our licence renewed unless we rewrote and reprinted our information pamphlet, leaving out any reference to these concerns. As we had no data to validate those concerns beyond doubt, this was duly done.

It may surprise some people that the HFEA, which collects all statistics from all clinics around the country, has not attempted to consider this and related issues seriously. Even today, the HFEA has not conducted long-term follow-ups on children born from IVF to see whether abnormalities after freezing, or indeed any other added procedure, might predispose to problems later in life. This does seem a seriously missed opportunity, in view of the fact that Britain is one of the few countries that have such a regulatory authority, and that therefore we in the United Kingdom are in an ideal situation to fill this gap in follow-up after IVF procedures.

There certainly seems to be no obviously increased serious risk to mothers during pregnancies and births resulting from embryo freezing. One American study produced the alarming claim that frozen-embryo implantation was seventeen times more likely to result in an ectopic pregnancy; but theirs was a sample of only nineteen cases. Contrary data from clinics in Aberdeen, Manchester and New Jersey suggest a very similar rate of ectopic pregnancies whether the embryo is frozen or not. Nor is there evidence clearly indicating that miscarriage is more likely after frozen embryo transfer.

There is one other repercussion which may pose a minor problem. The drug treatment that prepares the uterus for embryo implantation does alter ovarian function temporarily and changes the pattern of growth of the uterine lining. In someone who does not get pregnant after embryo transfer there is always the possibility that the next period may be delayed, early or quite heavy. Also, a few women experience irregular periods for a few months afterwards.

What happens when the clock strikes?

In addition to potential unknown physical risks associated with embryo freezing, there are some obvious ethical complications. As the unhappy story of the Davises proved, the created embryo is a 'hostage to fortune' while it exists in the limbo of the clinic deep-freeze. What becomes of this potential life if its parents should die, or end up disagreeing between themselves about its fate? There is also the very faint possibility that embryos stored for a long time may develop serious abnormalities, or simply be unable to produce a successful pregnancy. Whatever the realities, the UK legislation introduced in 1990 stipulated that frozen embryos could be stored for an apparently arbitrary maximum of five years before they had to be used, donated or allowed to thaw out and discarded. In certain circumstances, with consent from both parties who had created the embryo, an extension of a further five years can be granted.

I do not believe that many people had considered the implications of this ruling. They became clear in August 1996, when the first batch of embryos stored in Britain was due for destruction. News of this event spread around the world and feelings ran high, especially among certain Christians, to whom the law seemingly displayed a flagrant disregard for human life. Among those whose passions were aroused by the prospect of these embryos being discarded was an American Christian called Ron Stoddart, who ran a California-based adoption agency. Stoddart realized that he could not stop what was happening in the UK, but the issue turned his thoughts to what was happening to the multitude of surplus embryos created every year in his own country from IVF treatment. Where he had formerly specialized in matching Christian couples with suitable children from Russian orphanages, he now started to match them with embryos.

In 1997 Stoddart launched the 'Snowflakes Program', offering an almost win–win situation for infertile Christians. IVF treatment became ethically acceptable to them, on the grounds that any spare embryos would be frozen and then 'adopted' by like-minded couples. The like-minded couples, meanwhile, had the advantage of becoming parents without undergoing either IVF or the troublesome and sometimes unsuccessful procedure of adopting a child from another country. Both sets of people felt that they were 'saving lives'. Some liberal Catholic authorities took the line that, while a great sin had occurred in the in vitro creation and freezing of embryos, adopting them at least amounted to a positive act of good. There were also advantages to the procedure of a less theological nature. Since they were adopting, and not genetically related to the embryo, neither parent had a greater claim upon it. Adopting parents and donors also had the option of knowing and meeting one another. And the adopting mother got the all-important chance to give birth to the 'snowflake' itself.

By 2003, some thirty-one 'snowflake babies' had been born to carefully selected couples – who had to have been married for at least three years, and expressed a commitment to raising their child in a 'constructive, wholesome and spiritual home environment'. The federal government lent its backing, in the form of half a million dollars to promote the Snowflakes campaign. Yet despite government approval – and brochures that feature President George W. Bush holding one of the agency's most photogenic little adoptees – the Snowflake outfit seems to have remained a relatively small-scale concern, using the non-threatening terminology of adoption to appeal to a specific, strongly religious section of the American population.

Embryo adoption in the UK is handled by clinics in a neutral fashion – that is to say, without any particular religious overtones. It is likely that demand for it will remain fairly high in Britain, partly because of recent changes to the legislation. In a recent HFEA report, some 23 per cent of the clinics that responded to a survey reported that couples seeking a frozen embryo had to wait up to a year before they could find a suitable match. This wasn't helped by the fact that 11 per cent of potential donors who expressed an interest subsequently dropped out, after realizing the extent of the screening procedures. It's a situation that probably will not be helped, in the short term at least, by the lifting of donor anonymity (see chapter 5). Nor is the image of the whole freezing procedure enhanced by high-profile disputes like that of the Davises – or a more recent one in the UK.

In spring 2006 the European Court of Human Rights in Strasbourg arrived at a ruling after four years of legal wranglings and bitter recriminations between a British woman, Natalie Evans, and her former partner, Howard Johnston. The row, like the one between Mary Sue and Junior Davis, concerned the fate of a number of frozen embryos. These had been in storage ever since Ms Evans began exploratory treatment before IVF in 2001. Upon discovering that she had pre-cancerous growths in her ovaries, she and Howard agreed to the creation of six embryos. Her cancer treatment would leave her sterile. Subsequently, after an understandably stressful period, the couple parted. Natalie Evans then decided she wanted to have a baby.

Current UK law states that both partners must provide written consent to the usage, donation or destruction of embryos during the five-year storage limit. Either partner has the right to withdraw their consent, up to the point when embryo transfer takes place. Mr Johnston exercised that right, claiming that his former partner was effectively forcing him to start a family. Ms Evans argued that she had no intention of making him become a father, and in any case, since her ovaries had been removed, this was her only option for having a genetically related child. The case went to the High Court, and eventually the Court of Appeal, each of which found in favour of Mr Johnston. Whatever side one takes, Ms Evans's remarkable determination is eloquent testimony to the overpowering need many women feel to bear a child.

Like the plot of a suspense film, its tension inexorably mounting as the time-bomb ticks in the background, the Natalie Evans case became increasingly fraught as the five-year deadline approached after which, if no decision were reached, the embryos would be destroyed. In September 2005, seven months before this point, her legal team took the case to Strasbourg. They argued that the rulings of the British courts contravened the European Convention on Human Rights – both by denying Ms Evans the right to a family, and by denying the embryos a right to exist. The panel of seven judges expressed their sympathy, but in March 2006 a majority of them concluded that Ms Evans's rights under the convention could not outweigh Mr Johnston's right *not* to become a father. And they unanimously agreed that an embryo has no right to life.

As I write, Ms Evans plans to take her case to the Grand Jury of the European Court, while the destruction deadline looms ever nearer. She is quoted as saying that she still hopes her former partner will change his mind. Mr Johnston, meanwhile, remains steadfastly opposed, telling a BBC reporter, 'The key thing for me was just to be able to decide when, and if, I would start a family.'

One could argue that, by consenting to the creation of embryos with his sperm, Mr Johnson had already given that approval. A sperm donor donates sperm, for example, in the full assumption that this will go on to fertilize eggs and create human lives. And someone consenting to the creation of embryos with his partner's eggs is surely giving an even more solid version of his approval. It is difficult for me to sympathize with Mr Johnston's perspective, especially as he remains fertile, while his former partner has had her ovaries removed. He is to be commended, at least, for not seeing his former partner's defeat as a victory for himself. And it seems, from the rueful opinions expressed by almost everyone involved with the case, that this is a classic example of everyone getting what no-one wants. Perhaps the only victory has been for those at the cutting edge of research – for whom the European Court ruling confirmed the fact that the embryo has no intrinsic right to life.

But perhaps these painful issues surrounding the status and ownership of the frozen embryo will subside once a new technology becomes more widespread. For ever since the mid-1980s we have possessed the potential to control our fertility, and skirt around the ethical minefield of the unborn embryo, by freezing eggs and ovarian tissue.

Good in parts: a medical curate's egg

In June 2002 a couple in the midlands produced Britain's first 'ice baby', Emily Louise Perry. After seven years of failing to conceive naturally, Emily's mother Helen underwent surgery in 1992 to unblock her fallopian tubes, which may have been damaged as a result of appendicitis when she was just seven years old. The procedure failed to produce a successful birth, and the Perrys then began a four-year wait for NHS-funded IVF treatment. However, their strong religious views as Jehovah's Witnesses gave them concerns about the possible wastage of embryos that would result from in vitro fertilization.

The breakthrough for the Perrys came in 2001, when a nearby clinic obtained a licence for freezing and storing eggs, a procedure that bypassed their ethical objections to the use of embryos. In April that year some thirty-four eggs were harvested from Helen's ovaries. Rather than abandon the treatment when so many eggs had been produced, and to give her body a chance to recover from the fertility drugs that had been administered, Gillian Lockwood, Helen's doctor, decided on what was then a novel approach. Dr Lockwood has considerable expertise with egg freezing and all the mature eggs were frozen and stored for two months. In June eight eggs were thawed, of which five survived the procedure. Three of these were injected with Mr Perry's sperm, resulting in two embryos. Two weeks later, sadly, came the news that Helen had failed to conceive.

Three months later, the procedure was repeated. Six eggs were thawed, five survived and three were injected with sperm. Two fertilized and continued to develop to the blastocyst stage, at which point they were transferred to Helen's uterus. Two weeks later, on 19 September

2001, the Perrys were told that she was pregnant and subsequently their baby, Emily, was born.

Egg freezing offers potential to a range of infertile people, regardless of their ethical views. When we use frozen embryos, both a woman and a man have been involved in their creation – and both partners have a right to decide what happens to them. When we use eggs, frozen before they are fertilized in a Petri dish, the fate of those eggs is purely a matter for the woman concerned. There is also an additional 'window of scrutiny' after fertilization during which we can observe their development prior to implantation. Storing eggs gives cancer patients the potential to preserve their chances of having a genetically related baby further down the line. It also creates the potential for women to approach their reproductive lives with a new sense of self-determination. Healthy young women could have their eggs removed and stored, build up successful careers and then become pregnant when they decide the time is right, rather than being dependent upon their biological clocks. This could remove the pressure to start a family that sometimes leads people to form unwise relationships, creating instead the possibility of embarking on family life at the point where it is financially and emotionally most appropriate to do so.

Almost as soon as egg-freezing technology became viable, some entrepreneurs saw a niche in the market for it. Across the USA, and now in Britain too, egg banks began to tailor their services, and their prices, towards young, successful professionals wishing to delay their fertility. But while this treatment offers exciting possibilities, it remains like the dubious breakfast egg served to the curate in the famous *Punch* cartoon – good only in parts.

Egg freezing remains a relatively experimental technique, albeit one whose success rates are improving all the time. As a procedure, it presents a number of challenges. The egg is the largest cell in the human body, and as a result the cryoprotectant takes a relatively long time to penetrate it. The cytoskeleton, the structure that holds and organizes the chromosomes within the cell, is a very delicate object. Any damage caused by the formation of ice crystals could have serious consequences for the health of the child. In 1984, a team working with human eggs in Cambridge showed that cooling them at normal room temperatures carried risks of genetic damage.[6] And indeed, the first team to report a successful live birth from egg freezing,[7] in Singapore in 1986, also produced a fetus whose chromosomes were radically altered, and which miscarried early in the pregnancy.

In thawed eggs, the outer casing, the zona pellucida, becomes particularly tough, and for that reason ICSI (intra-cytoplasmic sperm injection) may be used to assist the sperm's penetration. But this procedure itself poses certain risks. By intervening in this way, we are effectively giving the sperm a helping hand – in some cases, allowing fertilization with sperm that would, for very good reasons, otherwise fail in their task. There is therefore a possibility of passing certain problems from the father to the unborn child (see page 168). The general picture is that there is a slightly elevated risk of abnormalities in babies conceived using ICSI. Around 3 per cent have some chromosomal abnormalities, whereas the figure for the general population is around 0.6 per cent. Around 20 per cent of ICSI babies have minor abnormalities, compared to 15 per cent in the general population. Admittedly, these abnormalities are likely to be associated with male infertility rather than the procedure of ICSI itself, but they are worrying. The major disadvantage to egg freezing remains the very low success rates – probably a live birth rate of around 10 per cent may be expected. I quoted the precise figures in the treatment of Helen Perry to underline the point that few eggs survive the thawing process in the first place – lengthening the already weighted odds of IVF treatment. Advances are certainly being made in this field. A joint Japanese–American team has developed a special storage device called a Cryotop, which reduces the risk of ice-crystal formation and boosts the survival rate of frozen-and-thawed eggs. When the team tried out their new piece of kit,[8] 58 out of a total of 64 thawed eggs survived; 52 were fertilized, 32 developed into embryos and 29 were implanted, resulting in 7 live births and 3 healthy pregnancies at the time they published their results. The success rate per embryo transfer was 34 per cent, impressively comparable with the best results obtained from using frozen embryos.

A Boston IVF clinic is experimenting with a sugar called trehalose,[9] found in the diet of people who eat mushrooms, honey, yeast, lobster and shrimps. Researchers noted that some creatures, such as brine shrimps and arctic frogs, use natural sugars to protect their cells from damage at low temperatures. They have now developed a technique for injecting eggs with trehalose, to see if this might produce the same benefits for humans. We have reason to believe that trehalose may be beneficial in other freezing procedures – mouse sperm that were cryopreserved using this sugar were more likely to be capable of fertilizing eggs.

It seems probable that expertise in egg freezing will become more sophisticated. But the low success rate still is a concern. A large number of people may be disappointed, particularly those who have frozen their eggs for 'social' reasons. A young woman who freezes her eggs, pursues a career until she is forty and then tries to have a child may be very bitter when she fails to become pregnant – especially given that, by then, her other options may be limited.

I also wonder about the advisability of people waiting for the 'best' time to become pregnant. Certainly there are times in all our lives when an unplanned pregnancy would be a disaster. But pregnancy is such a life-changing and significant event that, in my experience, nobody embarks upon it without experiencing misgivings at some point. Egg freezing creates the opportunity for us to listen to those misgivings rather than take the plunge. Women, in my view, are certainly no match for men in the procrastination stakes. But there is a risk that, unpressured by their biological clocks, some might repeatedly consider using the eggs 'after I've settled into the new job', 'once I've got the house sorted out' and so on, until their chances of a successful pregnancy may be quite seriously diminished.

Ovarian storage

Freezing and banking will, I hope, offer a long-term solution to the poor availability of donor eggs. There are also some remarkable alternatives within our grasp, involving the use of ovarian tissue and even whole ovaries. In the 1960s there were unconfirmed reports that Russian scientists had grafted strips of ovary into the abdomens of women and restored their hormonal function during the menopause – though these reports were admittedly quite difficult to believe.

I first got interested in this area when I was working in

San Antonio, Texas, in 1980. Here I was working not only with Ricardo Asch but with my good friend Carl Eddy, one of the most brilliant exponents of microsurgery in the world. I feel very affectionate towards Carl. While we were always quite productive when we worked in the laboratory together (the year I worked at San Antonio, I think I published no fewer than eighteen papers), Carl and I never worked hard. We arrived in the laboratory at 9.30 a.m., made coffee and went to chat to the monkeys. Then we would go to another part of the animal house and stroke rabbits for a bit. After that it was up to the lab for an hour and prepare an experiment – and then it would be time for lunch. If we didn't go to the local inn that served red snapper, we spent our lunch break playing Space Invaders on a very sophisticated arcade machine in the local inn. (In those days Space Invaders was absolutely the latest technology.) After returning to the lab, we might, some days, actually get round to doing an experiment; but nobody in this unit – then a mecca for the study of reproductive physiology – worked after 5.00 p.m. Occasionally, if I was really in luck, Carl would take me off for the afternoon to the local US airbase where we would watch the F16s taking off with their thunderous noise and the afterburners glowing. Carl was a very good artist and photographer, and he used to draw the planes and do portraits of the pilots. He frequently got to fly in the navigator's seat and the USAAF always seemed to welcome him. I still have his drawing of me in a Spitfire hanging on my office wall.

Carl was a biologist, not a medical doctor, but he was so much better at surgery than any of the expert doctors at Bexar County Hospital in San Antonio that he was always being called in by the surgeons to do the intricate, most complicated parts of tubal surgery in their human patients. He was amazingly gentle and modest about it; nowadays, of course, he would not be allowed near a patient for fear of litigation – much to the loss of the patients.

Carl and I used to do prolonged, tedious studies in monkeys, measuring the hormones their ovaries produced at different stages of the cycle. These experiments were usually designed by Ricardo Asch, who seldom appeared in the laboratory in person. The monkey would first be anaesthetized, and then we would pass a very fine catheter into the ovarian vein and artery on one side. Every twenty minutes we sampled the blood to assess how the ovaries responded to varying doses of pituitary hormones, often over four to six hours. At the end of the procedure the monkeys would wake up as if nothing had happened; but we would be crippled by cramp, having been sitting still, in sterile gowns, gloves and masks, for many hours, just taking regular tiny blood samples from the catheters. It was terribly boring, but the data we gained were extremely valuable much later in working out many important human treatments, including the best conditions and timing for egg production. In order to overcome the tedium, I bought a plastic chess set and a plastic chessboard. These we sterilized along with our surgical instruments, and we used to play chess across our little patient's abdomen. But one day one of the technicians stealthily borrowed my chess set, and in the boring hours that followed, Carl and I had to discuss science – and which objects we would throw if Ricardo's face appeared around the laboratory door.

Carl was already well aware of my attempts at transplantation of fresh rabbit ovaries in London in 1974, after which a surprising number of animals had conceived. During this particularly dull experiment, he asked me why we should not do something a bit more exciting. Why not modify the procedure to transplant an ovary alone (rather than together with its fallopian tube, which I had done)? Moreover, he said, it would be a real advance if we could learn how to cool it and store it. I pointed out that, in spite of the flimsy Russian reports, the problem was that nobody could freeze a whole ovary and therefore there would be limited practical value in exploring transplantation. This is because transplants would be most needed after a long time interval – say, for example, after a cancer treatment, during which the patient's ovaries might be temporarily stored outside the body.

The problem was, of course, that while freezing cells individually or in small clumps is relatively straightforward – anti-freeze can be used and will penetrate the cells to prevent ice formation – an ovary is very bulky and, although it might be possible to freeze the outside part, the inside would not be easily accessible to the anti-freeze. Moreover, cooling in the centre of the ovary would be a slower process, and deterioration and death of the tissue would be almost inevitable. (This, incidentally, is why attempts at freezing a whole human body after death seem so pointless.)

Carl pointed out that we could use our catheter technique to inject the cryopreservative into the ovarian blood vessels. By this means we might reach the deep parts of the ovary more effectively. Moreover, he argued that we should try a different cryoprotectant. He suggested dimethyl sulphoxide – which is quite a toxic substance but displaces water from tissues more rapidly. So, many years before anybody else attempted ovarian cryopreservation, we set about this procedure in a rabbit. We removed an ovary one afternoon, gently perfused its tiny vessels with dimethyl sulphoxide solution and placed it, in a polystyrene box, in an ordinary refrigerator at 4° Celsius. The following morning, Carl and I transplanted the ovary into another rabbit – another of the same inbred strain which would not reject the tissue very vigorously. Remarkably, one week later, the ovary was still reasonably healthy; and within another week, laparoscopy showed it had started to produce follicles. Of course, our rabbit (called Topsy) never got pregnant, and eventually the graft failed because of rejection – but we were stupid never to publish our experiment. We meant to repeat it, but got sidetracked on to other problems. If we had persisted, there is a faint possibility we might have ended up famous. But good science is very much about persevering and persisting – and in this instance we didn't!

Interest in ovarian freezing was widely stimulated in 1995. At that time I was working with Outi Hovatta from Finland (now a professor at the Karolinska Hospital in Sweden). She was very concerned that every effort should be made to preserve ovarian function in young women undergoing treatment for cancer. Using the chemicals we had first attempted to use in the rabbit fifteen years earlier, Outi was able to freeze small pieces of ovarian tissue very successfully. When they are thawed after using her process, they look completely undamaged under a microscope. Outi's work is now used in many parts of the world, and it is quite routine at Hammersmith to store small pieces of ovary for women undergoing chemotherapy.

Storing ovarian tissue like this offers various potential advantages. It is a rich source of eggs – every square millimetre of tissue contains hundreds of them. It may also be more robust than harvested eggs, meaning that there will be less wastage during the freezing and thawing process. And in order to harvest eggs, we need to stimulate the ovaries with the same dosage of drugs used during IVF; whereas by taking a piece of the field, rather than the crops themselves, we have the potential to obtain a bountiful supply later on, bypassing the impossibly long waiting lists for donor eggs. Of course, these pieces of ovary are too small to transplant back into the patient with any real hope of success; but the possibility remains that we may be able to extract the eggs from these pieces and mature them artificially in culture. Any drugs needed would be administered to the pieces of ovary in culture – at far lower doses than would be needed in a person. In vitro maturation like this is still only a pipe dream, but is an important area for research (see page 442–3).

Grafting away

But there has been much bolder work in the field. Quite radical surgery may offer a prospect of ovarian grafting restoring female fertility – at least for a while. In Edinburgh in about 1995, in a much more sophisticated experiment than the games that Carl Eddy and I had played back in San Antonio, Roger Gosden, a pioneer in ovarian preservation, and David Baird, one of Europe's most respected figures in reproductive biology, did the first serious work in the field. Gosden had already experimented with mice and then moved to sheep ovaries, which are roughly similar in bulk to the human's. In this new work they capitalized on the fact that all the eggs in the mammalian ovary are near the surface, in the surrounding skin, the cortex of the organ. They stripped the cortex away – rather like peeling a potato rather thickly – and froze the resulting strips of tissue in liquid nitrogen for storage. Weeks later, the strips were thawed and sewn onto the raw surface of the animal's uterus. The strips continued to produce hormones for two years after the operation. In 1999, furthermore, Gosden and Baird announced some live births. However, it was clear that the tissue did not work as well as a normal ovary. Grafting had reduced the number of good follicles that were formed, and many of the eggs that were produced did not yield normal embryos.

Somewhat similar work followed in France, when Bruno Salle's team in Lyon cut sheep ovaries in half before freezing and then regrafting. They hoped that, by cutting the ovaries in half, they would preserve many blood vessels and thus cause less disruption – but still have thin enough tissue for freezing. Four out of six ewes treated became pregnant but, for reasons that are not clear, three of the lambs died after birth. Not long after this, an Israeli team, led by Amir Arav, announced they had transplanted whole ovaries in sheep. While these animals did not get pregnant, the scientists were able to perform IVF successfully on the eggs they collected from the ovaries they had transplanted.

More recently still, it seems that other researchers may have been successful using similar procedures in humans. In September 2004 a woman in Brussels made history by giving birth to the first baby conceived following a transplant of her own ovarian tissue. Ouarda Touirat had undergone chemotherapy for a cancer, Hodgkin's disease, in 1997. Immediately before the treatment, doctors took five small strips of tissue from her left ovary and froze them in liquid nitrogen storage. Three months after the chemotherapy had successfully seen off the cancer, she had hot flushes and her periods stopped. It was confirmed that – as is usual after drug treatment of this sort – her ovaries had stopped working, at the early age of twenty-five. Then, in February 2003, Dr Jacques Donnez, with a team from the Catholic University of Louvain, replanted the stored pieces of tissue. I know Jacques Donnez well as a most able fertility surgeon – indeed, when I was working in Leuven in the 1970s (in the Flemish-speaking half of that ancient university) I had the pleasure of demonstrating some of our microsurgical techniques to him. Five months after Jacques' operation, Ouarda began to have periods again. And then, remarkably, she fell pregnant – apparently confirming that it is within our power to reverse the menopause.

Wulf Utian, the fashionable gynaecologist and executive director of the North American Menopause Society in suburban Cleveland, Ohio, is reported as saying of Donnez's work: 'This woman has essentially been restored to a normal reproductive state after having been made temporarily menopausal by chemotherapy.'[10] But he also warned there was a real, albeit small, chance that the pregnancy had occurred because her left ovary, the one that had not been damaged at the original operation, may have started working spontaneously. This is not a fanciful suggestion; I have seen a number of patients whose ovaries were left intact and who, seven or eight years after treatment for Hodgkin's, have had a spontaneous return of their periods and a pregnancy without any intervention at all.

It is interesting that other surgeons have repeatedly failed where Donnez apparently succeeded. It is undoubtedly difficult to ensure that, after freezing, the thawed ovarian tissue gets enough oxygen, for there are no blood vessels that can be rejoined and it takes time for a new blood supply to grow. And apart from the initial loss of oxygen and nutrients, it will always have a less than adequate blood supply once put back in the body.

There is another surgical strategy that would seem to

have a better chance of success. In some cases the bits of ovarian tissue, each of which can contain hundreds of immature eggs, have been transplanted to novel locations where there are many small blood vessels that might grow rapidly into the tissue. Conducted through these vessels, of course, the pituitary hormones can continue to do their normal work. The sites include the loose tissues around the bladder behind the pubis, areas in the abdominal fat and under the skin of the forearm. The advantage of these sites is that access is easy. Although there would be no chance of spontaneous pregnancy, eggs could be removed from these places without difficulty. And the approach might make it easy to tell when the woman was ovulating – in one woman's case, a tiny bump appeared under her skin, and the egg was then removed with a needle for IVF with her husband's sperm.

In March 2004 Dr Oktay in New York treated a thirty-year-old woman who had had breast cancer.[11] Her ovary had been frozen in liquid nitrogen at the time of chemotherapy. When she had fully recovered, he took fifteen small pieces of the surface of the thawed ovary and transplanted them under her abdominal skin. Within three months the pieces of ovary had started to respond to the woman's own hormones. Oktay boosted the response by giving her injections of FSH. Over the next eight months he was able to suck some twenty eggs from the pocket under her skin, some of them in fertilizable condition. Unfortunately the two embryos created by IVF and put into her uterus did not take. One of the problems with this approach is likely to be that, because the pieces of ovary are small (about the size of a fingernail), and because they have a compromised blood supply, they are not likely to remain functional for a long time. So if IVF does not work within a year at most after such an operation, it may be that a successful pregnancy will become increasingly unlikely.

There is another serious potential problem. This treatment may be unsuitable for many young women with different cancers. It could not be used where there was ovarian cancer, obviously, and might be dangerous for victims of leukaemia or Hodgkin's disease.[12] This is because the pieces of ovary would be quite likely to contain cancer cells from the blood or lymph glands. It would be horrible to contemplate that leukaemia might be cured by irradiation and chemotherapy only to recur years later because of a treatment for infertility. On the positive side, animal studies at the University of Leeds suggest that this risk may be quite low. In 2001 a research team there took ovarian tissue from women with particularly aggressive forms of cancer, and implanted it into mice. As a control, they also implanted diseased human lymph nodes into a second group of mice. The mice that received the ovarian tissue remained free of disease – although the lymph node group did not.

Testicular freezing

The freezing of testicular tissue is a relatively recent advance but has an increasingly important place in fertility treatments. The microscopic tubules in which sperm grow can be frozen, and sperm can be recovered from them later. It is also possible to take testicular tissue and literally homogenize it – rather like making a soup in the kitchen. A homogenized mixture, it turns out, is rather easy to freeze with more success. But until ICSI was well established this technology had little value, because there was no obvious way any sperm taken from the testicular tissue could be used to fertilize eggs.

There are essentially two main groups of men for whom these treatments may be appropriate. One is men who are very infertile and producing no sperm at all in their ejaculate – perhaps because the tubing in the testes is blocked, or because they have had a vasectomy, or because the testes are just functioning very poorly. The other is young men, or boys, who need radiation or chemotherapy for a cancer and whose sperm can be stored in this way for future use. Just as with the ovary, many cancer treatments upset sperm production permanently because the stem cells in the testes (which make the sperm) are quite sensitive to the therapeutic agents that kill cancers.

What is medically controversial about this treatment is that it is increasingly being used to achieve fertilization with sperm cells that are not fully formed. Some men have such poor testicular function that the testes do not manufacture mature sperm. In the UK the HFEA refuses to give a licence to use immature sperm from the testes because it is concerned that they may be at risk of causing an abnormality. The HFEA does not consider that people should run this risk in every throw of the dice to get a child; so it is currently banned. Consequently, a number of British patients have had this treatment in European or American clinics, with some limited success.

In my laboratory, Carol Readhead, a friend and colleague from California, Outi Hovatta and I have been trying to restore fertility in mice. We have first removed the stem cells that make sperm, and have then given the mice a treatment which prevents their testes making any new sperm. Subsequently, we have used very delicate surgery to replace these stem cells. If this treatment does eventually work in humans, as it does occasionally in mice, it will mean that boys having cancer treatment could have sperm stem cells stored for years and then have their fertility restored normally; for once the stem cells have been returned to the testes they should carry on making sperm indefinitely. Then there would be no need for IVF.

Religious views

Research into this area continues apace, and meanwhile a trickle of encouraging reports fire up the public imagination whenever a live birth provides the opportunity for a cute cover-photo. There is still a need for caution, though – there may be disappointment in store for some infertile women who read the over-confident media reports and imagine that the treatment can offer them instant hope. The specialists involved admit that they don't have the full picture at this stage.

The view of most religious authorities is that all these forms of tissue, egg and embryo freezing are permissible, but only within the context of a live marriage. In Islam, the concerns centre on the donation of these human materials to others, and whether that constitutes an act of adultery. Within this framework cases like those of Natalie Evans or Mary Sue Davis would never occur because, once a marriage ceases to be valid, so does the possibility of using the eggs or embryos. Islam reaches similar conclusions in cases where the husband has died, forbidding women to fertilize eggs with his sperm, or to implant embryos created with it. However, Egypt's spiritual head, the Grand Mufti, reversed this ruling in a particularly tragic case, where a man was killed while accompanying his wife to an IVF clinic in order to have their embryos implanted. Because all of his family gave their consent, and the man had demonstrated such obvious intention to create a pregnancy, his widow was allowed to implant. But the outlook remains less favourable for other women. I believe that in many Islamic countries, having children acts as an insurance policy for one's old age. Childless widows are thus at a distinct disadvantage, even though the technology may exist to help them.

In Judaism, there are many responses to the problems thrown up by egg and embryo freezing. For a living married couple there is no problem with either technique – although a very few extremely conservative authorities rule that, in any IVF procedure, masturbation to produce semen is wrong. But even in that case, couples can get round this ruling by using specially adapted sterile condoms, so that the semen is produced through a permitted act of intercourse.

In general, Jews consider that assisted reproduction is a good thing, because it fulfils the commandment or *mitzvah* to 'be fruitful and multiply'. But most authorities, like the former Israeli Chief Rabbi Yehoshua Bakshi Doron, point out that this commandment is meant to be fulfilled only within the setting of a marriage. On those grounds, it might be forbidden for an unmarried man to freeze his sperm before cancer treatment. Although this rabbi makes no specific ruling on the subject, this could possibly also be the case for an unmarried woman considering storage of some of her eggs.

But what about cases like those of the Davises, or Natalie Evans and her partner? There was, in fact, a similar dilemma in Israel in 1991, when Ruth and her husband Dani Nahmani divorced, after having frozen several embryos. The intention, following Ruth's hysterectomy, had been to implant them in a surrogate mother, who would bear them to term on Ruth's behalf. After the divorce, Dani withdrew his consent for the embryos to be used. In traditional Jewish law, the consent of both partners would be required to go ahead. But the Israeli Supreme Court eventually gave Ruth the right to use the embryos – in a judgment which, perplexingly enough, is also supported by some religious authorities. Among the minutiae of Jewish law are some clear precepts regarding how people should conduct themselves when they have entered into a contract. When two parties create an agreement with a common goal, neither side should withdraw arbitrarily.

Jewish law also classes frozen embryos and sperm together in a way that has consequences for cases where a husband has died, and his widow wishes to conceive his child. In both Britain and the United States, the law states that any resulting child is considered fatherless – a somewhat heartless position, largely arrived at to prevent complex legal battles over inheritance. Jewish law follows this position, considering such a child to be the same as one born to a Jewish mother and an unconverted non-Jewish father. But that doesn't mean the practice is outlawed.

Judaism has a number of strict laws surrounding the proper treatment of the dead. In general, the principle is that corpses are to be buried as swiftly as possible to minimize the possibility of disease. Accordingly, and also out of respect for the recently departed person, Jewish law forbids anyone to derive benefit from a corpse. This might include the use of their sperm or any embryos created with it – as the rule could be said to apply, not just to the body awaiting burial, but also to any parts of it stored somewhere else. However, the commandment to be fruitful and multiply may take a kind of precedence. The *mitzvot*, some rabbis note, were not created for people to derive any benefits for themselves, so fulfilling them can never be said to offer benefits. Additionally, the commandment to procreate is a rare example of a *mitzvah* whose fulfilment lies in the consequences of an act, rather than the act itself. So, provided the husband has consented, it is acceptable for his widow to bear his child, via stored embryos or fertilization with his sperm.

Steptoe and Edwards are often unwittingly given a bad press by people praising their achievements. Had they not tried to help Mrs Brown back in 1978, it is often noted, none of these tugs-of-war over embryos or agonizing quandaries over eggs and sperm would be troubling us. But the real significance of their breakthrough lies elsewhere. Until eggs were fertilized with sperm in a Petri dish, it was proper to speak of the 'mysteries of life'. That is exactly how life was – contained, beyond our view, in the secret depths of the womb. In vitro fertilization took life out of this biological black box, laid it out for our scrutiny and, in doing so, gave us the means to control it.

Donated Eggs and Sperm: The Most Generous Gift?

The matter of the bedsheets

One ancient Jewish tradition has it that Ben Sira, the grandson of the prophet Jeremiah, was conceived in somewhat unusual circumstances. Jeremiah, his grandfather, had fled to Egypt, apparently hoping to meet Plato. While relaxing in the pool where he was staying, he encountered some rather unpleasant men from Ephraim, from the northernmost part of Israel,[1] who were amusing themselves by ejaculating into the water. Not entirely unsurprisingly, he expostulated with them, whereupon they turned on him violently, offering to commit sodomy on him unless he did the same. Some hours later, his virgin daughter went for a swim. She, unhappily, was impregnated by her father's semen while bathing in the pool. Nine months later, she gave birth to a son, Ben Sira, who came out talking. Ben Sira's first words were to his mother: 'Do not be ashamed . . .'

Technically, then, Ben Sira was his mother's brother as well as her son, his grandfather's son as well as his grandson. He was the product of an incestuous union, strictly forbidden in Jewish law. Having no legitimate father, he should have been considered a *mamzer* – a bastard, illegitimate, unable to inherit, unable to father Jewish children. Yet he became a prophet in his own right, author of the book of Ecclesiasticus,[2] summoned to address the court of King Nebuchadnezzar. Why was a man sired in such strange circumstances able to attain such a respected position within his religious community?

For the Jews, a people of distinct cultural identity, whose history has been characterized by the need to live closely alongside other groups, the issue of 'who belongs' is of paramount importance. For that reason, Jewish legal minds have been considerably exercised by the business of considering who has the right to participate in the promises made to Israel, to call themselves a Jew, to marry another Jew, and to call their children Jewish. In the case of Ben Sira, the rabbis decided that he was legitimate. Adultery and incest, they noted, were crimes that proceeded from a forbidden act of intercourse. No intercourse had taken place (although his grandfather had presumably committed Onan's offence of spilling his seed on barren ground – or waters). So there was no crime, and Ben Sira was no pariah.

A related issue troubled the mind of a thirteenth-century French rabbi, Peretz ben Elijah of Corbeil. He wondered what might happen if a woman lay on a bed previously occupied by a man other than her husband. Should the sheets be saturated with sperm from that man, and should the woman subsequently become pregnant, what would be the status of the child? Drawing on the precedent set by the story of Ben Sira, the rabbi ruled that this child too would be legitimate, because no intercourse had occurred. What troubled him more was the possibility that the semen might have been left behind by an unknown, but Jewish man. If that was the case the resulting child, unaware of its origins, might one day – particularly in the small, self-contained Jewish communities of the day – marry its half-brother or sister, thereby committing incest. Better to lie in a bed previously occupied by a Gentile than by a fellow Jew. Better still, just change the sheets . . .

The musings of this medieval talmudic scholar may seem like an exercise in legal hair-splitting – especially as the likelihood of sperm surviving on a bedsheet, and getting far enough along the female reproductive system to fertilize an egg, would appear on the available evidence to be rather slim. But Rabbi Peretz's discussion touches upon issues that are deeply relevant today, as we come to grips with new technologies that allow us to create babies from donated genetic material.

A short history of sperm

The *Kathasaritsagar*, an Indian text written in the eleventh century CE (its title means 'Oceans of the streams of stories'), tells of a king who, wanting to make an offering to his ancestors, threw three rice balls into a river. Three hands reached up from the waters to seize the balls – those of a farmer, a priest and a warrior. The king had to consult the oracles to find out who was the rightful recipient. The oracles told him, 'The farmer is the man who married your mother, the priest is the man who made your mother pregnant and the warrior is the man who took care of you.' The king was advised to make the offering to the farmer – a judgment not unlike the one taken by the Jewish scholars. The man's relation to the mother through marriage was deemed more important than a genetic connection (or indeed, that of nurture).

A number of Hindu sources allude to the possibility of

children being sired by donated sperm. The epic poem *Mahabharata* tells of the death of a king called Vichitravirya – his name actually means 'Strange virility' – who left a number of widows. The king's mother invited a wise man called Vyasa to sleep with the widows, and the resulting children were considered to be the dead king's descendants. As we saw on page 105, this custom, in the form called Levirate marriage, was also prevalent among the ancient Israelites – though, of course, in the case of the Hindu wise man it was not practised to establish the genetic continuity of any resulting family.

The god Shiva, the Lord of Destruction, is said to have once spurted semen when he caught a glimpse of Vishnu in the guise of an arrestingly beautiful temptress. Sages collected Shiva's semen and placed it in the ear (a suspiciously womb-like structure) of a monkey, who then gave birth to Hanuman, the monkey-headed god. Mythically speaking at least, the Hindu imagination was clearly captivated by the possibility of creating new life without intercourse. In other cultures, people have long been interested in the practical possibilities. In 1322 a traveller's account referred to an Arab chieftain stealing sperm from the stallion of a neighbouring tribe and using it to fertilize one of his own mares. And in the eighteenth century Lazaro Spallanzani, as we have seen, was among the first to understand that fertilization depended on the presence of both eggs and sperm. Like a modern PhD student reliant on MRC funding, he was compelled to live on a modest priestly income while conducting his research. Nevertheless, Spallanzani's further interests were wide-ranging, covering the digestive tract, translations of the Greek classics and even a treatise on the science behind 'skimming' stones.

Despite his enormous intellect, Spallanzani could not,

at that stage, fully understand the fertilization process. Following earlier studies of sperm by the Dutch scientist Antonie van Leeuwenhoek, Spallanzani thought that these wriggling life-forms, visible under the microscope, were merely parasites that thrived within the nutritious substance of the semen. Nevertheless, as we saw in chapter 2, the Italian was the first to attempt artificial fertilization of eggs taken from frogs, fish and dogs – and the first to succeed.

The first clearly recorded example of human artificial insemination comes around the same time. In 1776 the Scottish physician John Hunter reported that he had successfully inseminated the wife of a draper who suffered from the condition known as hypospadias. As noted in chapter 3, this disorder, present in about 1 in every 350 males, causes genital malformations, so that the opening of the urethra, instead of being positioned at the end of the penis, can appear at points along the shaft, or even within the scrotum. Needless to say, this can make intercourse difficult and ejaculation into the vagina impossible. The draper gave Hunter some of his semen, which Hunter placed in a warm syringe and deposited inside the wife's vagina: the result was a pregnancy.

The history of donor insemination shines some illuminating sidelights on male values. In 1909 the American journal *Medical World* received an article from a Dr Addison Hard of Jefferson Medical College in Philadelphia. Apparently ignorant of John Hunter's achievement in the eighteenth century (but then, American doctors and scientists are prone to ignoring precedents in the European literature), Hard claimed to have been involved in the first case of human artificial insemination in 1884. It was, perhaps, the first use of donated sperm – and certainly the only recorded one to be carried out without full consent. It so happened that a rich Philadelphia merchant, concerned about his childlessness, consulted a professor at Jefferson College, Dr William Pancoast. His wife was a great deal younger than him and he had become convinced that he was responsible for their infertility. Dr Pancoast, eager to promote his reputation among well-to-do individuals in town, decided to help. The medical students in Dr Pancoast's class were approached and, after some interesting discussion among themselves, it was generally agreed that Addison Hard was the 'handsomest member of the class'. Accordingly, the appropriately named Mr Hard obliged in a suitable manner. The young lady in question was given a general anaesthetic and Dr Pancoast delivered the semen to her vagina. Though this turned out to be a successful fertility treatment, neither the husband nor his wife was aware of what precisely had transpired, and neither knew of Hard's existence. But nine months later the lady gave birth to a rollicking son. Although the whole class had been sworn to secrecy by Dr Pancoast, twenty-five years later the sperm donor – now Dr Hard – could not restrain his curiosity. He visited his son to see the fruits of his actions at first hand.

I have been unable to find any pictures or photographs that substantiate Dr Hard's claim to being the most handsome member of his class – but his article in *Medical World* suggests clearly that he was proud of his contribution and took some paternal interest in the outcome. There is nothing in its tone to indicate that he saw anything wrong with the procedure, even though it was undertaken without consent.

More recently, a few other physicians took a more cautious approach to artificial insemination by donor – conducting it in absolute secrecy. In 1954 the *British Medical Journal* published the first comprehensive account of such a procedure resulting in a live birth. Now artificial insemination could help couples, like the one described by John Hunter, for whom intercourse was difficult or impossible. But it also raised the possibility of fatherhood for men unable to produce fertile sperm – by using a donor. And this was where the problems lay.

Even though the possibilities had been under discussion for centuries, the reality of artificial insemination in humans did not arrive to a welcoming fanfare of approval. For the Roman Catholic Church, artificial insemination was another form of reproduction without sexual intercourse, and hence unholy. Pope Pius XII, in an address delivered in September 1949 at the Fourth International Conference of Catholic Doctors, declared it to be a sin and recommended gaol sentences for those practising it.

There was wariness among Anglicans, too. A commission chaired by the Archbishop of Canterbury in 1948 decided that insemination using the husband's sperm was acceptable, but that pregnancy achieved by donated sperm should be classified as a criminal offence. It wasn't until 1988, with the Family Law Reform Act, that a child born from sperm donation was considered legitimate. Nevertheless, donor insemination was undoubtedly being done, quietly, before this.

One of the real British pioneers was Dr Margaret Jackson, who lived in Devon. I was very friendly with her son, Mark, who was a medical student contemporary. I have a vivid memory of running up and down the hilly Jackson garden in Crediton with Mark's quite elderly father, passing a beer bottle as we practised a well-executed scissors movement against the imaginary opposition of a Welsh rugby three-quarter line. Margaret was equally eccentric and a truly independent thinker. She built up Britain's first semen bank and handled this very controversial issue in the 1950s and 1960s with great wisdom and compassion. Years later, when Margaret must have been in her eighties, I consulted her about a friend of mine who wanted to have insemination with a specific donor, to whom he was related. The given wisdom at that time was that arrangements of this kind were not ever to be contemplated. Margaret Jackson differed from the general view and, after asking me various questions about this family, advised me to go ahead. My friend and his wife now have two fine adult children, of whom they are justly proud. The children know who their biological father is and it does not constitute any source of worry.

Meanwhile, there were considerable legal concerns. In the same year as the original *BMJ* report, the Supreme Court of Cook County, Illinois, declared that, regardless of whether the husband had given consent, a pregnancy by donated sperm amounted to an act of adultery. The child was illegitimate, the husband had no rights over it, and the child had no rights regarding the inheritance of the husband's property. In the USA this view persisted until 1964, when the state of Georgia ruled that, provided written consent to the procedure was obtained by both husband and wife, the child was legitimate. Five years later, donor insemination embarked upon the path to true legitimacy in the USA, with an historic ruling in the case of *The People* v. *Sorenson*. Mr Sorenson had claimed that he was not obliged to provide support for his child because it had been conceived, albeit with his consent, by donated sperm. The court decided that a sperm donor had no more responsibility towards any resulting child than a blood donor did to the recipients of their blood.

Current figures from the HFEA state that in the UK some eight hundred pregnancies per year result from donation – but this figure includes embryos and eggs. Donor insemination success rates are not especially high, ranging from 13.5 per cent each month in women under thirty to just 2.5 per cent for those aged forty and over. Of course, it is unreasonable to expect a pregnancy immediately and most clinics would give a course of at least three treatments, bringing these success rates up to around 40 per cent at best. There are a number of circumstances where it can be a highly useful treatment. It offers a solution to the most severe male fertility problems – where the man produces very few or no sperm, or completely abnormal sperm. It is also used occasionally if a vasectomy has badly damaged the testicular production of sperm, or if an operation to reverse a vasectomy has failed. Donor insemination can also be of value in cases where a man is known to be the carrier of certain genetic conditions, or if he has undergone radiotherapy or chemotherapy that has compromised his fertility. More recently – to the accompaniment of a predictable outcry from some sections of the public – women have begun to use donated sperm in order to create one-parent families, and lesbian couples to create children 'of their own'. These latter treatments cannot be regulated by the HFEA because it hardly needs a doctor to inseminate somebody. Once a woman has persuaded a friend or acquaintance to give some semen, it can be done quite efficiently at the right time of the month using – for example – a basting syringe.

Although straightforward artificial insemination is one of the simplest treatments for infertility, freezing sperm and, more importantly, storing it when frozen, requires rather more complex technology (discussed in the previous chapter). One of the biggest players in the frozen sperm market is Cryos International Sperm Bank, based in

Aarhus, Denmark. In April 1991 the firm delivered its first samples – drawn mainly from the nearby university population – to a private local hospital; the subsequent inseminations resulted in five live births. Word spread that Cryos was offering extensively screened, high-quality sperm and, almost overnight, it became inundated with requests from abroad. By 2002 it was exporting to some fifty countries – exploiting, in many cases, people's preferences for tall, blond-haired, blue-eyed progeny. One of the reasons for its success is that Cryos guarantees anonymity for donors – a right protected under Danish law. As we shall see, this legal detail has profound implications for the availability and use of donor gametes, male and female. For many couples, it is the difference between misery and joy. The Cryos website claims to have in stock high-quality, frozen, tested semen from more than two hundred donors – mostly Scandinavians. Since 1991 more than ten thousand pregnancies have been achieved using sperm from this source. One cannot help wondering how many of those babies are related to one another, and what statistical likelihood there is of one of them meeting another and forming a fertile relationship when they are adults. In this context, it is interesting to observe that in Britain (see page 278) the law has been changed and identifying information concerning the donor is held on record so that children – if they know they were conceived by donor insemination – can learn who their genetic fathers were. This has resulted in very few men being prepared to donate sperm. I understand that it is said that the shortage of donors is so great in Scotland that there is only one sperm donor registered in that country. If this is indeed true, and if these children do not discover the true facts of their genetic background, this must surely increase the risk of consanguineous partnerships in these offspring.

Artificial insemination offers one effective solution to the problems that arise because of male infertility, or lack of a male partner. But, as we have seen in earlier chapters, a large – and, indeed, increasing – number of reproductive problems can be traced to the woman's store of eggs. In decline from well before she is born, a woman's ovarian stock is notoriously prone to shortages, stoppages and the effects of ageing. Even if your sperm comes at a premium price from a six-foot-tall Danish Nobel Prize winner, it won't lead to a live birth if there are problems with the ovulation cycle, the eggs, or the balance of hormones that mature them properly.

The egg race

Whereas sperm donation has a relatively long history, it is only in the past two decades that we have been able to take eggs from one woman, fertilize them in vitro, and give them to another woman. This technique was pioneered by the remarkably innovative Alan Trounson, and the first deliveries from its use occurred in the early 1980s.[3] It is now offered by many clinics, frequently to women experiencing early menopause or with ovaries damaged by cysts, to those who have failed to respond to high doses of ovarian stimulation, and to those carrying gene defects such as Duchenne muscular dystrophy.

Egg donation can be a useful solution to fertility problems. One advantage, of course, is that if the donated egg is fertilized by sperm from the male partner, both parents will have made a significant contribution to any children produced from the very beginning: the father is genetically related to the child, and the mother has nurtured the baby from conception and given birth. This is fundamentally different from sperm donation, where the 'adopting' father has no direct relationship with the child. Another advantage is that success rates achieved with egg donation are often slightly higher than those after straightforward IVF. This is partly because the eggs come from young women. In Britain, the HFEA stipulates that women over thirty-six may not be egg donors. The national statistics from some years ago show clearly that the ageing process hardly affects the efficiency of the uterus at all; the success rate depends mostly on the age of the egg donor.

Egg donors also tend to have been screened to rule out any conditions that might affect the quality of their eggs. And, of course, being younger, they tend to respond more readily to stimulation, producing more eggs.

Another slight difference is that, with routine IVF, the woman is exposed to high doses of stimulant hormones, which encourage her ovaries to produce eggs. But these hormones may also have some adverse effect on the uterus. Although a low dose of oestrogen needs to be given to the recipient of donor eggs before any embryos can be transferred, the dose is only enough to prepare the uterus to receive the fertilized egg. It is of interest that egg donation is actually more successful in women who are completely menopausal than in those who are still menstruating. This is because the hormones given to prepare the uterus for implantation are more effective, and embryo transfer is easier to time correctly.

For all these reasons, egg donation can be a highly efficient option for the recipient. However, egg donation is of major consequence for the donor. She has, effectively, to go through the equivalent of a complete IVF treatment, but without the last stage – embryo transfer. First, she will (at least in a good clinic – the HFEA does not stipulate any requirements in this respect) be thoroughly screened, and therefore faces the possibility of finding out that she may have acquired a serious medical condition herself. She and her partner (assuming she has one) need to have effective counselling to prepare them both for the possibility that they may have a genetic child about whose fate they will know nothing beyond the fact that it will be a sibling to her own children. And now, with the loss of anonymity recently stipulated by law, she needs to be prepared for the possibility that she may be traced eighteen years later by a child she did not even know she had. During the treatment, she will need to have regular hormone injections, with all their possible side-effects of hot flushes, headaches, bloating, depression and decreased libido. She could also run the risk of developing hyperstimulation and becoming unwell as a result. She will have to tolerate the inconvenience of attending hospital to undergo daily ultrasound scans and have frequent blood tests. Finally, she will undergo an operation to have her eggs removed. This is no different from the ordeal undergone by women seeking straightforward IVF treatment. But most women undergoing IVF are quite prepared to accept the discomfort and indignity involved, because after treatment there is the possibility of a baby. What motivates the average egg donor?

In some cases, of course, egg donors are genuinely altruistic. This is one of the greatest acts of charity that a woman can perform. Some donors, having experienced the life-changing experience of having their own children, strongly believe in giving the joy of motherhood to others. Sometimes the donor may have strong ties to the mother – perhaps she is a sister, cousin or close friend. Many women will go through all that trouble, and some risk, to make this awe-inspiring sacrifice.

But the nature and extent of the treatment remain a

serious disincentive. There are a number of partial solutions to this, but each raises difficulties. Eggs may be obtained from women undergoing hysterectomy or sterilization procedures. But the women undergoing these surgical procedures are usually well over forty – beyond the age limit set for egg donation by the HFEA, even if the recipient accepts a lower chance of pregnancy or a higher risk of Down's syndrome. The HFEA does not allow well-informed people to take risks of which it does not approve.

There is a massive shortage of egg donors. In most UK clinics the waiting list for donated eggs extends for several years, and there is little guarantee that one will be available at the end of the waiting period. There is evidence that this shortage could be overcome by paying donors. In the USA, egg donors – often young women at college – offer their services for at least $5,000 to $15,000 a donation. One of my colleagues on the production team for the BBC TV series that accompanies this book told me she was offered $30,000 to donate her eggs. Buying and selling is usually brokered by an egg-donor business, which negotiates with the clinic or undertakes the IVF treatment. But in the UK, selling eggs or sperm is not unreasonably regarded as undesirable – because there is a risk of exploitation. The HFEA stipulates that women donating eggs may be paid only their immediate expenses. So an option increasingly widely used is 'egg sharing'. This is effectively just another way of paying donors; but in this case the donors are women undergoing IVF themselves. In return for a free cycle, the woman donates some of her eggs to another, reducing her bill for this costly procedure. And the woman receiving the donated eggs pays double – the cost of her treatment and that of the donor's. As we shall see, there are many reasons to believe that egg sharing is a good deal more undesirable than selling eggs. I believe that these reasons cause sufficient concern to question the wisdom of the HFEA's decision to allow egg sharing under any circumstances.

Egg sharing, of course, does have practical advantages. First, no patient undergoes ovarian stimulation and egg harvesting without the *potential* reward of a child at the end of the process. As part of this, no woman is exposed to the risks of stimulant drugs without the possibility of benefit. And for women who are unable to afford IVF treatment, egg sharing seems to present an overwhelmingly attractive opportunity.

Another, related, option is so-called 'egg giving', where a woman undergoes two cycles of IVF. In the first, she donates all of the harvested eggs; in the second, she receives her own fertilized eggs. Egg giving, obviously, entails two cycles of stimulation – double the physical and emotional tolls, although most women undergoing IVF are likely to go through more than one cycle. One problem with this treatment is that some women have a reduced response each time they are stimulated, so it could be that they end up giving their best eggs away in the first treatment cycle.

Ever since the Cromwell IVF and Fertility Centre reported the first UK case of egg sharing in 1992, some people have raised objections. They noted that poorer women might be manipulated into giving their eggs away because they could not afford their own IVF treatment. It was also suggested that the harvest might be unfavourably split, so that the paying client received the 'better' eggs; and that egg donors undergoing this treatment might have their ovaries 'pushed' unusually hard to produce extra eggs. This is highly unethical, but it is very difficult to detect such abuses, and I have evidence from personal experience talking to a number of patients that this may have happened to them. All these women attended a London clinic which has now closed.

A study conducted in 1998 that assessed the feelings of egg-sharing donors and recipients found a degree of what might be called pragmatic altruism.[4] Participants cited among their motives a wish to help other infertile couples; 89 per cent said they were happy to participate, even if the outcome was unsuccessful for themselves, or for both women. My problem with many studies like this is that the clinic or clinicians doing the study will tend to have an interest in the research topic that is difficult to separate from the outcome of the research. No disrespect to my colleagues is intended, but such questionnaire-based research can lead to researchers receiving the answers that they most want to hear.

One reassuring study indicates that egg donors are just as likely to end up with a live birth as egg recipients – suggesting that the treatment does no harm and that the eggs are not being split unfairly.[5] While I am sure this is true in the Lister Hospital, which published this study by Dr Thum, one cannot escape the possibility that from time to time the donor's treatment will fail, but the recipient will get pregnant with the donated eggs. This is surely unacceptable, because the psychological consequences (unless the outcome is kept secret) can be profound. This possibility that a donor who longs for a child herself may not be successful, while the recipient of her eggs goes on to have a baby, must be potentially heartbreaking. I have been told this at first hand by women coming to my clinic after such failure. They seem to have been devastated by grief and are usually very angry. This is why I would never countenance this treatment in my own clinic when I was in active practice; and it illustrates the extreme importance of ensuring that people fully understand what the possible outcomes are before embarking on such treatment. And yet, while every UK clinic is obliged by the HFEA's code of practice to offer counselling, I am not convinced that all patients are fully prepared for the outcome. Social science research has recently indicated that counselling is not particularly helpful for a surprisingly large proportion of couples anyway (see page 149). The problem is that the need for a child is so strong that many people will take any risk to give themselves a chance, without, perhaps, giving themselves the time and space they truly need to decide whether egg sharing or giving is the right path to take. I have worked with a number of excellent, professional and highly trained counsellors over the years. But the fact remains that, even with the best guidance, people can give the impression of understanding all the issues and being ready for all possible outcomes, without this being so.

There is another, very controversial, source of human eggs. As I have already stated, a girl fetus, while still in her mother's uterus, has more eggs in her ovaries than she will have at any time after birth or during adult life. The fetal ovary is packed with millions of eggs, all of which have potential to mature and be fertilized. Roger Gosden is a reproductive scientist who has studied fetal ovaries. He has also always been deeply interested in the menopause: how and why it occurs, and whether it is predictable. In 1993 in Edinburgh, Roger Gosden took the ovaries from dead mouse fetuses and transplanted them into adult mice. If the normal female hormones were circulating, these transplanted ovaries worked. They produced eggs capable of being fertilized and developing into normal rodents. He published his results in a dry scientific journal – where they attracted the attention of the daily press.

Roger, a soft-spoken, ivory-tower academic, a purist who does not seek publicity, serious and earnest to the point of seeming almost unworldly, was dismayed by this surge of interest. During various interviews he found himself the centre of attention. When pressed, he answered that such a treatment ought to be equally possible in humans. And under further pressure he said that he would like to see such treatment at least discussed openly.

The media coverage included an international outpouring of unbelievable venom, leaving Roger blinking in the headlights. He was not remotely thinking of treating people himself – indeed, he is not a medical doctor, so this was not even a dream. I believe that Dr Gosden was simply pointing out that a huge amount of good ovarian tissue is wasted – usually incinerated – at the time of miscarriage or abortion. Given the prevalence of premature menopause, and given that there is a real shortage of eggs for donation, it ought to be reasonable, he felt, at least to debate the issues calmly and sanely. Once his quite moderate point was broadcast, the Health Secretary of the day, Virginia Bottomley, responded tartly that the National Health Service would certainly not pay for the effort. The Ethics Committee of the British Medical Association rapidly reviewed the idea – I do not think it took any evidence or discussed the feasibility of such a process – then publicly damned Gosden's ideas, running to denigrate them on television news. Within hours, the news of an 'intended transplant' was hawked around the world. George Annas, a lawyer at Boston University, was quoted as saying, 'the idea is so grotesque that it is unbelievable'. It was interesting that Mr Annas did not object to the potentially much more dangerous process of using fetal tissue to treat devastating conditions such as Parkinson's disease, and yet asserted vehemently that the fetal ovary should not be used as a source of eggs for donation. I suppose there is a difference between saving life and creating it, but the source is the same whatever is achieved. Other ethical 'experts' were not so critical. Dr John Fletcher from the University of Virginia mildly suggested that such an approach might do more harm than good.

There is no question that the pressure of publicity on Dr Gosden was intense, and the impression was firmly given that human treatments, if not clearly confirmed as done, were imminent. The *Daily Mail*, together with many other critics, reported that science was moving too fast to be controlled.[6] As it happens, thirteen years later, there is still not even a remote possibility that we could use eggs taken from a human fetus, even if we wanted to. We might be able to remove them and put them in culture, but mammalian eggs are useless unless they have gone through the subtle, complex and poorly understood process of maturation. Methods of maturing eggs outside the body are very primitive and far too crude at the present time to be used on humans. So there is no way that fetal eggs could be used for donation now, and I think it highly unlikely that this process will become feasible soon, if ever.

Of course, what was interesting about the wave of response to Roger Gosden's mildly made suggestion was that it showed, yet again, how the subject of human reproduction raises huge emotions. Stories about attempts to manipulate the reproductive process cause strong reaction in people who probably would not respond so vigorously to other potential scientific advances. Many newspapers felt that Dr Gosden had raised the 'yuk' factor. Admittedly, at the time, I too wondered whether using eggs from a dead fetus to create a live baby was a step too far. Looked at coldly, it does seem challenging to consider that a baby could have a mother who had been aborted as a fetus. What would one tell the child under these circumstances? Many other people with strong views about the ethics of abortion argued that if such an advance were possible, some women would inevitably seek abortions so that the ovaries of their fetuses might be used as a source of eggs.

Most religious sources that I read at this time seemed outraged by the possibility that the fetal ovary might be used in this way. So it may surprise people to learn that the Jewish position is somewhat permissive. Approximately eighty years ago, Rabbi Benjamin Weiss was asked theoretically about ovarian transplants. His reply was that it was hard to believe that this would ever be a possibility, but that it would be prohibited under Jewish law because it would involve removing an ovary – and therefore sterilizing the donor. But beyond this, he did not have particular concerns about such a transplant. The source would not necessarily matter because once an organ is transplanted, it becomes an integral part of the body of the recipient. Following the transplantation of the ovary, he considered, the nourishment and maturation of the eggs within the ovary would depend upon the recipient and not upon the donor. So, under the Jewish law,[7] there is a body of orthodox opinion arguing that the legal mother of the child is the person in whose body the egg is fertilized.

One particularly sad outcome of this controversy is the effect it had on Roger Gosden, who was aghast at the outcry and felt very threatened by it all. Roger is a delightful, warm individual, a modest scientist who was simply making a 'blue skies' observation. I think he was horrified at the vituperative nature of much of the criticism expressed, and it may have been this that led him to decide to leave the British Isles and settle in Canada, where he currently conducts his research.

Leaving aside the fraught issue of 'robbing' the aborted fetus, the biggest issue about egg donation is that it involves one woman giving something that is uniquely hers, which is in limited supply in her ovaries. There are related problems for sperm donors, and I address these later, but the issue is heightened for women. Women have a smaller number of reproductive years compared to their lifespan than men. Moreover, ovarian stimulation can never be seen as anything resembling a pleasurable experience.

Whose baby? Definitions of parenthood

When an egg is successfully fertilized and placed into another person's uterus, and a live birth results, whose is the child? Under UK law, the mother who bears the child is in all legal respects its parent – the genetic mother has no rights over or duties towards it. But the situation differs from country to country, and even within different branches of the same religion. In Judaism, for example, most rabbis rule that egg donation is permissible. But arriving at that decision is by no means straightforward. Membership of the Jewish religion is *matrilineal* – that is, one is Jewish if one's mother is Jewish. But egg donation raises the question of what precisely determines 'motherhood' – the process of gestation and birth, or the common ownership of genes.

In many other aspects of Jewish law, decisions rest upon facts that are observable – one of the reasons why Judaism is generally informed by scientific knowledge – and on this point too it seems that it is the act of birth that confirms motherhood. Meanwhile, under the principles established by the facts of Ben Sira's miraculous conception, if the egg is fertilized with the birth mother's partner's sperm, no 'forbidden act of intercourse' has taken place and so the child is considered legitimate. Moreover, as we have seen, in Jewish law an ovarian transplant becomes part of the mother's body. Rabbi David Bleich, in his excellent work *Contemporary Halakhic Problems*,[8] cites Rabbi Eliezer Waldenberg,[9] who took the most rigid view of IVF. To Rabbi Waldenberg, the IVF baby has neither mother nor father, (a) because fertilization occurs through the mediation of a third power, i.e. the Petri dish, (b) because conception occurs in a manner which has no relationship to genealogy, and (c) because the removal of an egg from a mother's body, whether it is her egg or not, destroys any genealogical relationship. Rabbi Bleich, like nearly all other senior rabbinical voices today, does not view these objections as valid. Indeed, Bleich cites the Talmud, pointing out that a Jewish fraternal relationship exists between twins born to a woman who converts to Judaism during her pregnancy.[10] As a proselyte is regarded as a 'new-born child', all previous relationships are severed. The situation of the pregnant convert is analogous to that of a woman who receives a donated egg. It is the process of parturition which defines the mother in Jewish law.

So, if a non-Jewish mother has donated the eggs, the child becomes Jewish. Indeed, some authorities argue that *only* a non-Jewish donor is, in fact, permissible – because of the potential risk of incest if the resultant child does not know its origins and marries a half-sibling. Others argue that this applies only if the donation is anonymous – and a very few authorities tentatively suggest that the only permissible form of egg donation should be between unrelated Jews who are fully acquainted with each other's family trees. But this latter is not the general orthodox view.

The issue is more cloudy in Islam, the fastest-growing faith. Across its traditional homeland, the Middle East, there are large numbers of infertile couples and a thriving business has emerged to treat them. Egypt boasts some fifty IVF clinics for its population of seventy million, while tiny Lebanon (pop. five million) has fifteen. Though Saudi Arabia and Iran permit IVF only for married couples, there are numerous centres even in these countries. Also, the western press often overlooks the fact that, although many countries adhere to the one religion, there tends to be a wide diversity of cultural beliefs and practices across and within them. Islam itself is divided into Sunni and Shi'a sects, which have differing positions regarding infertility treatment.

In Sunni Islam, widely and perhaps mistakenly seen as the more tolerant branch of the faith, egg donation is strictly forbidden because of problems relating to inheritance. A child resulting from a donated egg is regarded as the issue of a union between its father and a woman to whom he is not married, and therefore is illegitimate. In Shi'a Islam, the majority religion in Iran and Iraq, egg donation has been rendered possible by rulings from the spiritual head in Iran, Ayatollah Ali Hussein Khameini, in the 1990s. He declared it permissible, provided that both parents and the donor adhered to Muslim codes of parenting. However, the child could inherit only from the donor, since the infertile parents were regarded as only its caretakers. Another possibility exploited by some Shi'a couples is to engage in a *muta* – a temporary marriage – so that the husband is married to both donor and recipient at the time the egg is fertilized. Other Shi'a authorities accept egg donation because they say it is akin to polygyny – the practice of having more than one wife – which is widely permitted in Islam. But this view also leads them to forbid sperm donation as being too close to polyandry – having more than one husband – which is forbidden.

Having rejected all acts of procreation that do not result from intercourse between married couples, the Roman Catholic Church cannot, of course, permit egg donation. Accordingly, the Vatican made the following pronouncement in 1987, relating to both sperm and egg donation:

> [It] violates the rights of the child. It deprives him of his filial relationship with his parental origins and can hinder the maturity of his personal identity. Consequently, fertilization of a married woman with the sperm of a donor different from her husband and fertilization with the husband's sperm of an ovum not coming from his wife are morally illicit.11

I find the Roman Catholic objections more salient than those from Islam or my own faith, perhaps because they show a concern for the feelings of the child, rather than its rights to property or membership of a race. They refer also to the powerful and important notion of protecting the value of the family. But inheritance of a parent's genes is no guarantee that a child will experience a fulfilling filial relationship with that person. There are some appalling fathers and mothers in the world, and some who, though unrelated by blood to their children, have set Olympic standards in parenting. And, as we shall see, the problems associated with a child's feeling of personal identity and knowledge or ignorance of his or her origins are complex.

When a man or a woman has donated genetic material,

it is natural for him or her then to have strong feelings towards the resulting child. This is particularly likely to be the case when siblings or close friends have provided the gametes. In watching the child growing up resembling themselves and exhibiting many of their traits, they may experience a strong bond. In some cases, this can be entirely positive, with the child growing up with the luxury of 'an extra parent', someone to whom they can go with their problems, with whom they can experience the freedom of a close relationship uncluttered by the pressures of daily life. But what happens when the child's parents make decisions with which the donor disagrees? They may wish to send the child to boarding school, hot-house him or her with extra tuition to create a tennis star, or adopt a *laissez-faire* attitude to teenage rebellions; the donor parent may disagree, but has no right to interfere. The child itself may experience divided and conflicting loyalties – or play one parental influence against the other. It is easy to imagine such a situation leading to considerable conflicts.

Having said that, I have been very impressed by many Asian families with whom I have come into contact, working as I did in west London where many people of Asian origin have settled. Among that community, donation of an egg is often regarded as a very positive way of helping a family member. It does not seem, as far as I can tell, to blur family values or relationships, and I have not been aware of long-term problems after such arrangements. In the next chapter I deal with the issue of surrogacy, and I have seen a number of Asian women bear a child for an infertile sister or cousin – as far as I know without any pressure being exerted. But although I have seen and treated many such couples, it is always difficult to get an accurate view from the outside of a relatively closed community. It is of interest, too, that I have not seen much enthusiasm for the use of donated sperm within the Asian families I have treated.

The three-parent child

The ethical complications are even more fraught with another, rather curious, form of donation. *Ooplasmic transfer or cytoplasmic transfer* involves combining parts of eggs from two sources. It has been claimed to be potentially useful when older women are attempting to conceive. It was pioneered by the Dutch scientist Jacques Cohen, a brilliant researcher who has also been actively involved in treating infertile couples by assisted reproductive technology for over thirty years. He is now director of Tyho–Galileo Research Laboratories and vice-president of a private institution, Reprogenetics, in the USA. After leaving the Netherlands he came to England to join the team of Drs Edwards and Steptoe, where – following the technology described by Trounson – he was the first to freeze and successfully thaw the human blastocyst.[12] He has developed a number of techniques that have revolutionized IVF, and certainly his work on improving the culture of embryos is interesting – though, I have to say, not widely adopted. He claims to have been the first to undertake 'assisted hatching' (see page 170), which is intended to promote pregnancy by drilling the egg's outer 'shell' following fertilization. As it happens, this was a technology started by my group at Hammersmith in 1989–90, and soon abandoned by us as unhelpful.

Dr Cohen is certainly not reticent in trying new things, and he first considered cytoplasmic transfer in about

1996. The idea was not his – Dr Audrey Muggleton-Harris had tried it in the mouse in 1982 – but it is fair to say that very little animal work had been done between the time she published her results and the time Dr Cohen subjected human eggs to the technique. The procedure involves taking the ooplasm, the fluid surrounding the egg nucleus taken from a younger woman's egg, and injecting it into the egg of an older patient whose IVF treatment has previously failed. This fluid contains *mitochondria*, inside which are little pieces of DNA that are responsible for energy production in the cell. Of course, 99.99 per cent of the DNA is contained inside the nucleus of the cell, but a tiny proportion is in this fluid (see page 397). The idea behind the procedure is to rejuvenate the 'older' egg and make IVF more likely to be successful. Once the egg has had the mitochondrial transfer, it is fertilized and then transferred to the uterus. So it involves not egg donation, but the insertion of just a very few genes from a donor egg. Technically speaking, a resultant child would carry some DNA from two mothers. But the characteristics inherited from the donor, via the ooplasm, would be minimal: only around forty of the twenty-five thousand genes in the body. And even those genes are not those connected with any obvious characteristics – like appearance or personality – but are important only in controlling how the cell uses energy.

Dr Cohen's first reports of this technique were highly encouraging.[13] It seemed that older women who were hopelessly infertile and in whom IVF had failed to produce fertilized eggs could now be helped. I got interested in Dr Cohen's work at first hand when I made the BBC Television series *Superhuman*. I interviewed a desperate couple from New Jersey who had talked themselves into trying this procedure. We filmed them undergoing treatment twice without conception – and although they were in much distress at the failure, I remember thinking privately at the time that they were lucky. Dr Cohen was undoubtedly getting pregnancies – though of course, there is no way of knowing how many of these successes occurred through chance and how many because of the modification of the eggs – and he published his first successes in the journal *Human Reproduction*, writing, 'This report is the first case of human germline genetic modification resulting in normal healthy children.'[14] However, what was made a little less clear – at least in the press reports – was that some of the conceptions had chromosomal abnormalities. I cannot help feeling that this was a situation where scientists were too ready to experiment in humans before doing more detailed animal work. This is an argument that I have had with Dr Cohen. In any event, the US government decided to take no chances and banned the procedure in 2001. Since then, the formidable Alan Trounson in Australia has published work with animal eggs suggesting that this technique could be problematic.

Another variant is *autoplasmic transfer*. This involves injecting a woman's eggs with ooplasm taken from her own eggs – removing the donor element, and thus the mixing of genetic material, from the equation entirely. Scientists in China have also pioneered a kind of reverse procedure, whereby the nucleus of the mother's egg was transferred into a donor egg before fertilization. In one instance, multiple embryo transfer subsequently resulted in the birth of triplets. This might just be useful in cases where a mother risks passing to her children certain rare genetic conditions that are carried in the mitochondria. She can have her 'own' babies, but 'hosted' inside someone else's egg.

In my view, the great danger with all of these

procedures is that the whole technology risks being out-lawed because of the impatience of a few pioneers. Research these days is done so often in a male-dominated, testosterone-driven environment where first publication is everything. Reproductive technology is not only big busi-ness, it is also a high-profile area given huge publicity by the media. Many of my colleagues undoubtedly get a rush of adrenalin from being in such a field. But before any of these techniques are tried out on humans, a great deal more careful research needs to performed on animals. This is especially the case where we are modifying DNA. We already know that mutations of DNA can cause children to be born with appalling, agonizing mal-formations and defects, condemning them to a short life of misery. The need for caution is paramount.

I began this section with the suggestion that it was now possible for a child to have three parents – but of course, it is possible to have at least six. Consider the scenario of a donor egg, some donor sperm and cytoplasm from another donor egg all combined to produce an embryo that is then placed in the uterus of a surrogate mother and adopted after birth by a couple unrelated and unknown to any of the previous contributors.

Embryo donation: adoption before birth

Some couples undergoing IVF produce more embryos than can be transferred during one treatment cycle. These 'surplus' embryos may be stored in liquid nitrogen. This resource has made it possible, with consent, to offer frozen embryos that are 'surplus' to the requirements of the couple who generated them as a donation to women who cannot conceive. It is an alternative to egg donation, but with the potential disadvantage that neither parent of the child that is born has any genetic relationship with it. Technically, the procedure is precisely the same as egg donation, and in both cases the outcome somewhat resembles adoption.

There is a curious and interesting story about an experience I had in 1983 at a medical congress in Los Angeles. At the time, there was argument about whether natural-cycle IVF was potentially better than using stimulated cycles. Stephen Hillier and I had lingering doubts that eggs obtained after hefty ovarian stimulation might be less likely to be normal and produce viable embryos. But there seemed no way of getting enough naturally fertilized eggs to compare with IVF embryos.

At this congress, a tall, slightly shambling man, looking incredibly benign and smiley, went to the podium to deliver a paper that seemed sensational.[15] John Buster had timed the cycles of a number of women who were not using contraception and asked them to have sex on a specific night. Any embryo that had been produced should have arrived in the uterus about one hundred hours later. So, five days after he had detected ovulation, he inserted a fine catheter through the cervix and flushed out the contents of the uterus. Remarkably, he obtained fertilized eggs from some eleven of these patients, and microscopic evidence suggested that quite a number of these were incompletely developed, or abnormal. Five of the embryos looked reasonably developed and were transferred to the uteri of infertile patients. Two of them implanted successfully and resulted in live births.

Buster's paper was mesmerizing for many reasons. First, it was extraordinarily interesting because it gave further evidence that humans produce a large proportion of abnormal embryos. Second, at that time a pregnancy rate of two out of five seemed extraordinarily good. But third, nobody in the entire audience seemed in the slightest bit concerned about the ethical issues all this raised. Trembling, I got to my feet. How did they get consent to do this, I asked? And then, rather more boldly, how much were these donor women paid? And then, had each of them been told that if an embryo had been formed but was not extracted in the flushing out, she could become pregnant herself? And finally, had the donors realized that there might be a risk of ending up with an ectopic pregnancy, potentially a serious surgical emergency? If the flushing pushed the embryos back into either fallopian tube, they could lodge there and implant.

John Buster was not remotely put out, though out of the corner of my eye I could see one of his colleagues who had done the experiment with him, Maria Bustillo, getting angry. Consent was no problem,[16] he said, still smiling beatifically. Payment was trivial, apparently. And finally he casually opined that any of these women could always have a pregnancy terminated if she wished, and the risk of ectopic pregnancy was probably non-existent.

I have to admit I desperately wanted to repeat John Buster's experiment. It seemed really interesting, and it was clear that we could have gained some extraordinarily valuable data about early human development from such an approach. But I did not think there was a serious prospect of getting ethical approval from a British ethics committee, and probably quite rightly too. Nevertheless, the experiment has since been repeated in Milan.[17] Dr Formigli and his colleagues flushed the uteri of fifty-six women, obtaining a total of twenty-three embryos, of which seventeen were considered to appear sufficiently normal to be transferred to recipients. Eight of these women became pregnant. One of them was the infertile sister of a woman having her uterus flushed; she gave birth to a daughter.

Who 'deserves' a child?

The shared experience of infertility can create enormous and very gratifying amounts of goodwill between donors and recipients. Knowing that your eggs are going to help someone in the same boat as yourself seems to be a powerful incentive. But whether they feel they are parting with a baby, or a small quantity of eggs invisible to the naked eye, there is a sense in which many women feel a responsibility towards the results of their donation. That means they may have quite strong views about the sort of people they wish to help.

This is particularly so in cases of altruistic egg donation. In my experience, younger, fertile women tend to be more comfortable when they know that their eggs are going to someone in roughly the same age bracket as themselves. Some donors also have strong religious or cultural views, which motivate them, on the one hand, to want to assist others, but on the other hand to impose limits on the types of people and families they want to benefit. It is as if their 'payment' is the knowledge that their eggs are going to the 'right sort' of home. But this feeling can often clash with laws and changing cultural values.

In the USA, the commercial operators of fertility clinics were quick to see the potential of egg and sperm donation for creating children outside the 'traditional' family unit headed by married parents. From 1970 onwards, many marketed their services directly at lesbian couples, or at women who, for whatever reason, wanted babies outside marriage. In 1978 the news that lesbians were attending UK fertility clinics for donor insemination provoked an outcry in the press. Rhodes Boyson MP, renowned for his outspokenness, was quoted as saying: 'This evil must stop for the sake of the potential children and society, which both have enough problems without the extension of this horrific practice. Children have a right to be born into a natural family with a father and a mother. Anything less will cause lifelong deprivation of the most acute kind for the child.'

Although it was more cautiously worded, a clause in the 1990 Act that led to the creation of the HFEA drew on some similar sentiments. 'Centres [offering IVF treatment] are required to have regard to the child's need for a father and should pay particular attention to the prospective mother's ability to meet the child's needs throughout their childhood.' This was not only a reminder to consider, and use as a criterion for deciding whether to embark on treatment, the standard of parenting a resulting child might receive; also, by mentioning the 'need for a father', the Act reflected a viewpoint that was already becoming outdated. I believe both considerations lie beyond the remit of governments or physicians.

There is no scientific evidence to support the claim that children who grow up with lesbian mothers are likely to experience anything akin to the 'lifelong deprivation' referred to by Rhodes Boyson. A number of studies were conducted in the 1970s and 1980s,[18] and the results were strikingly consistent: the children of lesbian mothers did not show any increased likelihood of psychological disorder or difficulties in forming relationships with their peers. More recently, two UK social scientists followed up participants of a 1992 survey, now in their early or mid-twenties. Similar results came through – there was no increased likelihood of psychological harm or abuse from peers. Those who did report hostility and anxiety tended to come from poorer backgrounds, where there was perhaps an increased likelihood of neighbours and classmates objecting to homosexuality.

One argument frequently raised against allowing lesbian couples to have children is that their offspring may grow up with some sort of skewed view of sexuality, and are more likely to be influenced into becoming homosexual themselves. But of the twenty-five young people assessed in the recent study,[19] only two were homosexual. To argue that the sexuality of the parents is an overriding factor in sexual choice is, in any case, rather specious in my view. How do we explain the large numbers of homosexuals who grew up in heterosexual homes?

Large-scale studies of single-parent families have suggested an increased likelihood of psychological problems and poor educational prospects. In the UK, local councils are creating 'mentoring schemes', recognizing that, particularly in the case of boys, there may be a link between the lack of a father and truancy, crime, drugtaking, etc. But do we know that the child's need for a father is the overriding factor? The bulk of the work done on single-parent families concerns people from poorer economic backgrounds, where a whole raft of problems may have an effect on the child's welfare. Other pressures on single mothers, such as social stigma and lack of support from the wider family, may be affecting the quality of the parenting given. Single women seeking donor insemination are more likely to be affluent and economically self-supporting. In addition, the children born to these mothers will not have experienced the emotional turmoil of separation and divorce.

A study conducted between 1997 and 2005 examined

twenty-seven single mothers who opted for donor insemination, and compared them to some fifty married women who conceived in this way.[20] The main reason for conceiving voiced by most of the single women was that they felt that time was running out, and that this offered them the easiest way to have a child. They didn't want to have casual sex, or hoodwink a man into getting them pregnant. The results suggested no difference between them and the married women in terms of anxiety, depression or parenting-related stress. Interestingly, the children of the solo mothers, when assessed at the age of two, actually showed fewer emotional and behavioural problems than those growing up in a two-parent setting. And the majority of these mothers, some 87 per cent, hoped to have a relationship with a man in the future.

So we have little concrete evidence that growing up in a lesbian or single-parent family has *itself* any damaging effects on children. In any case, it is objectionable that infertile people are subjected to a disproportionate level of scrutiny about the kind of life they can offer their offspring. When a fertile woman receives confirmation of her pregnancy, her GP does not then ask a number of awkward and unnecessary questions about her private life. No hospital would refuse to allow her to give birth on its premises because of her low earnings, lack of a partner or sexuality. I believe my overriding duty, as a fertility specialist, was to assist pregnancies with maximum concern for the welfare of mother and child. I might have refused treatment if I thought that the health of either would be seriously at risk. I might have refused treatment if I had grounds to believe that the well-being of any resulting child would be seriously at risk. In all other circumstances, I would respect the right of women to try to have children.

The responsible donor

In 1991 America reacted with horror to a court case concerning Dr Cecil Jacobson, who had run a fertility clinic in the state of Virginia. Throughout the 1980s Jacobson had offered hope to childless couples with costly hormone injections and the chance to try the latest pioneering treatments in artificial insemination. Many were delighted with the results, going on to give birth to healthy sons and daughters. Others, after being shown their developing babies on an ultrasound scan, were told that there had been a problem. Their babies had died, and would be reabsorbed into the womb. One woman, who had gone through this twice already, sought another opinion. She was told that she was not pregnant; in fact, she never had been. It subsequently came to light that the hormones Dr Jacobson was dishing out could mimic the symptoms of pregnancy – up to a point. In fact, the dosages were so strong that even a man might have produced a positive pregnancy test result if he had taken them. But this was no mere medical accident. Jacobson deliberately pretended that these phantom babies were visible on ultrasound scans, taking great pains to point out heads and limbs and hearts, when in fact he was indicating bits of intestine and even faeces.

An American news crew exposed Jacobson's fraudulent practices and his licence was revoked. An inquiry was held, during which numerous supporters of the rogue doctor came forward – in some cases with the children he'd helped them safely to conceive. But at that point, people started to notice something odd about these children. They all bore a strange resemblance to one another – and to Dr Jacobson himself, an overweight and short-sighted individual. A whistle-blowing secretary eventually revealed that Jacobson did not – as he claimed – use sperm donated by medical students. Fifteen minutes before a patient came in for insemination treatment, the doctor would disappear to the lavatory and masturbate. When he was accused of having fathered at least seventy-five children himself, he quite openly admitted it – although he added, worryingly, that he could never know how many he had actually been responsible for. Even when he was sentenced to five years in prison, he maintained he had done nothing wrong.

There were all sorts of theories as to why Jacobson acted as he did. One of his victims suggested he was in it for the money – but that did not seem to hold water, as he hardly lived a life of luxury. It was also suggested that, as a devout Mormon, Jacobson thought his duty was to fill the earth with children. But that seems equally unlikely, since his church was quick to condemn his actions. Perhaps we will never know what motivated him. But it is interesting that, though all the women who had received treatment at the clinic were offered DNA testing to determine whether their children were Jacobson's, only seven came forward. The rest remained silent, presumably preferring not to know.

Using gametes from known donors can be fraught with difficulties. A ready solution is to adopt the policy we use with all other donated human material: strict anonymity. This has the advantage that donors – particularly those who have given early in their adult lives – will not run the risk of being confronted by their offspring one day, perhaps at a point when it could be disastrous for them. The man who donated sperm in his university days, and subsequently married a woman who is infertile, is one example: the sudden appearance of hitherto unknown offspring could place great strain on his marriage. Another might be a woman who donated eggs and then produced a large family in straitened circumstances. How might her children feel when a new sibling walks into their lives who has enjoyed all the benefits of being the only child in a wealthy family?

On the other hand, ever since sperm and egg donation became reality, it has been accepted that it may be helpful for parents and potential children to possess some information about the donors. The amount of information has differed considerably in degree and content. In the USA, where egg and sperm donation became major business concerns, clinics capitalized on offering infertile parents the maximum degree of choice. In 1990 Shelly Smith, a former actress working as a family counsellor in California, decided to set up as an 'egg broker' between willing donors and infertile couples. She hand-picked her egg suppliers, many of whom were aspiring Hollywood starlets, on the basis of looks, health and a rigorous battery of psychological testing. Other suppliers joined the fray, using their transparency as a marketing tool. 'Your child will grow up without secrets,' pledged one brochure for a sperm bank. 'They will not grow up fantasizing that their father is the lost King of Bavaria or Charles Manson.'

The comments of that particular enterprise, Rainbow Flag Health Services, became particularly apposite in 2003, when it emerged that a murderer had been on the books of the legendary Cryos International Sperm Bank in Denmark. Heine Nielsen was gaoled for life after killing his baby daughter.[21] Recipients of sperm from Cryos subsequently learned that Nielsen had served six years for killing another daughter, prior to donating seminal fluid. Fortunately, Cryos was able to assure anxious parents that none of Nielsen's donations had been used, because he had never completed their exhaustive psychological testing procedures.

Cases like Nielsen's and that of Dr Jacobson heightened the need felt by many parents to know as much as possible about the source of donated sperm and eggs. Many clinics will offer an impressive booklet, complete with photos, qualifications, even a tape-recording of the donor's voice. In the UK, every effort is made to ensure that the child matches the hair and eye colour and ethnic background of its parents. But in the USA, even more specialized matching is on offer, so that people can select donors who are athletic, musical, good linguists or even rapacious businessmen.

The right to remain invisible

In the UK, it has long been considered that parents and children have certain limited rights to information about donors. When the HFEA established its register of donors in 1991, certain facts were enshrined in law. The husband or partner of the woman giving birth was the legal father, provided he had consented to the treatment. The sperm donor had no legal rights over any resulting children, and there was no obligation on the parents to tell a child how it was conceived. In 2005, a modification of the Human Fertilization and Embryology Act of 1990 also allowed children to find out at the age of eighteen whether they were born from donation, and at the age of sixteen whether they might be related to anyone they wished to marry. Therein, of course, lies an important social and legal conundrum, since these provisions entitle individuals to know about their genetic origins before they are entitled to know that they have reason to question them!

In addition, clinics were given a degree of leeway as to the non-identifying information they could hold on record about donors – for example, year of birth, medical history, marital status and number of children.

In 1991 a survey at the Radcliffe Hospital in Oxford showed that the majority of men and women donating sperm or eggs were prepared to allow non-identifying information about their background to be released.[22] But 85 per cent of sperm donors made it clear that they would not consider donating sperm if their anonymity were not protected. The Oxford group warned that if the convention protecting donor anonymity were changed, recruitment of new donors would be significantly reduced – to the detriment of infertile couples requesting this treatment.

However, since the year 2000 there has been increasing pressure to change the status of egg and sperm donors in British law. One of the problems has been the existence of a mass of poorly researched information, which has undoubtedly influenced the debate and, sadly, the Department of Health. One article published from the Department of Psychology at the University of Surrey is fairly typical.[23] The researchers, Amanda Turner and Adrian Coyle, give the impression that donor insemination produces widespread unhappiness in the offspring so conceived. It may well do, of course, but their paper is not convincing. Their 'evidence' is taken from only sixteen participants in a questionnaire study; and those sixteen come variously from the UK, USA, Canada and Australia. The authors state that two respondents were British – though we know only one was living in England and is recorded as being known to the investigators – possibly a friend? So whether these respondents are remotely representative of a wider problem in the UK is highly dubious.

Whatever the case, they would most likely be a self-selecting group, which is how they came to respond to a questionnaire. One of the respondents tells of the effect of the eventual disclosure of her paternal relationship. 'My initial reaction was to laugh . . . The man I thought was my dad was such a creep that it was nice to know that I wasn't genetically related to him.' Another respondent says: 'I always felt like I didn't belong to these people – I searched for evidence of my "adoption" for many years as a child.' A third is reported as remarking: 'My father did not like me and this [disclosure] made it worse. He disinherited me. I was outraged by my father's malevolence.' The only conclusion one could come to is that these individuals have been scarred by being reared in dysfunctional families. But the authors of the paper imply that many individuals born as a result of donor insemination experienced problems of personal identity, abandonment and mistrust. This kind of very poor science is not helpful in informing difficult decisions about a highly complex area of medical practice.

The Department of Health announced a review of its own policy in 2001. The next year, Baroness Warnock, who had chaired the 1984 government inquiry into fertility research that eventually led to the HFEA's formation, proclaimed that she had changed her mind. In 1984 she had supported anonymity for donors. She now said that her beliefs had altered, though whether she had any experience of meeting children resulting from donor insemination is unclear. Her views came at a politically sensitive time, during which a high-profile court case contested individuals' rights to information about their genetic origins. Joanna Rose, a 29-year-old Australian woman living in Britain, sought non-identifying information about her biological father, and argued that there should be a voluntary contact register for all people in her situation. But because the 1990 Act creating the HFEA was not retrospective, Ms Rose was not entitled to this information and discovered that records had actually been destroyed. Her lawyers argued that this contravened the European Human Rights Act on the ground of discrimination – since children conceived by donor insemination ought to have the same rights to information as adopted children. In July 2002 the High Court granted a judicial review of the legislation.

Meanwhile, the British Fertility Society (BFS) – the main professional body in the UK concerned with the issues surrounding fertility treatment and embryo research – started to see the writing on the wall. It published its recommendations regarding donor anonymity, stating: 'Anonymous donation is encouraged in accordance with sperm donation practice. The implications of donor anonymity are different according to the parties involved. It has been shown that 85 per cent of potential donors would not enter a sperm donation programme unless anonymity is maintained.'

Quoting the 1991 Oxford study, the BFS held the line that sperm donation would virtually dry up if anonymity was not maintained. This point was repeatedly chorused by a large number of my colleagues in the press and on television. It seemed to be the only point they could think of making – and it must have sounded to people listening that they were against change because their practice would be damaged if nobody wanted to donate sperm any longer. It was, I believe, a very unwise argument because, apart from anything else, if any practice causes damage to individuals, it is right to ban it. If donor insemination is causing misery to thousands of children who are at a loss to find out more about their genetic origins, then it is difficult to justify it on any grounds unless the alternative is worse still.

The information being published by the HFEA was not much better. They mostly sounded out public opinion about egg sharing and anonymity through the mechanism of a website. Inevitably, consultation done in this way is likely to emphasize problems and unlikely to elicit a broad or balanced view from the public. It is more likely to record polarized attitudes: concern from a few people who feel very damaged and defence of the status quo by people with an interest in maintaining it.

In April 2003 the government launched UK DonorLink, offering genetic testing to match offspring with donors who register with the service. In the next year, the Department of Health wrote to all assisted reproduction clinics in the UK, asking for the opinions of their staff and donors on altering the anonymity provisions. Responses came from 140 donors and 42 clinics. Many voiced concerns about the likely decline in the supply of donated sperm and eggs if anonymity were lifted. Others pointed out that removal of anonymity would have no effect, since many parents would still choose not to tell their children how they had been conceived.

I put my objections to removing anonymity in writing. I pointed out that the issue of the number of available donors being reduced was not relevant in itself. My main concern was that far more parents would simply keep the nature of the conception a secret from the child. Under these circumstances, a child finding out by accident later in life could be caused horrendous damage. And with fewer sperm donors there was undoubtedly more likelihood of one person fathering more children, potentially raising the chance of consanguineous relationships when these 'secret' children became adults. I was also concerned about the implications for egg donation. If the HFEA continued to allow egg sharing there was always the probability that some donating women, who had not got pregnant through their own IVF treatment, would find out – eighteen years later – that they had a child after all and that child had traced them. This seems a very cruel, and all too likely, scenario.

In January 2004 the Department of Health announced its final decision at the HFEA's annual conference. From 1 April 2005, donors of sperm, eggs and embryos would lose the right to anonymity. Donor offspring born after that date would, once they were eighteen, be able to have access to identifying information, including name, last known address and date of birth. The decision still needed parliamentary approval, because it constituted an amendment of the 1990 Human Fertilization and Embryology Act. Unfortunately, the legislation proceeded through the House of Lords hurriedly. The Chief Whip begged me not to vote against the new proposal, but in the event the legislation was put through on a day when I could not attend to speak or vote. The anxieties that one or two of my medical colleagues expressed in that debate were effectively ignored.

People who donated sperm or eggs before the cut-off date are not affected, since the Act is not retrospective. If a woman had had a child using sperm deposited before April 2005, and then wanted to create a sibling using the same sperm, she would be entitled to do so, bypassing the dilemma of one of her children being entitled to know its father's identity while its sibling is not. The timing of the Act means that the first requests for information will not occur before 2023. But its consequences are being felt immediately, in a manner which leads me to deplore the way the decision was taken.

The most obvious effect of this legislation is to make

potential donors of eggs and sperm increasingly wary about taking the plunge. Since financial remuneration, in this country, is limited to a tiny sum to cover travelling expenses, there seems very little incentive for people to donate. All of this might sound like scaremongering if it were not for the fact that the results are becoming immediately, and poignantly, obvious to those in the profession.

After Sweden lifted anonymity rights for donors in 1984, the number of clinics offering artificial insemination dropped by 70 per cent. In 2003, when the government of New South Wales began to consider the same measures, donor recruitment became so difficult that one clinic advertised 'free holidays Down Under' to entice Canadian donors into making the trip. In Britain, many clinics have reported similar shortages since the measure passed in 2005 first came under consideration. A study undertaken in the same year asked former egg donors whether they would consider giving again now that their anonymity had been lifted. Some 36 per cent said they would never have given, had they known that this was going to happen, and only 54.4 per cent said they would now consider donating again.[24]

This reluctance is making itself felt in a very real, observable manner. Cardiff Assisted Reproduction Unit announced last year that it would have to start turning patients away. In Sheffield, the Centre for Reproductive Medicine and Fertility has limited treatments to those living in the immediate area. It has also begun to recommend IVF – costlier and more medically demanding upon the woman's body – rather than donor insemination, because it uses fewer sperm. From Glasgow, Newcastle and Edinburgh, there have been reports of dwindling shortages, five-year waiting lists and spiralling costs caused by the need to import sperm from European suppliers, such as Cryos in Denmark.

The net result of all this is more misery for infertile women and couples. As supplies of sperm and eggs dwindle, health authorities will become increasingly picky about whom they choose to treat, rejecting people for reasons that would not come into the equation if there were no shortage. As the UK fails to offer hope to so many of its citizens, the bulk of whom are infertile through no fault of their own, increasing numbers of people – including many who can ill afford it – will travel abroad to seek treatment.

Seeking alternatives to legislation

There is no question that secrecy, in general, is bad. It is very unwise to withhold information about their conception from children, and this information needs the most sensitive handling. I remember how my mother, in her role in managing adoption (see page 90) invariably took great pains to ensure that couples were properly advised. She always recommended them to tell adopted children that they were indeed adopted, and to start giving this information in a simple way from the earliest years. Apart from the general need for openness, if the truth comes out all at once later it can do so in a very destructive way. I know of cases where children have found out that they were adopted at the time of the break-up of their parents' relationship just before divorce. If a child starts to rebel as a teenager, learning from a disgruntled or estranged parent that they are not genetically related can be extraordinarily damaging. I remember how, when I was at St Paul's School, aged about fifteen, a boy who consistently failed to conform in and out of class was told by our history master that he 'bore no resemblance to' his studious and conscientious elder brother, a pupil who had recently gained a scholarship to Cambridge. This prompted the boy to confront his parents and for the truth to emerge in a most upsetting fashion.

It may be that some of the fears being voiced currently are unfounded. We will know better in 2023; but it is possible that, for many donors, the apparent disruption of meeting a child to whose genes they contributed may in fact be a rewarding experience. A study published in *Human Reproduction* in 2003 indicated that, while an overwhelming number of donor-conceived children were curious about their biological parents, they did not necessarily want a great deal of contact.[25] In the study, conducted by doctors at the University of California at Davis and a California-based sperm bank, young people with an average age of fourteen were consulted about their feelings towards the facts of their parentage. They had all known about their origins from an early age, and some 44 per cent said that this had had no impact, or a positive impact, upon their lives. Eighty per cent said they wanted some contact with the donor, but in most cases this boiled down to meeting them, so that they knew what they looked like.

Perhaps, over time, as the perception of the risks diminishes, there may be a cultural shift back towards donation. And maybe, as some argue, a guarantee of anonymity means very little in an age of increasingly sophisticated global communications. You might be inclined to agree after hearing the story of a fifteen-year-old American boy who traced his biological father on the internet. He had sent a sample of his cheek cells to a popular genealogy website, which led to his discovering two men with very similar DNA to his own. Given reasonable odds that they were related, the boy sent details to a tracing website, along with some information about the donor's date and place of birth that his mother had been given when he was conceived. The search revealed only one possible person – and he and the boy made contact some ten days later. Nobody could have known at the time the boy was conceived that such powerful research tools would be widely available fifteen years later.

The advantages of telling a donor-conceived child about his or her origins, in a measured and gentle way, at the earliest possible age, are obvious. But Acts of Parliament, whose consequences could affect hundreds of existing and future lives, should not be made on the basis of a few reports. These grave decisions, where they involve important matters of public policy, should be made only after thorough and rigorous scrutiny of the arguments and evidence, using the same criteria that we use to assess the value of a medical or surgical treatment, or the safety and benefits of a new drug therapy.

It is also pointless to make comparison, as many in favour of lifting anonymity have done, with the feelings experienced by adopted children. Many of these do feel abandoned and rejected, even inclined to blame themselves, surmising that they must have been somehow imperfect because their mothers gave them away. The same can hardly be said to apply to most donor-conceived children, whose biological parents did not give them away, but donated gametes in order to fulfil the wishes of parents who longed for a child. If anything, donor-conceived children are likely to feel more loved, more wanted, more special, because the parents who have brought them up were prepared to go to such lengths to conceive them.

It is almost certain that the consequence of lifting anonymity will be more, rather than less, secrecy within families. We have already seen this trend in a number of British clinics. Parents, especially those who have gone through a great deal to produce a child, may feel especially protective towards their offspring. They may worry that, at some stage in the future, their child may reject them in favour of its biological father or mother. This could only increase the likelihood of their keeping the facts a secret, and of the child finding out in a way that is less than ideal. It would seem that this has been the case in Sweden, where, since donor anonymity was lifted, some 80 per cent of parents who conceived by donation have not told their children about their biological origins.[26] The principle behind the changes to the law is based upon a well-meaning perception that it is healthier for children to know about their origins than not. But lifting donor anonymity is not going to ensure that children necessarily *do* know about their origins – unless we start printing the details on their birth certificates.

One solution might be to introduce a 'double-track' system of donation, which has been tried at a number of clinics. Donors can choose to be anonymous or not, and prospective parents can select eggs and sperm from either of the two lists. This leaves the choice within the hands of individuals, rather than government.

Another remedy might be to ameliorate concerns about the effects of lifting anonymity by allowing more realistic payments to donors in the UK. While I baulk at the idea of creating a commerce in human gametes, it is perhaps inevitable, given that it is now so easy for people to travel abroad for treatment if they cannot find it here. Certainly, the current HFEA rulings, arbitrarily restricting payment to a nominal fee of £15 and a maximum of £50 a per day for expenses, are a further obstacle to donation – particularly as they have not increased in line with the costs of living. Women donating eggs may have to travel long distances for daily check-ups and miss work as a result of that and the egg collection, and also, in some cases, pay for child-care. It is thoroughly hypocritical to allow what is happening in practice at present: namely that, through the device of egg sharing, women are being paid several thousand pounds for donating their eggs – only they do not receive money, simply treatment in kind. Surely the HFEA has taken the worst option of all?

It is not unknown for a woman to come to an IVF clinic with a 'dear friend', who is actually someone she contacted through placing an advertisement in a magazine, and who is secretly receiving a sizeable cash sum for her efforts. It would be far better for the well-being of everyone concerned if these transactions were conducted openly, and placed under more reasonable constraints. Also, we do not bypass the ethical problems just because money is removed from the equation. Not all unpaid egg donors are doing it for reasons of sheer altruism. Some young women are motivated to donate eggs out of guilty feelings resulting from a previous abortion. While this is an understandable motive, it can still lead to considerable anxiety and regret when the recipient goes on to produce a baby.

Despite the possible advantages of egg donation, there are no quick fixes to the dilemmas that can arise. Without doubt, the most perfect arrangement is when a young, healthy woman donates her eggs for uncomplicated reasons. But arrangements like these are in very short supply. A good example comes from my own experience during filming for the BBC series *Making Babies* in 1996. After one programme we received 403 enquiries from women who wished to donate eggs. Following the usual extensive counselling and screening procedures, we were left with just eight – less than 5 per cent of the original body of volunteers. Given the large numbers of people who would benefit from the treatment, and the very small numbers of donors coming through, the difficulties of egg donation are always likely to persist.

Until very recently I have felt strongly that the removal of donor anonymity was wrong. But in the making of the BBC series which is associated with this book, I met the remarkable artist Stuart Pearson Wright. Stuart's work is meticulous, arresting and poignant, as well as brilliantly executed. I had a long conversation with him which was filmed in his studio when I went to see some of his very fine portraits. We both have a deep interest in the theatre, and much of his work has a powerful theatrical element.

Stuart does not know who his father was. He was conceived at King's College Hospital by donor insemination, and managed eventually to get a rather battered photograph of the limited record they have at the hospital, which he showed me. From this it seems he has two siblings, half-brothers or sisters whom he has never met. Stuart is deeply troubled by what he feels is his personal loss of identity. He showed me some remarkable self-portraits – including a particularly moving one, painted with half his face missing. I asked him whether his loss of a father would present difficulties when he had his own children. He replied:

One of the things that would sadden me about being a father would be … passing on the question to them of who their grandfather was. But I think it's true … that there will be some form of a sense of resolution through having children. I certainly feel that my very strong desire to have children is to, to quite a large degree I think, informed by my predicament. My situation is desperate in that there, there's so little offered by way of solution – I know for a fact that I have two half-siblings, assuming that the pregnancies functioned and were successful, I have two half-siblings who would probably be roughly my age, who may or may not realize that they've been born by artificial insemination, which makes it difficult. But I have to say that every time I take a journey on the underground or the bus, whenever I'm in the street, I look at every single face I pass or I see on the tube and I'm always looking for them and I've always had a very strong feeling that if I ever see either my father or my half-siblings, I would recognize them because I know my own face so well through drawing it hundreds and hundreds of times...I'm looking for them constantly and I'm hoping that one of them will see this [television interview] and, and recognize me in them or them in me and make the connection because...I think that would fill in so much of the, the jigsaw puzzle...it would be ...probably very difficult and very painful as well but maybe wonderful, who knows. But it's certainly almost as strong a desire as, as the desire to find my father...

I felt very close to tears during this interview. I wondered what damage I had done during the course of my career in pursuing treatment by donor insemination.

Fertility tourism

The brochure for one particular fertility clinic in Greece is an impressive product – adorned with attractive photographs of the Mediterranean coastline, picturesque villages and state-of-the-art facilities. It offers infertile women and couples a complete 'IVF holiday package', including flights, accommodation in a range of hotels, and ample opportunity to pass the inevitable and tense waiting periods associated with treatment in the Mediterranean sunshine.

On the face of it, there should be nothing wrong with going abroad to seek medical treatment. For years we have been hearing stories about people travelling to destinations as diverse as India, South Africa and Russia – either because the costs are lower than being treated at home, or because those countries have expertise in a particular procedure. I certainly object to the xenophobic point, raised in some quarters, that the 'standards of care' are likely to be poorer in a foreign country. Where IVF is concerned, clinics in Hungary and Slovenia, for example, boast extremely good success rates.

But fertility tourism has a number of pitfalls that desperate women and couples may be inclined to overlook, particularly if they have been unable to find treatment in their own countries. In the first place, IVF treatment is a demanding and potentially bewildering procedure, fraught with the risk of complications and disappointment at many stages. In some foreign clinics, the staff speak exemplary English and every effort is made to explain what is going on to the patients. In others, that is not the case. And even with the most solicitous care in the world, I believe most women would prefer to be in their home country if the treatment goes wrong. Second, the success rates advertised in some brochures may be misleading, because data are not collected in the same way as in UK clinics. And there may be other differences, such as less stringent procedures concerning the screening of donors for HIV, hepatitis, other sexually transmitted diseases and genetic conditions. Finally, no benefit comes without cost. In the case of many foreign IVF clinics, the reason why treatment is so cheap is that standards of living in those countries are very low. Women and men may be donating sperm and eggs out of sheer poverty rather than altruism and a dispassionate consideration of the implications. There is, therefore, the issue of whether your cut-price treatment is exploiting somebody who is poor and extremely vulnerable.

The story of Svetlana, recently told in the *Observer*,[27] is an object lesson. This 27-year-old married woman with two children regularly sneaked out of her flat in a Soviet-era concrete block in Kiev and, without telling her husband, went to the local infertility clinic, where she was given large amounts of FSH by injection each day for two weeks. She was told that the process was simple and she would be paid $300 – to her a small fortune – if she produced enough eggs. Svetlana went through this process five times, yielding up to forty eggs a cycle – perhaps four times more than the normal response to a lower (and less dangerous) dose of drugs. As a reward she was given another $200 bonus. The women who were receiving her eggs were coming from outside Ukraine, including some from Germany and the UK. Some of the eggs were frozen and shipped to other countries, again including the UK. Unofficial figures suggest that around four hundred embryos were imported into Britain last year, but it could be more. The women who went to Kiev for treatment paid around £2,000–3,000 for it. Even allowing for the cost of the treatment they received, the clinic would net ten or fifteen times what Svetlana was being paid. Svetlana was given 375 units of FSH – a huge dose for an averagely fertile woman. She was lucky; she experienced only mild hyperstimulation and could still get home after each treatment to look after her children. But she could easily have had a massive reaction. Ovarian hyperstimulation in its full form can cause blood clots, permanent damage to the ovaries, stroke, renal failure and, of course, death.

Some clinics in regulated countries have arrangements with clinics like the one in Kiev. It is said that there are several practitioners in London who are prepared to refer patients to clinics overseas for procedures that are banned in the UK. Some appear to have either an arrangement like this with an independent clinic in another country, or a satellite clinic where they can offer such banned treatments. These treatments include egg donation where cash changes hands and donors are paid (and exploited), multiple embryo transfer and sex selection. Fertility tourism is sometimes a scandalous business and has hazards for some couples. But it will continue to be promoted so long as we impose restrictions in Britain that encourage it, and do not deal effectively with those of our colleagues who are profiting from it.

British regulation: the Human Fertilization and Embryology Authority

In this book I offer what may seem some harsh judgements on how reproductive medicine is regulated in Britain. When the HFEA, the regulatory body, was first mooted, most clinicians like myself felt that this was an extremely promising development. At the time, it seemed that it would certainly ensure better information for patients, that it would reduce the risk of patients being exploited by unscrupulous practitioners, that there would be minimum standards of good laboratory and clinical practice, and that research that threatened the public interest or offended public morality would be prevented. It was also hoped that the existence of such an authority would deflect the vigorous attacks that were being made on this branch of medical practice by the 'moral right'.

In fact, while the establishment of the HFEA produced minor benefits, there is little evidence that British citizens are better off as a result of its existence. The only exception to this is that the practice of multiple embryo transfer has been largely curtailed in Britain,[28] with the excellent effect of reducing the number of multiple births, many of them premature. But the clinical results of IVF in Britain are certainly no better than in most other western countries,[29] and there is little evidence that IVF patients here are less exploited than elsewhere. And it is of great concern that the UK's pre-eminence in research in the field has taken a nosedive.

Although members of the HFEA sometimes boast that our legislation – and our regulatory body – is the envy of the world, it is clear that the deficiencies in the British system are so great that very few advanced countries (with the exception of Canada) have copied it.

First, it is difficult to justify the singling out of one specific clinical treatment, namely IVF, for regulation. No other form of medical practice is similarly regulated; medical practice does not suffer unduly in consequence. It has been said that the reason for regulating IVF so strictly is because it deals with the beginnings of life and with small babies. But it is worth reflecting that we do not have a similar regulatory body for the practice of clinical obstetrics and midwifery, nor for fetal medicine. Nor is neonatal paediatrics regulated in this way. Practice in all these areas has consequences as least as weighty as the practice of in vitro fertilization. It has also been said that the need for special regulation is to prevent 'evil practices' such as human cloning. But human cloning is a criminal offence, and it seems unthinkable that it could be conducted in the UK without discovery; clearly there is no need for a regulatory authority to prevent such procedures.

This singular regulatory focus on IVF has perpetuated its image as a 'dubious' area of medicine where disreputable things may happen. As a consequence, assisted reproduction in Britain is often perceived as unsuitable for public funding – 'fringe medicine' rather than life-giving and life-enhancing treatment; and as a consequence of *this*, IVF has been pushed increasingly into the private sector, where patients may pay high fees for treatments that often, in other similar countries, are considerably cheaper. How can it be justifiable to charge £1,000 for embryo screening for chromosome defects in the UK, when the same service is performed at least as effectively in parts of Australia for one-seventh of the price?

The 1990 Act of Parliament also demands that clinicians 'must take account of the welfare of any unborn child' that might be born as a result of treatment. This position is unique to IVF procedures; it doesn't apply to other infertility therapy. It is undesirable in principle because it risks discriminating against patients who are deemed to be 'unsuitable', and unworkable in practice because no doctor can forecast the future. The HFEA has controlled this aspect of treatment over-cautiously; its rigorous insistence on trying to ensure that clinics take the welfare of the unborn child into account adds a huge amount of extra paperwork and bureaucracy, which is generally unproductive.

It has been asserted that the presence of a regulatory body allays public fear that devious practices are going on behind closed laboratory doors. But there is little evidence that the HFEA does allay public fear. On the contrary, the presence of such a regulatory body may raise public anxieties by focusing on issues that are seen to be more threatening than they really are.

In my view, the corporate failings of the HFEA are also

of considerable concern. The HFEA was put in place to ensure good data collection so that sound and adequate information was available for public scrutiny. But the HFEA's data collection mechanisms are inadequate and the data are published approximately three years out-of-date – so it does not give patients the information they need to make decisions about which clinic to visit. Moreover, it seems that it has not always been able to collect accurate data. A perusal of some recent annual reports shows that it has not always given national success rates for treatments. It is said that this is because it has had repeated difficulties with the way data have been collected, collated and placed on the computer. To quote from the HFEA's annual report of 2002, 'During 2002–03 a modern flexible database system was successfully developed and implemented and transfer of data from previous systems will take place shortly. A programme is planned to check previous data and resolve any inaccuracies. This will ensure the register is able to meet the HFEA's statutory duties.' As I write it is the middle of 2006, and this does not yet seem to have happened.

More importantly, the most critical data are ignored. Treatment results are certainly important, but information about the long-term consequences of different IVF procedures such as gonadotrophin therapy, changes in culture media, long-term culture to the blastocyst stage, sperm microinjection, embryo freezing and embryo biopsy is not recorded. All these procedures, done in either humans or animals, have been occasionally associated with risks. In Britain, we have virtually no knowledge about the potential hazards of different IVF procedures, nor any record of the abnormalities produced by them. Given that in the past ten years reports of a high incidence of fetal abnormality after treatment in some IVF centres have emanated sporadically from several other countries, such as Sweden, Australia and the USA, this is a potentially serious deficiency.

The quality of much other information given out by the HFEA is poor. Quite indifferent material is optionally offered to prospective patients. *The Patients' Guide to Infertility* is in many places inadequate and inaccurate. It focuses very strongly on in vitro fertilization and not nearly enough on other equally successful alternative treatments. Another criticism concerns the quality of the information the HFEA gives to the press. It is acknowledged by many journalists that the HFEA is not an easy body from which to get information or comment.

The HFEA has claimed that it has for some years been conducting vigorous audits of the results it receives from various clinics. If this is true, it is surprising that there is very little feedback about these audits. To date, there has been no explanation from the HFEA of why one or two clinics are getting extraordinarily good success rates, often more than double the average of other clinics. Is this because some clinics are doctoring their results, or because they are surprisingly skilled? Conversely, one or two clinics appear to achieve far less success than average. These differences are of critical importance to patients and should also be a matter for more public scrutiny.

There is some evidence that research in IVF may have been inhibited. If clinical practice is to improve, research is essential. Yet though for many years Britain led the way in reproductive medicine and physiology, recently our research output has been rather poor, particularly where it concerns human embryos. Even though Parliament voted overwhelmingly in favour of liberal regulation of research in the field of embryonic stem cells, our publication record and citation record are less good than those of the USA, China, Australia and the Scandinavian countries, and no better than that of Israel, which has a population half the size of London's. Currently, researchers have to apply for ethical approval from the regional ethics committees set up by government – quite a rigorous procedure – and thereafter for ethical approval from the HFEA. This extra barrier is time-consuming but pointless. It also tends to be ludicrous when the HFEA requires even minor modification of consent forms that have already been approved by the regional ethics committees. Such modified consent forms have been returned to the regional committee for their approval, and the consequent games of bureaucratic 'ping-pong' can prevent a particular research project from being started for months. All these barriers tend to inhibit researchers, who are inclined to be diverted into areas more likely to be easily pursued but possibly less valuable scientifically. One leading British embryologist with a significant international reputation left the field, demoralized, because she found the HFEA and its overbearing approach to her research intolerably invasive, making her feel that her work was in some way unethical.

The HFEA is a very bureaucratic body and is expensive for patients, to whom the charges that the HFEA makes on practitioners have inevitably been passed on. IVF must be one of very few medical procedures where the patients themselves pay for the regulation of their treatment. Hammersmith Hospital – a medium-sized clinic – paid approximately £135,000 for its clinical licence in 2005, and in addition spent money on at least two full-time salaries to service the administrative requirements of the authority. Inevitably, this adds to the considerable cost of the treatment. Moreover, the HFEA recently increased the charges made for applications for research licences. To apply for a new licence to research on, say, embryonic stem cells costs £700. Modification and renewal of licences may also be heavily charged. In no other field of medicine is such a charge made for application for ethical approval, either in this country or in the United States.

At the time of writing, it seems that the British government is considering some adjustments to the legislation concerning reproductive technologies. In chapter 7 I discuss the area of IVF that, at least on the surface, most requires regulation. But even in this area, the detection of genetic defects in embryos, it is very dubious whether the impact of the HFEA to date has been useful. As we shall see, the prevention of genetic diseases by screening potentially affected embryos is so tightly regulated that this treatment, sixteen years after the first births from IVF, is still very hard to get in Britain. It is to be hoped that as society matures in its attitudes we may take a more constructive view of how to regulate with a lighter touch, enabling this much-needed medical treatment to be offered more widely.

6

Womb to Rent

While they were on this discourse and pleasant tattle of drinking, Gargamelle began to be a little unwell in her lower parts... Whereupon an old ugly trot in the company, who had the repute of an expert she-physician, and was come from Brisepaille, near to Saint Genou, three score years before, made her so horrible a restrictive and binding medicine, and whereby all her larris, arse-pipes, and conduits were so oppilated, stopped, obstructed, and contracted, that you could hardly have opened and enlarged them with your teeth, which is a terrible thing to think upon; seeing the Devil at the mass at Saint Martin's was puzzled with the like task, when with his teeth he had lengthened out the parchment whereon he wrote the tittle-tattle of two young mangy whores.

By this inconvenient the cotyledons of her matrix were presently loosed, through which the child sprang up and leaped, and so, entering into the hollow vein, did climb by the diaphragm even above her shoulders, where the vein divides itself into two, and from thence taking his way towards the left side, issued forth at her left ear. As soon as he was born, he cried not as other babes use to do, Miez, miez, miez, miez, but with a high, sturdy, and big voice shouted about, Some drink, some drink, some drink, as inviting all the world to drink with him. The noise hereof was so extremely great, that it was heard in both the countries at once of Beauce and Bibarois . . .¹

296

So, in 1532, the physician and Franciscan monk Rabelais described the birth of Pantagruel, after labour was induced by his mother ingesting sixteen quarters, two bushels, three pecks and a pipkin full of tripes. His book was rapidly banned by the church. Although not exactly a surrogate arrangement, this birth is painted in a grotesque way that is highly reminiscent of much of the comment about surrogacy in the 1980s. And when the news hit the headlines, the noise around the world was possibly equally as outraged and outrageous.

The ludicrous preoccupation with surrogacy, and the continued prolonged adverse publicity that surrounded the news stories (echoed in the reaction to the idea of cloning twenty years later), had a serious effect on how the new reproductive technologies were perceived in Britain. The outspoken condemnation of surrogacy by some members of the Warnock Committee, and the extra-ordinary attention given to it in some newspapers – broadsheet and tabloid – skewed public opinion and sullied the reputation of fertility medicine. What virtually no commentator ever bothered to point out was that surrogacy, though not quite as rare as Pantagruel's mode of delivery through his mother's ear, was always going to be a very uncommon practice.

Surrogacy: medicine, morals and politics

Surrogacy is essentially of two different kinds. So-called 'partial surrogacy' involves insemination of the host mother – either, as tends to be usual (at least in those cases that are publicly documented), by artificial means, or naturally, that is, by sexual intercourse. Partial surrogacy by natural conception, incidentally, has probably always gone on, and even now is possibly much more widely practised than people imagine; but any records kept tend to be sketchy. After partial surrogacy, the child is genetically the host mother's, but the father is usually part of the adopting partnership. So-called 'full surrogacy' has to include IVF treatment because both the eggs and the sperm come from the adopting couple. The adopting mother goes through the whole panoply of IVF treatment, her eggs are fertilized with her partner's sperm and the embryos are placed in the uterus of the host mother. Thus the host mother has no genetic connection with the child.

The massive focus on the allegedly aberrant and immoral nature of surrogacy certainly made the medical community anxious; and when Parliament did not immediately produce legislation in this area, many members of the medical establishment came to favour the creation of some kind of regulatory authority. So, as I have related in chapter 2, we as a profession established our own voluntary licensing authority under the excellent chairmanship of Dame Mary Donaldson.

It was against this disappointing and difficult background that we at Hammersmith made two attempts at IVF surrogacy. I think I may have been the first doctor anywhere to try this procedure – and this was in the very early days of IVF, when only a handful of babies had been born in this way around the world. Even so, I was concerned that we should preserve strict confidentiality. Early in 1983, May C,[2] a 42-year-old woman who had had a hysterectomy, consulted me at Hammersmith. She had never had a child. My immediate impression was of a rather funny, highly focused, quite feisty woman. She was not employed at the time, having left the baker's shop where she had worked for nearly twelve years. Although it was now impossible for her to become pregnant, she was still somewhat obsessed with trying to have her own baby. The adoption authorities, to whom she had repeatedly applied, had turned her down – mostly, it seems, on the grounds of age.

May, though she had limited formal education and no science background, was very intelligent and had worked out (and this was before the newspapers got excited about IVF surrogacy) that we might provide her with a solution. Her ovaries had been left intact at the hysterectomy, so she intended to find a way of going through IVF, and had come to me, after a great deal of discussion with her immediate relatives, to request treatment. She wanted her husband to produce sperm to fertilize one of her own eggs in vitro, with any resulting embryos being 'hosted' in the uterus of her sister, Elisabeth, who had agreed to the idea.

May was querulous when I was slow on the uptake and very concerned when, having eventually understood what she wanted, I initially raised numerous objections to attempting this treatment. But she returned three or four times – in those days, there were no substantial waiting lists to see NHS patients, and no rigidly kept appointments (we just stopped working quite late in the evening when the last patient had been seen) – and eventually I began to realize that she ought to be taken seriously. So I asked to see her sister. Elisabeth was rather a mild, quiet, unassertive person – it was obvious that the genes for being pushy did not run throughout the family. May's husband also seemed a bit mouse-like. So obviously I wondered whether Elisabeth had been railroaded into agreeing to co-operate. I continued to agonize over what I should do for almost four months. Both women went for counselling with their respective husbands. It is often said that people working in human reproduction are anxious to break new boundaries and test new limits. I believe that this is a very unfair assessment of our thinking. On the contrary, in this case I think we were very cautious; we spent a long time temporizing, trying to dissuade our importunate would-be patient from considering this kind of treatment.

When May continued to be very persistent, I discussed the situation with the chairman of my department, Professor Murdo Elder. He felt that, provided properly informed consent had been obtained, this would be a reasonable procedure to undertake, and suggested that I go to various senior colleagues in confidence and gauge their reactions. Although most people think that issues of this kind get presented to ethics committees in the health service, there is actually no mechanism for this to happen. Ethics committees are concerned with the regulation of *research*; medical *treatment* is, as a matter of principle, not regulated in this way, but is traditionally considered to be a matter of individual clinical judgement. I think that this is a sound way of handling difficult issues. But clinicians need to be sufficiently responsible to seek advice from other colleagues and to take enough time in making up their minds about such issues, except when there is real urgency. In the past two or three years, many hospitals have set up clinical governance committees. I suppose such matters could be raised before one of these bodies, though I doubt whether formal regulation will reduce the chances of making errors of judgement.

Eventually we decided to treat May. We were satisfied that all four people involved, the women and their respective husbands, were fully aware of the problems that might arise, particularly the potential for conflict if a child was born. Above all, I was concerned that the bearing mother, Elisabeth, might feel very possessive of the child that she delivered and might wish to interfere in

May's nurture of her baby. It turns out that I was right to consider this a potential difficulty, for other surrogate arrangements since that time have shown that this can be a real problem. I was also increasingly worried that the press might come to hear about what we intended. It began to dawn on all of us what a massive story this could be. At that time, anything involved with in vitro fertilization was big news, and it seemed obvious there would be risks to both our patients if the newspapers came to hear about our projected treatment. I therefore insisted that if May was to be treated, this would be a strictly confidential arrangement and that no journalists, under any circumstances, would be informed. And there must be no selling of a story to a newspaper. Both wives and their husbands readily agreed. My reliable secretary kept the clinical notes of both women in a locked cabinet in my inner office and as few staff as possible were put in the picture.

We started by synchronizing the menstrual cycles of both women. At the time there was no precedent for placing a donor embryo in another woman's uterus. I knew that Alan Trounson in Australia was attempting to do this with donated eggs, as was Peter Kemeter in Vienna. But at this time no cases in which pregnancy had been achieved had been published in the literature. At Hammersmith at that time there was no way we could freeze embryos. So it seemed vital that if we were to get a successful implantation we would need carefully to synchronize the preparation of Elisabeth's uterus with egg collection from May. Both women were placed on the contraceptive pill, and then stimulation commenced using a natural cycle with the addition of clomiphene tablets. This helped to keep the two women in synchrony as far as possible. Unfortunately, after a month or two, it became obvious that May's ovaries were so badly scarred after the hysterectomy that they were not responding very well. So we abandoned these cycles and waited three months. We then gave clomiphene tablets again, but this time added some injections of FSH. The response was poor, but a follicle did seem to develop and we hoped we might be able to go ahead to egg collection. But just eighteen hours before the egg collection the follicle collapsed, and we had to abandon this cycle.

Sadly, at this stage the whole situation suddenly became very complex. That Friday afternoon I received a phone call from a Sunday newspaper. The journalist, who had quite a reputation for reporting sensationally, claimed aggressively that she knew we were undertaking a new IVF treatment. Although she had a number of facts quite wrong and it was clear she couldn't identify the patients concerned, I was horrified. I decided to deny all knowledge of what the journalist was talking about – my only option if I were to protect my patients. But this added complication inevitably meant that it would probably be wise to delay a further treatment cycle. To continue at this point would clearly risk the loss of my patients' privacy. To this day, I do not know where the leak came from. Only three or four other people on the unit knew that the treatment was being conducted, and I do not believe that either May or her sister had blurted out this information to a friend.

So we waited four months, and then offered May another attempt. But by this time she felt that the situation was too stressful, and wanted longer to think about it. A few months on, she came back for review and was keen to try again. We attempted another treatment cycle, obtained three follicles and fertilized the eggs – but, to May's extreme distress, no pregnancy resulted. A few months later, it seemed that May wanted to go ahead with a further attempt but that her sister was becoming increasingly doubtful; and in the event, although May came back for repeated consultations, we never tried again. As time went on, she became increasingly bitter about her situation and, for a while, was very angry with her sister and angry with us. I lost contact with her about a year and a half later.

This whole experience greatly influenced my attitude to surrogacy treatment. I began to feel quite strongly that it posed too many risks for the women involved, and even to feel that perhaps we had been lucky not to have been successful. I never attempted another surrogacy treatment. I still think that this episode was a very good example of the disadvantages that can attach to pursuing the 'latest' medical advance. Sometimes raising false hopes by trying every possible treatment can result in a great deal more harm than good; even to this day, I wonder what happened to that unusual, gutsy woman, and how her relationships with the rest of her family eventually turned out.

We were certainly prudent in being discreet about this treatment. When surrogacy eventually did hit the news, the British authorities and media reacted with extreme distaste. In 1984 Mary Warnock and certain other members of her committee recommended – though the committee was seriously split over the issue – in the strongest terms that the practice should be outlawed.[3] Then, in 1985, it emerged that a woman called Kim Cotton had received some £6,500 in order to give birth to a baby for a childless couple. The baby was made a ward of court for seven days until hasty legal rulings made it permissible for the couple to adopt it. This was a powerful stimulus for many MPs to call for legislation to regulate all IVF. A Bill was introduced in Parliament, but it was directed solely at commercial surrogacy, which was made a criminal offence in the UK.

Meanwhile in the USA, public opinion was being similarly exercised by the celebrated case of 'Baby M', born in March 1986 to a 26-year-old woman called Mary Beth Whitehead on behalf of Bill and Betsy Stern. Mary Beth had signed a witnessed contract that was fully legal in the state of New York. But despite having received $10,000 for her efforts, she was unable to deal with the difficult feelings she experienced after parting with the baby to whom she had given birth. She visited the Sterns' home and abducted the baby, subsequently telling the courts, 'I signed on an egg. I didn't sign on a baby.' As well as being swift-footed, Mary Beth was obviously pretty nifty at aphorisms. Of Bill Stern she is reported as saying, 'He just considers me a uterus on legs,' and of her baby, 'I gave her life, I can take it away.' A lengthy legal battle eventually ensued in the neighbouring state, New Jersey, before Bill Stern was finally awarded custody of the child. There this sad affair – or at least its public aspect, played out in the full glare of press attention – ended.

In 1987 the surrogacy debate was widened still further when a South African grandmother living in Tzaneen, a predominantly Afrikaans-speaking town in the Transvaal, became a surrogate mother for her own daughter, Karen Ferreira-Jorge. Matters were made more sensational by the fact that this 48-year-old, Pat Anthony, carried triplets.

Karen had borne one child already, but following this had had a hysterectomy at the age of twenty-three. Both she and her husband were deeply concerned about having an only child and wanted more children. So, after discussion with her mother, Karen's eggs were fertilized in vitro with her husband's sperm, and four embryos were transferred to Pat Anthony's uterus.

While the world's press attitudinized, kicking up a huge fuss, there was surprisingly little condemnation of this treatment in South Africa – though some religious authorities and members of the medical establishment were quick to deplore it. Representatives of the Roman Catholic Church (both Mrs Anthony and her daughter, Karen, are Catholics) were appalled; theologians belonging to the three main branches of the Reformed Church, to which most Afrikaners in Tzaneen belong, indicated strong disapproval, calling the arrangement 'unsavory'; and the attempt was declared 'undesirable' by the Medical Association of South Africa. Perhaps unsurprisingly, Dr Hein Odendal of Tygerberg Hospital in Cape Town,[4] the head of the country's leading in vitro fertilization laboratory, and presumably associated with a competing clinic, was fairly vitriolic in his opposition.

But the South African press seemed pleased that their country, in spite of its growing problems, could compete with others in wielding the latest technology, and they gave a mainly positive spin to the story. I remember when the pregnancy was announced in the *Daily Mail* in London my own reaction was very critical, until my wife pointed out what an extraordinarily generous thing it was for a mother to have done. But perhaps the most important aspect of this story was missed by nearly all commentators: namely, that it was surely very irresponsible to have transferred four embryos. In the event Pat Anthony was lucky, and the babies were born prematurely as healthy triplets. But a complex pregnancy of this sort in a woman in her late forties could have ended in disaster.

Religious views on surrogacy

As we have already seen in the tale of Sarah and Hagar, discussed in the prologue, there has since ancient times been a feeling that surrogacy arrangements can be fraught. In general, the Jewish attitude to surrogacy is clear-cut – and it centres, as we have seen, on the issue of who should be regarded as the child's true parents. Jewish law has almost invariably regarded the birth mother as the mother of the child – even if she has been implanted with an egg from someone else. But while there is concern about the risks the various parties might run in bringing about such a birth, there would be no objection to a legal adoption afterwards. For that reason, the few Jews who have sought surrogacy treatment have also tended to try to find a surrogate mother of their own religious background, so that the child is Jewish, too.

Nevertheless, a number of Jewish sources have come out strongly against surrogacy. Notable among them was the late Chief Rabbi Dr Immanuel Jakobovits, who officiated at my own wedding. Dr Jakobovits, a very great teacher and scholar, was a member of the House of Lords when I first entered that chamber in 1995, and I was very moved by his warm welcome and his continued friendship. I think Margaret Thatcher, who was Prime Minister when he was made a peer, said of him that 'he is my Chief Rabbi'. Certainly, although he sat on the cross benches, he was notable for his sometimes strong conservative views. Nevertheless, though we quite frequently disagreed politically, he was a man for whom I had greatest respect; he was regarded internationally as an expert on medical ethics and he was a leader known for taking many courageous decisions. It was invariably a privilege to discuss with him the many moral issues which were constantly raised during parliamentary debates. He had an undoubted gut reaction against surrogacy, saying on one occasion that 'to use another person as an "incubator" and then take from her the child she carried and delivered for a fee is a revolting degradation of maternity and an affront to human dignity'. But I cannot help feeling that Lord Jakobovits, on this occasion, was speaking from a profound sense of emotional conviction rather than from reflection on the finer points of Jewish law.

In the Arabic language, in which the Qur'an is written, the word for parents – *walidan* – comes from the verb *walada*, to give birth to. Accordingly, the holy book of the Muslims is replete with references that make Islam's position very clear. To give just two examples:

None can be their mothers except those who gave them birth.

We have enjoined on man's kindness to his parents – in pain did his mother bear him and in pain did she give him birth.

So Islam also considers the birth mother to be the true mother. But it outlaws surrogacy on the grounds that it involves insemination outside of marriage, which is forbidden.

The baby brokers

Even if the Sterns had no religious misgivings, they must certainly have been aware that surrogate arrangements had been clouded by controversy ever since they had first attracted the attention of the American media. Back in 1976, a Michigan-based lawyer called Noel Keane had been approached by a number of couples asking him to find them surrogates, who would agree to be impregnated by insemination using the husband's sperm.[5] Keane began to act as a womb broker, placing discreet adverts in Michigan papers, offering a fee for women who would be prepared to assist. But some newspapers refused to place Keane's advertisements, voicing serious misgivings about the ethics of his new business. Others ran critical pieces denouncing him. Matters became worse when Keane became concerned that, under state law, the sale of babies was illegal. That could mean payment for surrogacy was also illegal, because in this form of the transaction the mother paying for the service bore no genetic relation to the child, and could thus be deemed to be buying it. As Keane had no wish to act outside the law, he removed the element of payment from his brokering scheme. This, unfortunately, had the immediate effect of discouraging most potential surrogates to come forward. He set up in Florida, where more lenient laws may have meant that paid surrogacy was no problem.

I met and debated with Keane on television when the issue of surrogacy became big news in the UK and Keane flew across the Atlantic, I imagine to promote his work. I was impressed – and not favourably – by his apparent unconcern about the bearing mother. It seemed to me that, as far as he was concerned, the women who had been commissioned to get pregnant had signed a contract, and he made it quite clear that he would take all steps to enforce it. He may have had qualms, but I doubt it. Over time, Keane was joined in the brokering business by a number of other pioneers, among them Bill Handel, based in Beverly Hills. Handel did make some attempts to assess the psychological state of the surrogate mothers he hired, but mostly, I think, with a view to matching surrogates with parents using a sophisticated battery of psychological tests. I had the dubious pleasure of meeting Handel as well as Keane. I asked them both the same questions. What would be the situation if the bearing mother became ill as a result of being pregnant – for example, if she developed toxaemia of pregnancy, or liver failure? These objections were rapidly dismissed: all problems of this kind would simply be dealt with by the medical insurance cover that was invariably taken out first. The answer to another concern – what if the toxaemia became so serious that the mother had a stroke and suffered brain damage, or died as a result of an embolus or haemorrhage? – was more blurred. Both men were at pains to insist that the best possible medical treatment would be provided, and there was provision for life insurance cover too; but it seemed to be the case that if the woman was being paid for her 'work' then ultimately she had to accept the risks it entailed.

Neither of these practitioners, nor indeed any other person involved with this trade whom I met, ever gave me what seemed completely satisfactory answers to questions concerned with the baby. What will happen if the baby is born prematurely and possibly brain-damaged – or if the surrogate mother gives birth to a Down's syndrome child and the commissioning parents reject it? And supposing this baby is then also rejected by the bearing mother? The answer seemed to be the Victorian solution of the orphanage. Then there was the question of twins: supposing, after IVF, a single embryo splits into two and the commissioning parents do not want their life disrupted by caring for the consequences of a multiple pregnancy; what will happen then? Again, how will the commissioning parents react if the surrogate mother needs an abortion for the sake of her own health – will her needs outweigh those of the unborn child? And does a surrogate mother have the right to choose a certain diet, or an exercise routine, or a method of birth that the commissioning parents feel may put the unborn child at risk?

Another of these gentlemen, and one not troubled by reticence, is Dr Richard L. Levin of Kentucky. From his website, www.babies-by-levin.com, I learn that in the 1970s 'Dr. Levin traveled to Europe to collaborate with Winston firsthand. After working together for a brief period, Winston confirmed that Dr. Levin has indeed reproduced his microsurgical Tubal Ligation Reversal technique.' For some reason, I have no recollection of Dr Levin's visit to Hammersmith. I am extremely sorry to have forgotten it – but it was a long time ago, and I was probably quite busy peering down my operating microscope. When he espoused the cause of surrogacy back in the USA, he explained how pleased he was with his personalized car number-plate, which read 'Baby 4 U'. He made the rounds of local and national talk shows, explaining that he paid surrogates several thousand dollars for their service. Womb rental had become a market, rather than a time-honoured biblical custom.

As it became a market, of course, others were swift to condemn it, and the 'Baby M' case provided the initial spark for an explosion of condemnation. At this stage, critics tended to focus on the wider ethical implications of commercializing reproduction, rather than on the feelings of the individuals concerned. Some feminists saw it as the further degradation of women's bodies by men. Ginea Corea wrote, with some justification, that 'a woman is again seen as the vessel for men's seed', just as she was under Aristotelian biology: now again, women were 'serv[ing] as the passive incubator'.[6] But others said it empowered women – enabling them to use their biology to earn money, just as a man's muscles might allow him to make a living from hewing logs or building houses. Perhaps both arguments were equally exaggerated. In the USA, a country founded on the principles of freedom and opportunity, many argued that legislating against surrogacy was a breach of a person's most basic rights.

As noted at the beginning of this chapter, in partial surrogacy the host mother is genetically the mother of the child, while in full surrogacy she has no genetic connection with the baby to whom she gives birth. It seems to have been the troublesome matter of the genetic link between the surrogate and the child that created the problem in the earliest disputes. If there was a disagreement, it was the bearing mother who had the rights over the child, matched only by the rights of the man who had given the sperm. The woman who had commissioned the surrogacy had none at all – only the rights that may have existed in whatever contract she might have drawn up prior to making the arrangement. And courts were very reluctant to uphold those rights. In the 'Baby M' case, even though the court of New Jersey found in favour of the Sterns, it ruled that 'a contractual agreement to abandon one's parental rights . . . will not be enforced in our courts. There are, in a civilized society, some things that money cannot buy.'

Ethics, choice and the law

IVF technology should have meant that, because the surrogate mother was only the host, and not the genetic parent, of the child, the potential for problems would be diminished. It would seem on the face of it less likely that a full surrogate mother, having borne a child that was not hers, would feel as distraught as Mary Beth Whitehead at giving it away, or would wish to press her right to be acknowledged as its mother. From the parents' point of view, all that was required was a healthy woman of proven fertility, who would agree to conduct her pregnancy in the best interests of the child. This, in turn, meant that the numbers of women willing to undergo the procedure shot up. In 1981, when non-IVF surrogacy was the only kind available, there were around a hundred such pregnancies in the USA. By 2000, there were 1,210 per year.

But the possibilities raised by IVF did not mean that controversy subsided. On the contrary, if anything it was heightened, because surrogacy now contained such a strong element of dispassionate commerce. People argued that it amounted to no more than another way of trafficking in women – in particular for the poorest, who would be coerced into renting out their bodies by the promise of money. In the USA, it was claimed that black women would be exploited the most, not just because of poverty, but because their skin colour made it so appealingly obvious to a commissioning couple that the white child she bore was theirs and not hers.

This kind of feeling trickled into the way countries formulated their legal provisions about surrogacy. Germany, as so often with issues of human reproduction, took a hard line and banned commercial surrogacy outright. The German approach to reproductive technologies owes much, I think, to the scars left by the political events there preceding and during the Second World War. France, though generally much more liberal about matters related to IVF, followed suit. In Canada, the government permitted non-financial surrogacy arrangements, but in law the birth mother remained the legal parent of the child, regardless of who had contributed the egg. In the UK, after the 'Baby Cotton' case, the government took a similar line, outlawing payments beyond necessary expenses for the surrogate mother. Today, when a surrogate mother gives birth to a child, she remains its legal mother until the commissioning parents formally adopt it – and adoption is routinely permitted only when the parents are married, and when one or both has contributed some of the child's genes.

The picture in the USA remained a patchwork of varying laws, with the federal government steering well clear of the debate. It seems that, particularly in these and like matters, the states value their independence from Washington – and this attitude has undoubtedly affected the way IVF is researched and practised there. Some states, like Kentucky and California, make partial, paid-for surrogacy extremely straightforward; others, like Michigan and New Jersey – where the earliest controversies erupted – remain opposed. In California, a state for the most part associated with liberal attitudes (except, perhaps, where smoking is concerned), a landmark ruling in 1993 defined motherhood in an entirely new way. In the case of *Johnson* v. *Calvert*, an economically disadvantaged black surrogate signed a contract and then changed her mind after giving birth, filing for custody of the child. Media opinion swiftly made this a case of poor blacks versus wealthy whites, even though the married woman who had sought the surrogacy was Filipino. The Supreme Court in California decided that the natural mother was the woman who had intended to bring about the birth of the child and raise it as her own. It enforced the contract. Motherhood, in their definition, became largely a question of intention, less about genetics and not about birth.

Because governments reached such differing conclusions about the ethics of surrogacy, the potential has
opened for an international trade in rented wombs.
Australians, frustrated by restrictive laws in their own
country, came to California. Americans with less money
crossed the border to Mexico. Gay couples in the UK
circumvented the rigid adoption laws by travelling to
liberal American states to commission a surrogacy
agreement. In German, Belgian and Dutch newspapers
Polish women discreetly offered their services, for the
equivalent of two years' salary. The trade continues to this
day.

My impression is that only very few British couples consider surrogacy. Moreover, the numbers who seek
surrogacy arrangements abroad are likely to remain small,
because of a legal loophole in the regulations. Commercial
surrogacy is forbidden, but, in a typically British compromise, 'reasonable expenses' are permitted. This prevents
anyone from setting up a 'Baby 4 U'-type brokering service,
but it leaves the question of fair remuneration comfortably
in the hands of the individuals making the transaction.
Carole Horlock, a British mother of two who has produced
nine surrogate babies, is estimated to have earned around
£100,000 from her unusual career choice. She told the
Sunday Times that her principal motivation was not money,
but helping childless couples to experience the joys of parenthood. 'If I didn't like a couple, I wouldn't do it,' she told
a journalist in 2003. Yet one news report suggested that her
family are unhappy about her actions and motives. ITV
reported that her father, a committed Christian, had condemned her actions, and apparently her ex-husband,
Stephen Horlock, has been quoted as saying: 'Money is
definitely what drives Carole. She will never reveal
how much she makes from it. She has never really held

down a proper job. I find the whole surrogacy thing bizarre.'

But if Ms Horlock enjoys her freedom, some argue that thousands of poor and underprivileged women may be sucked into the surrogacy trade because they have few other options – and this is a tangled ethical issue. It is possible to argue that the whole economic order from which we in rich countries benefit rests upon the availability of cheap labour in poorer countries. A pair of trainers, a mobile phone, a bicycle – all of these things come conveniently to us because someone in another part of the globe is earning a tiny amount to produce them; but to stop buying such products, arguably, will only make the poor 'producers' poorer. There is also the question of where one draws the line. Why is it deemed 'all right' for a western woman to rent out her womb, but not all right for one from Mexico or India to do the same thing?

If surrogacy remains fraught with difficult issues on a global scale, it is no less problematic when we consider the feelings of the individuals who engage in it. Sarah's response to Hagar's pregnancy provides a good model for the sort of emotions that can arise when an infertile woman sees another carrying her husband's child – even though the use of artificial insemination or IVF has at least taken sex out of the equation. But these are only the tip of the iceberg. What happens when the surrogate has to hand over the baby? It could be psychologically very damaging for her to have to part with the child. Even if the baby is not genetically related to her, she has carried it for nine months and gone through the intensely moving and hormonally disruptive experience of childbirth. While it was generally assumed when IVF surrogacy became available that a mother who is not genetically related to the child would be less concerned about giving it up, I know of no good social science research clearly indicating that this experience is potentially less damaging to the host mother. We also now know that the environment she unconsciously provides in the womb may have profound effects on the child's physical health later in life and may well influence the personality of the baby. This, of course, is one of the reasons why most of the American surrogacy contracts include clauses that make it quite clear that the host mother may not undertake risky activities during pregnancy – such as smoking or drinking any alcohol.

The above issues may loom larger depending on whether the surrogate is a friend or relative of the commissioning couple, or someone unknown to them. Arguably, a friend or relative may act more responsibly during the pregnancy – but if it fails, there may be a serious rupture in a previously close relationship. Those close to an infertile couple might be more easily persuaded to help them out, but then be wary of backing out, even if they have privately realized that they have strong misgivings about the procedure. On the other hand, using a previously unknown surrogate entails trusting a stranger with one of the most precious and delicate transactions any humans could ever make.

The difficult feelings surrounding a surrogacy extend beyond the two women involved. They may be able to create strong bonds because of a shared understanding of a woman's need to bear a child. The infertile woman can sometimes 'enjoy' her pregnancy by proxy, discussing it with the surrogate, accompanying her to antenatal appointments and generally feeling involved. But her male partner may find himself out of the loop – standing in an extremely uncertain relationship to the surrogate mother. And what of the child itself, when it is born? What will be its relationship to the surrogate? If she is involved in its life, could it become confused about its true identity, and will it feel abandoned if she remains absent?

All of these questions and possibilities were posed when our government first began to discuss the ethics of surrogacy in the 1980s, and they continue to be raised today. But perhaps the issue is not so problematic as it appears. Public and press opinion tends to home in on controversial cases, such as that of 'Baby M', where the surrogate mother experienced agonizing feelings of loss upon parting with the baby – and to overlook the great majority of surrogacy arrangements that proceed quietly, smoothly and happily, and end with all-round satisfaction.

In 2003 a major British study by Susan Golombok and her group at City University looked at the experiences of thirty-four surrogates and forty-two parents at around a year after birth.[7] What they discovered – in contrast to the soap-opera tangle of emotions envisaged by many of surrogacy's critics – was a body of largely positive experience. Moreover, the participants' feelings differed very little according to whether full or partial surrogacy was the method used, or whether the surrogate was a previously unknown person, or a friend or relative of the couple. Some of the surrogate mothers reported feeling unhappy at the time when the baby was handed over – but a year later, only two of them said that they felt occasionally upset. None showed any discernible signs of depression. Nearly two-thirds (59 per cent) said that there was no 'special bond' between themselves and the child, and none reported feeling that the child was like her own.

Across the sample of people studied, relationships between parents and surrogates were untroubled during the pregnancies and remained cordial thereafter, even if they had decided not to maintain regular contact. There were no reported cases of the parents refusing contact. In many cases, the surrogates were and intended to remain a low-key presence in the children's lives, visiting, for instance, at birthday parties, but very few expressed any misgivings, or wished for a more active involvement.

There were certain drawbacks to this study, of course. A possible weakness is that perhaps dissatisfied individuals did not readily come forward. Moreover, asking questions a year later may have allowed some of those involved to forget any difficult feelings that may have arisen during and immediately after the pregnancy. It could also be the case that people's feelings and relationships may change, sometimes for the worse, as the children grow older. Perhaps the greatest weakness is that the surrogates and parents studied were all drawn from one surrogacy charity (set up, in fact, by Kim Cotton, whose giving birth as a surrogate had stirred such controversy in 1985), but not all the individuals involved in surrogate arrangements responded to the enquiry. And, as members of such an organization, they all tended to be people with at least a moderate degree of interest and involvement in the subject, well-informed as to the possible pitfalls, and not necessarily representative of all the individuals who might try a surrogacy arrangement. Nevertheless, the results of this study do seem more encouraging than I for one would have thought likely.

Since this study was published the same group, headed by Professor Susan Golombok (but now in Cambridge), has published an important follow-up.[8] Their work has continued to grow in stature and they have done what is widely considered to be the most valuable and carefully controlled studies in this difficult field. In their 2006 survey they compared their thirty-four surrogacy families with forty-one donor insemination families, forty-one oocyte donation families, and a control group of normally conceived children and their parents. They found no difference in psychological well-being between the various groups and no disadvantage in not having a gestational link with a child – that is to say, mothers who had commissioned another woman to give birth did not have a less good relationship with their children. It is interesting that 44 per cent of these surrogating couples had already told their three-year-olds that they had a different birth mother, and another 53 per cent said they planned to do so. All of them had told other people outside the family unit, and in most cases both paternal and maternal grandparents were aware of the arrangement. It is just worth mentioning that secrecy was more common among families who had undergone sperm donation – about half of them had no plans to tell the child of its origins. Oocyte donation families were rather more open: two-thirds had already told the child by age three or were planning to do so.

Another recent event has also caused me to wonder whether, in my originally hard line about surrogacy, and in particular my concern about the bearing mother, I over-reacted. In April 2005 Teresa Anderson, a 25-year-old student nurse from Mesa, Arizona, gave birth to quintuplets in a hospital in Phoenix. Her five baby boys were all the biological children of Luisa Gonzalez and her husband, Enrique Moreno. This infertile couple had been trying for over ten years to start a family. The pregnancy lasted thirty-three weeks, during which time the host mother gained nearly 6 stones in weight. It must have been horrendous, and the delivery was, of course, by Caesarean section. Though Teresa had decided to act as a surrogate having had four of her own children already, and knowing how fulfilling the experience had been, there is no question that there was also a strong pecuniary motive – it was reported that she hoped to earn $15,000 to help her own family.[9] However, after an ultrasound scan had shown that she was pregnant with five fetuses, she decided not to accept any money from the commissioning parents. They were clearly not particularly wealthy, and she was concerned that they would need all the money they could earn to nurture their new family. Of course, Teresa may have considered that in some way this was her family, too. But I do think she was being genuinely altruistic, and her reaction suggests that, although this must have been a demanding and unpleasant pregnancy, she did not feel at all damaged by giving up these children.

Perhaps the real point to be noted here is the shocking irresponsibility of the medical team who, in my view, jeopardized Teresa's life and health, the safety of these children and the wellbeing of the parents by transferring five embryos simultaneously. In Britain, the regulatory authority banned the transfer of more than three embryos as dangerous some years ago. It turns out that one of the little boys had a heart defect. This is the kind of malformation that is sometimes associated with multiple pregnancy. It surely must be only a matter of time before somebody sues a doctor for negligence for trying to boost success rates in this manner.

Parenthood, whether natural or assisted, is a life-changing event accompanied by strong and conflicting feelings, and it would be irresponsible for us to ignore them. But the generally positive reports – along with the willingness of women like Carole Horlock to repeat the process many times – suggest that surrogacy can offer hope to parents and fulfilment to those who help them. Clearly, there is a real potential for people to be exploited through commercial trading in surrogate mothers. But it just may be that this potential is one that is fed and nurtured by governments who try to restrict individuals' choices.

7

Pre-implantation Diagnosis

I go on as usual suffering frightfully, at this moment I am in agonies
of pain; my knee gets worse daily and I get more desperate daily. If
this continues for long I shall be driven to Bedlam or to Hanwell,
where I shall be fortunately able to terminate a wretched existence
by knocking out my brains.

Thus wrote Prince Leopold to his elder sister Louise in
June 1870. Queen Victoria had nine children, and
Leopold, possibly her most intellectually able offspring,
was crippled by haemophilia, a serious genetic disease,
one of whose manifestations is the failure of the blood to
clot. After various horrifying episodes, he finally died at
the age of thirty of a brain haemorrhage after a light fall
on a hotel staircase in Cannes. Fortunately, only one of his
descendants, his grandson Viscount Trematon, was a
haemophiliac; and he, too, died a very young man.

Haemophilia is a sex-linked disease. It is carried on the
X-chromosome, but does not affect females who carry it
because they also have a normal X-chromosome, with a
normal version of the blood-clotting gene to compensate.
However, boys who inherit the 'wrong' X-chromosome
from their carrier mothers do not have the compensating

normal X-chromosome and in consequence suffer from the disease. On average, 50 per cent of daughters will inherit the damaged chromosome from their mothers, and 50 per cent of sons will be affected. Men with haemophilia who live long enough to have children will pass the affected chromosome to all their daughters, who will be carriers.

Prince Leopold's faulty X-chromosome was, of course, inherited from Queen Victoria. We know that because she had two daughters who turned out to be carriers as well as the affected Leopold. Before Victoria, there is no record of anyone having haemophilia in any branch of the royal family, so it seems that the abnormal gene arose as a new mutation in either the egg or sperm that made the embryo that became the Queen.[1]

Princess Alice, Victoria's third child, was one of the carriers who, through no fault of her own, caused havoc throughout Europe. Her haemophiliac son, Frederick William, died aged three, but two of her five daughters carried Victoria's fatal X-chromosome – and it was Alexandra, Alice's sixth child, who married Nicholas, the Tsar of all the Russias. It was in a desperate attempt to control the constant bleeding of Alexei, their spoilt haemophiliac only son and youngest child, that Alexandra sacked all the doctors. In their place she called in Gregory Rasputin, who exploited his friendship with the Tsarina grossly. So you could argue that the consequences of that illness, and how the Tsar and Tsarina dealt with it, were a significant factor in the deaths of the entire family at Yekaterinburg in 1918, as well as for the popular unrest which eventually erupted in the Russian Revolution.

Victoria's gene on the X-chromosome also affected members of the German royal family and, more importantly – inherited through Beatrice, her youngest child – afflicted some of the royal house of Spain. Alfonso, the heir apparent and Victoria's haemophiliac great-grandson, was excluded from the succession because of his disease, and the weakening of the monarchy in Spain contributed to events leading to the Spanish Civil War.

Exactly one hundred years after Queen Victoria's death, our team at Hammersmith made the first attempt at screening for that haemophilia gene which had such profound consequences in European history. At the time we could not detect the actual gene, but we could identify the X- and Y-chromosomes, and at least prevent affected males from being born.

Towards an early warning system

The story of how I got involved with pre-implantation diagnosis really began about ten years earlier. In 1982, just a very few scientific colleagues were considering the potential that embryonic cells had for further development into other tissues. They have a remarkable characteristic. Any embryonic cell in the first three days or so after fertilization can theoretically develop into any one of the many cell types in the human body. This area of study has much more recently led to the explosive development of stem-cell biology, which is discussed in chapter 9. In the early 1980s this line of enquiry didn't seem particularly relevant or practical. But one name I had noticed in the scientific papers that appeared at that time was that of Alan Handyside. I had never met him, but he had published an intriguing paper on mouse embryonic stem cells in 1982, and then a riveting one in February 1983 on the ability of cells from mouse eggs to produce a whole variety of different tissues – skin, muscle, gut and glands – in an experimentally formed cancer. Vague enquiries told me that he was unusually bright, having obtained a double first in natural sciences at Cambridge, and that he now worked at the Medical Research Council unit in Carshalton. So I made a mental note to meet him; but Carshalton seemed a trifle inaccessible – so I forgot all about it.

In 1985 we faced change on the IVF unit at Hammersmith. My colleague and friend Stephen Hillier told me that he was going to leave us. I was not entirely surprised. Stephen and I had struggled together to establish IVF in the public sector, with very little support, and we had fought many difficulties while he had published some good experimental work. Yet while Stephen's outstanding published contributions had made a real difference to medical practice, I had known for some time that he felt that he had done all he wanted to do with in vitro fertilization; he wanted to move on to more basic science and a purer form of research. And he had had a very good offer from the MRC Unit in Edinburgh – then by far the leading centre for ovarian physiology. As he jokingly pointed out to me: 'You can never get pure data from studying human patients. Rats are the only way of doing science.'

But IVF was growing well at Hammersmith, and we now had a highly successful programme (admittedly conducted in inadequate premises: our patients used to queue on the stairs between two wards where they had blood taken, egg collections were done on the labour ward and our counselling room was a disused broom cupboard). The clinical work was a significant academic resource, and it was obvious that I should look for another senior scientist to help develop our research. Several people suggested that I should meet Alan Handyside. At our first meeting, I was deeply impressed by this scholarly, quiet individual who was clearly outstandingly intelligent with a rather shy sense of humour. As it turned out, he was without doubt the brightest person with whom I have ever had the privilege of working. As we talked, we got very eager about the opportunities there were in pursuing the subject of stem cells. It seemed that the faint possibility of growing these from human embryos could offer solutions to some fundamental problems and just might make a massive impact on human medicine.

After various negotiations over the next few weeks, Alan eventually agreed to join forces with me at Hammersmith Hospital. But we never did get involved in stem-cell biology together, and I look back at this as one of my many great missed opportunities. It was almost fifteen years later when the whole medical world woke up, and research into human embryonic stem cells started to hit headlines. Since then an entirely new field of human biology has been developing in a most exciting way, and I have entered this field very belatedly.

At the time I hoped to persuade Alan to join us, the clinical facilities may have been poor but most of my laboratories were still housed in derelict huts, built as emergency facilities during the Second World War and condemned in the 1950s. Yet these dilapidated structures had hosted some key scientific developments. Ian Donald had messed around there with wires, valves, bits of radio sets and oscilloscope screens, and built one of the first ultrasound machines. Nobody thought that his badly blurred images were the start of a twentieth-century medical revolution. Erica Wachtel, a refugee from Nazi Austria (who started work in Britain as a housemaid), radically changed gynaecology with her brilliant work which led to the cervical cancer smear test; John McClure Browne had pioneered his work on toxaemia of pregnancy and localization of the placenta; and Harry Gordon had developed treatments for rhesus disease that saved babies all over the world. Literally thousands of students from across the globe had been housed in these huts during the unique intensive training programmes run at Hammersmith. Even in the 1960s and 1970s the facilities were inadequate. Now, the roof leaked, draughty windows didn't fit, sterile working was virtually impossible, and our closest colleagues were cockroaches. So Edinburgh was a fantastic opportunity for Stephen, and it didn't surprise me that a gifted scientist like him should choose to leave this environment and move to more glistening quarters.

One of the young research workers who had been seconded to me from Northwick Park Hospital,[2] where the prestigious Clinical Research Centre was then housed, was Richard Penketh. He was a junior doctor who had been working at University College London, researching antenatal diagnosis. At that time considerable improvements were being made in screening the fetus for abnormalities. The use of amniocentesis to aspirate cells that had been shed by the baby in the womb was a significant advance, giving doctors access to the fetus to make refined diagnoses about its health. These techniques sometimes get a bad press nowadays, but antenatal screening has been a boon to many women and their families. Few injuries are worse than carrying a child to term with increasing doubt about its health and its ability to survive. Throughout my career in obstetrics I saw huge shock and grief in both men and women who found out late that their babies were badly damaged, very unlikely to survive, or likely to be severely handicapped and in pain. The ethics of termination of pregnancy are beyond the scope of this book, but antenatal screening, used with humanity, gentleness and responsibility, has been of vital importance to countless humans. Nevertheless, amniocentesis had its problems. Not only did it carry a small risk of miscarriage, but it could not be done until relatively late in pregnancy; so if, as a result of what it revealed, an abortion was requested, it was for an advanced pregnancy, which was very distressing for the mother.

Richard Penketh wanted to improve the 'early warning system'. He was working on a revolutionary new technique to remove cells from the placenta using a very fine tube inserted from below through the cervix. The promise of chorion villus sampling – as it was called – was that it could be done earlier in pregnancy than amniocentesis. In the mid-1980s it was still experimental, but Richard had learnt at the feet of Bernadette Modell at University College Hospital, so was very much in the forefront of that field.

Richard was a tall, earnest, bespectacled, rather gangly individual who used to jump around the laboratory with huge enthusiasm and massive energy. He had the delightful habit of flitting from idea to idea, and among his most endearing features were his eternal optimism and his winning smile. And he seemed brimming with joy to be working at Hammersmith in our shabby surroundings. Whenever I work with people, I always think of them in terms of those animals in *Winnie the Pooh*. So it was appropriate to think of Richard as Tigger. As it happened, Alan arrived to run our research programme at Hammersmith on the same day that Richard started. They were very different. I hope Alan will not take offence if I describe his cautious, thorough, self-contained and quiet method of serious analysis in A. A. Milne terms, too. If Richard was an obvious Tigger, Alan was the archetypal Eeyore.

Soon after his arrival, Richard asked me whether it might be possible to get access to very early embryos. He wondered if we could examine the chromosomes in individual cells of the embryo, possibly to make a diagnosis of its genetic health before a pregnancy was started, immediately after fertilizing the egg in vitro. This was the first serious glimmering in our minds of pre-implantation genetic diagnosis.

It so happened that at the point when my two new colleagues joined me, a political event of importance had recently happened. On 15 February 1985 the House of Commons had staged the second reading debate on the Unborn Children (Protection) Bill put forward by Enoch Powell, the Member of Parliament for South Down in Northern Ireland (which he held with the most slender majority). I have treated this Bill in more detail in chapter 1; we were certain that, if it became law, in vitro fertilization in Britain would effectively be snuffed out. Even if treatments staggered on, the proposed legislation would prevent any improvements being made. A few scientists and doctors from across the UK became involved in fighting the Bill, and I was sucked into the activity, partly because of our research and the large IVF programme we ran at Hammersmith. One of the key arguments against this Bill, put repeatedly by certain scientists, was that if research were allowed to continue we might very well acquire the ability to prevent some crippling and fatal genetic diseases. This was a very powerful argument; but at the time there was little, if any, objective evidence that we could deliver on such a promise.

During a brave speech in the second reading debate, Dafydd Wigley MP rose to defend embryo research. He asked Enoch Powell:

Does the Rt Hon. Gentleman accept that the Bill will prevent medical research into in vitro embryos of up to 14 days, research which is central to the overcoming of many genetic disorders which are transmitted down the female line, and in doing so, will end the hopes and aspirations of countless thousands of families of disabled children and prevent doctors from undertaking research which can be accomplished successfully within the next few years?

Dafydd Wigley spoke with deep passion and with tears in his eyes. Only six weeks earlier, his son had died of a rare, fatal genetic disorder, and he was totally committed to the idea of protecting responsible, ethical medical research. I remember the occasion well, as I was sitting in the gallery under the stairs in the House of Commons, a seat as close as possible to being in the actual Chamber without taking part in the debate. After Dafydd Wigley sat down, his remarks were rejected or denied by speaker after speaker. Scorn was poured on the idea that our research could ever prevent genetic disorders. I remember wanting to jump over the Bar of the House to join the debate. It was one of the most frustrating mornings of my life. Little did I imagine that, almost exactly ten years later, I would join Parliament and take part in similar debates in another Chamber.

In one sense, it was not so unreasonable for Members of Parliament to be dismissive of what Mr Wigley had said. There was no evidence, in practice, that we could prevent genetic disease from the beginning of pregnancy, and it seemed much more likely that the need for abortion would continue. But many of us were deeply aware of the agony that women experience in deciding on abortion, and I had always been conscious of the psychological injury it inflicted. It seemed right to question whether there might be a better way.

Sex and the single gene

So Richard Penketh started to examine mouse embryos to see if he could look at their chromosomes and diagnose their sex. Sex selection may seem unconnected with the problem, but it was not. There are around three hundred serious gene disorders that are sex-linked.[3] Like haemophilia, most of these diseases are often carried by women but affect boys. Kipling was not far wrong in stating that the 'female of the species is more deadly than the male'. These diseases affecting only males include Duchenne muscular dystrophy, a wasting disease which commits young boys to a wheelchair – and usually to an early death; few survive much beyond their teens and twenties.

Richard's thinking was that if we could diagnose the sex of an embryo, then, when giving IVF treatment to a woman whose family carried one of these defects, we could transfer only female embryos to the uterus, ensuring that she would not bear a boy affected by the condition. In due course the Enoch Powell Bill failed, as I have described in chapter 2, even though there had been a massive majority in its favour. But the issue rumbled on in a very menacing way, and it was obvious that eventually there would be legislation that could be very damaging to our research and clinical practice. So when Alan got his feet well under the laboratory bench that year, we stopped talking about stem cells and started to question whether or not we could show that the screening of embryos for genetic disorders was feasible. Up to then, IVF had been only an infertility treatment. The idea that there were wider and very important contributions to be gained from it was entirely novel.

Our target was the group of conditions known as

single-gene disorders. These should not be confused with all genetic diseases in general. There are a vast number of genetic disorders – indeed, most diseases have some genetic basis, including diabetes, coronary vessel disease, Alzheimer's and polycystic ovary syndrome. But a number of genetic disorders – nearly all of them serious, most fatal – are the result of a change in just one gene; so their inheritance is more specific, and the abnormality in the DNA which causes them is sometimes easier to identify.

We did consider not using IVF at all. But I doubted the safety of flushing the embryo out of the uterus to examine it. I knew Dr Buster had flushed embryos in California a couple of years earlier for other reasons (see page 264), but to me it seemed far too dangerous. So we discussed a number of possible methods for couples carrying these disorders using IVF. The most fruitful approach seemed to be to stimulate the ovaries to get a good number of eggs, produce several embryos, remove cells from each, and stain the chromosomes in the cells we removed to see if they were male or female.

With Alan's help, and under his expert supervision, Richard carried on with his mouse work. It was rather haphazard, and in retrospect I have to say I didn't fully accept how revolutionary his approach was. I now feel I should have given him rather more support than I did. Alan, who was a purist academic scientist, was somewhat doubtful too, and once he had decided to look into embryo biopsy he tended to move in his own direction. Shortly after this, Alan asked me to recruit Kate Hardy, who had worked with him at Carshalton. She is a scientist with whom I have subsequently had the closest and fondest collaboration for almost twenty years. Another Cambridge graduate, Kate very much wanted to do a PhD; so she joined us as a doctoral student, starting her outstanding work on the development of the human embryo. Having gained her doctorate, over the next two decades she published a series of key papers on human embryology, bringing her to the forefront of her subject internationally. Over time, she became the most supportive person in our laboratory, and organized many of our activities meticulously – a perfect Rabbit.

There were numerous difficulties to overcome. One of the first things to do was to establish safe ways of taking cells from the human embryo. Some work on this had already gone on in a number of animal species. Robert Edwards had done it in the rabbit in the 1960s; others had operated on mouse embryos. After it had been discovered how to freeze cattle embryos, some work had also been done in agricultural animals. Within months, Alan and Kate had a good system. Using a microscope I bought for the purpose, and micromanipulators to control unwanted movement, they were able to deposit a tiny microscopic drop of acid on the zona pellucida, the outside shell of the embryo, to make a small hole. A fine pipette, so delicate and thin that its tip could not be seen with the naked eye, could then be inserted through this hole and a cell sucked out for analysis.

The next problem was that the cell we had sucked out was invisible. How could we reliably stick it to a glass slide, fix it, wash it, stain it, and wash it again and again, without losing it down the sink? If we were going to deal with human patients, this was going to have to be 100 per cent reliable. So Alan, with the help of Kate, worked on methods to glue the cell onto glass surfaces without damaging it.

It became increasingly apparent that a very important development had happened in the United States. In 1986 Kary B. Mullis and his colleagues published work in

Nature describing the polymerase chain reaction (PCR).[4] This was a method for examining short sequences of the DNA inside genes, and Mullis had been working on it for a year or two. It was one of the great advances in modern biology and Mullis subsequently won the Nobel Prize. PCR is now an absolutely standard diagnostic technique used in a vast range of contexts: for human disease, to identify bacteria, to identify dead or extinct animal species, and in forensic investigation when a criminal may have left traces of his DNA at the scene of a crime. So Dr Mullis's publication was very exciting; but it seemed that the PCR method would not be sensitive enough to test the DNA in just one single cell. Mullis needed several hundred cells – still a tiny amount of DNA – to get a result. So his technique could not be used for just one or two cells from an embryo.

Over the next year or two, Alan concentrated much of his time on refining PCR. Ideally, we hoped to be able to detect a specific DNA sequence in just one cell. And it had to be utterly reliable. There would be nothing worse than transferring an affected embryo which we had wrongly asserted to be free of serious disease. While Alan was refining the PCR process, we also had to find ways of liberating the DNA from individual cells. Painstakingly, Kate used to collect single cells and then treat them, looking for ways of encouraging the cell to disintegrate while keeping the DNA intact. Rapid freezing and cooling was one of the most effective processes, but sometimes chemical methods were chosen as well.

There was another critical issue. Diagnosis not only had to be accurate – it had to be fast. In attempting to make a diagnosis in a human embryo, we would be racing against the clock. Ability to grow embryos in culture for more than three or four days really successfully was limited at that time. We considered freezing the embryo after taking a cell away, to give more time to make a diagnosis, but this seemed a very poor option. In the 1980s freezing destroyed more embryos than it saved, and we were also concerned that the process might damage any embryo with a hole in its shell from where the biopsy had been taken. So we felt we would need to take our cells from the embryo on the second or third day after fertilization and, ideally, place it into the uterus within the next twelve hours at most. But I was concerned that even the refined PCR process could not be done that quickly. Also, the standard method of measuring DNA needed a radio-activity 'stain' attached to it; and to detect this radioactivity in the very tiny amounts emitted required a considerable exposure time for it to register on photographic plates or light-sensitive material. With the biopsy taking considerable time as well, the whole diagnostic process would just take too long.

Alan set about solving these various problems. He collaborated with a number of scientists, of whom three stand out. First was Ted Tuddenham. Ted is a haematologist of considerable distinction and an internationally acknowledged expert in the various genes causing haemophilia. He recognized the value of trying to identify the sex of the embryo by DNA amplification, realizing that this could be a good way of preventing children having haemophilia. Many years later, as it happens, in collaboration with our team, Ted refined these techniques very considerably. We can now make a specific diagnosis for haemophilia and transfer either a male or a female embryo to the uterus with safety, in the knowledge that the baby will not carry the disease.

Another scientist was Bob Williamson at St Mary's Hospital. Bob ran one of the best DNA laboratories in

Britain and also had a formidable intellect. I always found him slightly intimidating. On one occasion I went to visit him in his scruffy laboratory office in Paddington – I have memories of manuscripts and papers piled everywhere. He had just dashed off a paper describing our work together which he wanted to publish. His way of showing it to me was to use the tiny screen of his brand new word processor. This was, remember, in the mid-1980s when most word processors had green fuzzy lettering on a dark background. Bob was as always a wild mixture of enthusiasm and impatience. The paper was about nine pages long – over three thousand words in all. He flashed the first page onto the little screen and then pressed the scroll button, so that in the next thirty-five seconds the entire electronic manuscript flashed in front of my eyes. I rather think I caught my name among those somewhere near the top of the paper, but the rest was a blur. At the end of the scrolling, he looked up at me and said: 'Well – what do you think? Sure to be good enough to be accepted by *Nature*, don't you think?'

The last of this trio, another brilliant man, was Charles Coutelle. Charles now lives in London, but in those days, before the Berlin Wall was destroyed, he was an East German citizen, a committed Marxist living in East Berlin. He worked at the Central Institute of Molecular Biology, where he was doing outstanding work as the director of that unit with a research budget of perhaps around $10,000 a year. I had first met Charles when I visited Berlin to attend a conference on ethics and biology. I found him to be an enchanting and interesting individual, with his round, rather old-fashioned spectacles slightly resembling Wol. Wanting to expand his work and collaborate with scientists beyond the Iron Curtain, in 1988 he visited St Mary's Hospital and joined forces with

Bob Williamson. At that time he was interested in the idea of making a specific diagnosis of gene defects using the DNA from a single cell. Bob was a world expert on cystic fibrosis, so Charles wanted to look for the actual mutation in the gene that causes that disease. Charles had little time in the west and a limited budget. During one frenetic three-week period while he was working in London we supplied him with every unfertilized human egg we could obtain – not an easy matter. These came from a number of altruistic patients, some of whom were known carriers of cystic fibrosis. Charles, absolutely committed, worked night and day for those three weeks. He hardly ever left the hospital, preferring to sleep in Bob Williamson's draughty and unprepossessing laboratory in Praed Street.

Examining eggs and refining PCR

Alan and I decided that biopsy of the embryo two or three days after the egg had fertilized was the best strategy. One advantage of this approach was that there was a good deal of animal work to show that this was safe – it did not seem to damage further development. Another advantage of examining an embryo (rather than an egg) was that by the time the embryo is formed the complete genome is present, with both sets of genes from mother and father. But in the USA, many scientists were frightened of the 'moral majority' or could not get ethical approval to experiment with embryos. Around the same time we were starting to think about transferring biopsied embryos, Yuri Verlinsky and his colleagues in Chicago were advocating screening eggs. The procedure they favoured was to analyse the polar body – the part of the egg that is discarded when one-half of the paired set of chromosomes is extruded. Like all cells, the egg starts with paired chromosomes, but in order for fertilization to take place without increasing the number of chromosomes, the egg has to get rid of one half of the pair. Sperm do this as well, but at an earlier stage of development.

So Verlinsky was removing the polar body, which appears at the side of the egg just as it matures at the time of ovulation. He and his colleagues found it was quite easy to remove this little blob of material – easier than removing a cell from a fertilized embryo. But its potential for analysis was limited. It could be used only to identify genetic disorders that were carried by the mother. For example, if the mother was a cystic fibrosis carrier, then finding a mutation for cystic fibrosis in the polar body meant that the egg was free of the problem. But this technique could not detect any genes inherited from the father. Nor would the technique be any use for sexing the embryo, because the sperm at fertilization is responsible for determining sex. Another big drawback was that a large number of eggs would need to be biopsied because, of course, the embryologist could never tell which ones would fertilize. This made polar body biopsy quite labour-intensive. It did have one potentially useful advantage, though: if the chromosomes were abnormal in the polar body – if, say, there were too many Y-chromosomes or one was missing – this would indicate that the egg, too, was abnormal.

All in all, the drawbacks of polar body biopsy seemed to outweigh strongly any advantages. There was, however, another method of embryo biopsy that was being talked about. In a number of animal species, scientists had taken cells from the blastocyst. At this stage there might be 100–250 cells, and therefore there would be much more material available for PCR. This had the real advantage that the PCR diagnosis would be more soundly based: it is a huge responsibility to carry out these complex tests and make a diagnosis on the basis of a single cell alone. Having ten or twenty cells would be more reassuring, a luxury. But biopsy of the blastocyst often caused it to collapse and we were not sure that, in the human, that would be safe. Also, at that time, no human culture system had ever successfully maintained blastocysts in culture for more than a few hours – and no documented pregnancies had resulted. So Alan and I decided to concentrate on the original plan, taking cells from the embryo two or three days after fertilization. It meant we had to be absolutely accurate and meticulous in everything we did. But it was, we felt, the right decision. The key had to be to refine the PCR process – and this, in time, turned out to be one of Alan's great achievements.

Refining PCR was quite problematic. With each experiment, we first mixed the DNA from our precious cell in a little tube, and, given that the identification of a single normal cell and the extraction of its DNA was time-consuming, the resulting carefully measured mixture could take quite a long time to prepare. Then we meticulously added our precisely manufactured DNA primers, which had the exact sequence needed to identify the mutation of interest. Next came the expensive polymerase enzyme, which at that time was very difficult to obtain. Once the sealed magic tube was prepared, we were ready to start PCR amplification. The procedure began with a single molecule which would be present from just one cell. This was one double strand of the DNA, taken from the fluid in the tube. This was heated up to 93 degrees Celsius in a hot water bath for around one minute; this would 'melt' the DNA, dividing the double helix into its two separate strands. When the tube was then rapidly cooled to around 54 degrees Celsius for two minutes, the polymerase enzyme would start working, and each of the two strands (moving around with Brownian motion[5] in the fluid) provided a template for a new double helix to form. After warming to 72 degrees over two minutes, reheating to 94 degrees would halt the working of the enzyme and split the new double helices into separate strands – so now there would be four strands. Cooling would cause doubling up again. This alternate heating and cooling, repeated over thirty or forty times, would eventually produce around one million copies of the helix. With one million molecules there would be enough DNA to measure its molecular weight on an electrophoretic gel.[6]

But it was a laborious process, and we did it by hand. In practice it meant plunging the tube into a hot water bath, just below boiling point, for a minute, then plunging it into the 54-degree bath for two minutes, then moving it back again. Nowadays automated machines do this. But doing it by hand meant standing by the water baths and endlessly repeating the same motions over several hours. And because we did not know the best temperatures for the reaction, there was a lot of trial and error in getting the conditions right. I used to come in on a Sunday and occasionally help Alan – and once or twice he trusted me to get on with it. Alan was highly musical, so he agreed that the best way to avoid tedium was to clap on the headphones from a Walkman. He recommended Mozart or Bach as being appropriately rhythmical and soothing. One Sunday he gave me a particularly precious PCR mixture in a tube containing a single, dissolved, invisible egg from a patient who was a carrier of a rare gene defect. It had taken weeks to get approval and obtain this material. I, impatient as ever, decided that Sunday to play Shostakovich's Tenth Symphony. I became engrossed, and on a particularly loud chord in the third movement plunged my hand, holding the precious tube, into the 94-degree water bath. There was no way the tube could be rescued from the bottom of the tank of almost boiling water.

Around this time it became apparent that we really could detect specific DNA sequences in single cells. And various improvements in our procedures sped up the PCR amplification process, enabling us potentially to make a diagnosis on the same day as the biopsy. In the meantime, other work in our laboratory was demonstrating that removing one or two cells from a four- or an eight-cell embryo did not seem in any way to damage it. The animal embryos and the human embryos that we biopsied in this way continued to grow normally afterwards and, in a few days, had made up the missing cells. A large number of normal mice were born and the human embryos looked quite promising. And as other workers were beginning to biopsy monkey embryos safely, we were reassured.

Into the real world

Meanwhile, Richard Penketh persisted with his attempts to try to stain mouse and then human embryos. Eventually he was able to get a reasonably good stain which showed up both the X- and the Y-chromosome in individual cells; but usually the diagnosis tended to take more than a day and was not as reliable as DNA amplification. Sadly for him, the ideal probe – a correctly prepared artificial piece of DNA which would stick to the chromosome – was not yet available; so by this time I had decided that we should concentrate on

PCR, which was now where we put all our resources.

But Richard contributed substantially in another way. He had met a desperate patient, Janet D.[7] Janet was the carrier of a horrible disease called Lesch-Nyhan syndrome. It is carried by females on one of their X-chromosomes and affects all boys who carry the damaged chromosome. It is a very unpleasant disorder indeed. Janet had a son, Brian, who at the age of ten was totally spastic. He was unable to make any kind of co-ordinated movement, could not feed or clean himself, could not speak clearly, had considerable brain damage, was incontinent and needed to be nursed constantly, twenty-four hours a day. To make matters worse, like other sufferers from this disorder he repeatedly tried to mutilate himself. He had had to have all his teeth from the front of his mouth removed to prevent him from biting off his lips and tongue. Many of these children die as a result of self-inflicted injury, when they get an overwhelming infection. In addition, Brian had to be strapped into a wheelchair throughout the day. His arms and legs were tethered by straps so that he could not harm himself. In spite of this, on one occasion he attempted to move his body around in the chair so as to throw himself down the stairs. Janet and her husband had tried patiently and selflessly to do their best for their child. It meant never having a break, except on very rare occasions when carers could take over – which distressed Brian massively; a holiday, of course, was impossible. But to make matters worse, Janet had had a series of other pregnancies. On four occasions she had had miscarriages; on another, she had had a stillbirth. In her remaining pregnancies, antenatal diagnosis had revealed a male fetus, and Janet – totally reasonably, in my view – did not feel that she could cope with another child who might be affected like this, so all these male fetuses were aborted without Janet knowing whether they were affected or healthy. The injury to her was massive and the strain on her marriage severe.

Occasionally, major medical advances are made because of the needs of an individual patient. In Janet's case, it seemed imperative to try to help her by screening her embryos. She was quite happy to go through IVF and it was our hope and prayer that we might be able to find some female embryos to put into her uterus. By this time, Alan was working on a DNA primer that could detect the Y-chromosome, using the PCR process. Using this technique we could identify the males. It seemed very reliable and Janet was delighted at this possibility. She recognized that anything we did was taking a gamble, but she trusted us to have made the tests as accurate as possible. The thought of having a healthy child filled her with joy and she felt that she had nothing to lose. From our point of view, Janet was rather a good bet. Her options were stark: she could play Russian roulette, continuing to getting pregnant in the hope of having a normal baby; she could continue to have repeated abortions; she could have another desperately sick child who would eventually die of this horrible illness; or, of course, she could stop trying to get pregnant.

A number of people would feel that this last option – to abstain from sex or use effective contraception – was the only ethical decision. Some religious people with whom I have debated argued this. But there was the question of what Janet herself felt was in her best interests – irrespective of the fact that most of the contraceptives she had tried either made her feel ill or were not totally reliable. I am unconvinced that any member of the public, any politician, or indeed any regulatory authority has the right to interfere with that kind of decision, made between a patient and a responsible team of doctors who have been licensed to practise medicine by the General Medical Council. Moreover, it was clear from many soundings, and from our own research ethics committee, that most people in Britain see sex selection in this situation as ethically sound. Janet was a good bet for another reason. She was clearly very fertile, and therefore there was every reasonable hope that IVF might succeed.

So in 1989 she went through a treatment using IVF. Thirteen eggs were sucked out of her follicles; from these, eight embryos formed. On the second day, early in the morning, Kate and Alan worked together painstakingly to remove a single cell from each of five embryos that were growing well. To make the biopsy possible, they used our microscope equipped with the micromanipulators placed on a massive marble table top, supported on heavy legs cushioned on squash balls. This killed virtually all vibration – provided people nearby walked around on tip-toes. Even so, a lorry drawing up outside 30 yards from our derelict hut caused such movement that removing a cell safely was temporarily impossible. The biopsies took about three hours; and then Kate had to place each cell into an analysis tube, before disintegrating it for PCR.[8]

This was a difficult job. First of all the cell, invisible to the naked eye, had to be identified, and then pipetted into the tube. But the cell had to be examined really carefully as it went in – first of all to make certain no sperm got into the tube (which might have given a false DNA result), and second, to make sure as far as possible that the cell was normal, and contained a nucleus. If it was an abnormal cell – as many embryonic cells were – and had no nucleus, it might give a false negative result. As we were attempting to sex the embryo by PCR, Kate had a hugely stressful responsibility. If she failed to notice that the cell was abnormal, or if the cell got stuck on the end of the pipette when she removed it from the tube, we would get a negative result. With no signal from a Y-chromosome on the PCR later, we would be likely to assume wrongly that the embryo was female, and transfer it. Kate told me that, every time she tubed a cell, her hands shook violently. It was surprising, and very much to her credit, we were able to start the PCR process as soon as we did.

Some five hours later, Alan was ready to run the gels (see note 6, page 516). The final stage was when he retired to the photographic dark room wearing a plastic visor – looking like an unusually studious Cyberman from *Dr Who*. (The visor was to protect his eyes from the ultra-violet light with which he would illuminate the gel.) I joined him in the dark room, wearing another visor. The light shone. We peered hard through the gloom. Very silent, we switched the light off and then on again. But there was no little band of coloured light on the gel. The PCR hadn't worked.

Janet was very robust, though her husband kept complaining about the inconvenience. A few months passed and we repeated the treatment. This time the biopsy was quicker, and the PCR actually seemed to work. Some embryos did not give us a signal from the Y-chromosome, others did. As all our control cells clearly worked,[9] giving us the expected indications of a Y-chromosome either present or absent, we assumed that those embryonic cells not showing the presence of a Y-chromosome were female.[10] A number of embryos that appeared to be female were obtained and two were transferred. But Janet did not get pregnant. By this time, Janet's husband was reluctant for us to try a third time, so there was a delay of several months.

Meanwhile, I had come across two wonderful women in my clinic who seemed very deserving of embryonic sex selection. One had a mutation on one X-chromosome that had resulted in her having two severely mentally retarded boys; the other had a sex-linked mutation that had caused the death of her four-year-old son from adrenoleuko-dystrophy. Both of these brave women and their husbands had cared for their young children during serious illness. Both women were adamant that they did not want to consider abortion. They both felt that it was ethically wrong to terminate a pregnancy, particularly as all they could do was to abort any pregnancy with a male child, not knowing whether it had the mutation. At that time there was no way of doing a PCR for the specific diseases they carried. Both families were utterly in favour of our trying to put female embryos back into the uterus. They fully understood that these girls might be carriers and in time give birth to children that might have the illness; but they all felt that, twenty years or more down the line, it was likely that there would be improved ways of dealing with the problems they now faced. Both wanted to go through treatment, and both agreed to have an amniocentesis if they became pregnant, to confirm we had made a 'correct' diagnosis.

We proceeded to line up both these women for treatment, together with Janet for her third attempt. By now we had a fairly efficient system, though the winter of 1989 was approaching and there were pools of water on the floor of our hut. But the radiators were fairly efficient, except when the hospital steam generator shut down.

Janet's treatment was first; we transferred two embryos – and for a third time pregnancy eluded her. Elizabeth E. and Mary C. went through treatment very smoothly a couple of days apart.[11] The PCR worked quickly – we had a result at six-thirty in the evening, and I was able to explain to them both that each had two female embryos to be transferred. Although there were more embryos than this, we reserved the remaining 'spare' embryos for tests to confirm that the PCR had indeed worked correctly. Examining these 'spare' embryos over a few days would give us much more time to do more secure tests to validate our results. We felt we should put only two embryos back into each woman, because these were fertile women and with more embryos there would be a risk of multiple births. We were not prepared to attempt to achieve a pregnancy 'at all costs'.

I vividly remember the excitement when these two women came, within a day of each other, to have their embryos transferred. I do not know why, but I felt sure we would be successful with one of them. My hand is pretty steady on the whole; years of delicate microsurgery have helped me to train myself against tremor. But inserting the catheter through the cervix, though completely painless for both patients, was for me a matter of some trepidation. The catheter seemed finally to be in place, and I gently squirted the syringe. The catheter was removed, and was checked under the microscope to make certain that it was now empty and the embryos had been released. All we could all do now was wait.

Twelve days later, we received blood samples from both women and they were analysed in our laboratory. Dina Packham, our biochemist, rang me at about two o'clock in the afternoon to report, in a very matter-of-fact voice, that both pregnancy tests were strongly positive. I looked at the ceiling and thought 'Whoops'. This was not a moment of great elation, but rather a deep feeling of responsibility towards our hugely brave patients and the wider implications of what might be possible. When I walked into Alan's office, a few minutes later, to tell him the news first, we were both pleased but there was no real sense of euphoria. If these children were born healthy, it would confirm that the technique worked. It was a small piece of medical history; but I was well aware of what faced us. Every test that we had done and numerous tests on different animal species had shown very clearly that biopsy had not harmed any embryos. Normal offspring had been born in at least six or seven species. Not a single abnormal fetus had ever been formed. But there was also a strong element of unspoken tension in both our minds as we considered the potential enormity of having caused an abnormality by our meddling with these embryos.

Within a month we were able to scan Elizabeth with ultrasound. We saw that the embryos were growing normally – and we could see two hearts. A few days later we found Mary was carrying twins as well. Within a few weeks, the amniocentesis results from both women confirmed that each was carrying two female babies. We felt that this news justified submitting our results to *Nature*, and Alan, Kate and I wrote the account. Within a few days, John Maddox, the journal's editor, had agreed to publish the work.

There followed what seemed to be an unconscionable wait. Maddox was not prepared to say when he intended to publish our paper. When he did finally publish it, by some curious coincidence he happened to choose an important week. It seemed that the whole world's journalists attended the press conference that he organized. Alan and I answered the usual questions about the ethics of our approach. The following morning, the *Daily Telegraph* published four pages on this 'breakthrough' with fetching photographs of our two patients – now very large with their twin pregnancies. And this feature appeared the day before the government Bill on Human Fertilization and Embryology was to be debated in the House of Lords. It was far from certain that the Bill would be approved – Margaret Thatcher had decided on a free vote – so it was still by no means evident that IVF could continue and that the research we were undertaking would prosper.

I think that John Maddox's decision on the timing of publication (despite his protestations that he would never try to influence a political process) contributed to the very reasoned and tolerant debate on the Bill's second reading that followed. Over the next few weeks we were able to see that the massive majority that Enoch Powell had marshalled in favour of his Bill some five years earlier – around 75 per cent against liberal legislation and embryo research – was totally reversed in both Houses, with a two-thirds majority in favour.

But there is a rather sad end to the story. I delivered both women in the early autumn by Caesarean section. Elizabeth had two bouncing babies, Lisa and Harriet – but Mary's pregnancy ran into trouble at the very last minute. For unaccountable reasons, one of her twins died just twenty-four hours before the Caesarean section was done. I bitterly regretted not intervening sooner, but there was very little indication of any reason to do so. The death seemed to be due to interruption of the baby's blood supply. A post-mortem was done and there was not the slightest indication that the problem was related in any way to the pre-implantation diagnosis. But we delivered both women with a poignant mixture of joy and sadness.

The kindness of colleagues: Raul Margara

Alan and I took most of the credit for the development of pre-implantation diagnosis, but none of this work would have been possible without my closest colleague, Raul Margara. Raul had joined me from Argentina in 1977. I first met him at the big World Congress of Fertility and Sterility in 1974. Immediately after I had very proudly given my plenary session paper on microsurgery, this very youthful and rather bedraggled-looking young man came up to me and said he wanted to work with me in London. I just smiled, thinking (a) that this was impossible, because I did not have any kind of establishment to employ anybody, and (b) that I could not imagine anybody being foolish enough to want to work with me – even for a week. So I just continued smiling vaguely, and Raul said that he did not want to come yet. He wanted to leave Buenos Aires and move to Paris to complete his surgical training. Remarkably, he said he would see me in three years' time in the New Year of 1977. I just shook his hand and waved him goodbye – never expecting to hear from him again.

In the first week of January 1977 I was told that a very polite, well-spoken man had arrived at Hammersmith out of the blue, saying he had come from Paris. It was Raul Margara. True to his word he had arrived, expecting to start work immediately. I pointed out that I had no money and could not pay him. This, he insisted, was no problem; he simply wanted to work as a volunteer and train in microsurgery. After six months he had done so many tedious tasks in the laboratory that I thought I ought to take him to an international workshop on microsurgery in Belgium, where I was due to speak. My wife wanted to come because the meeting was in Leuven, where I had worked for a year, and we both wanted to revisit old friends there. So our two children, Tanya, aged three, and Joel, who was still a baby, came with us. I could not afford the air fare, so we drove off to the Dover ferry in my battered Peugeot, with Raul in the back. The crossing was so rough the ferry hove to in the middle of the Channel. Tanya was tearful and my wife, Lira, was quite queasy. At this point Raul just took Joel the baby and cuddled him for the rest of the trip.

That was the moment when my conscience was pricked. He was one of the family. I clearly had to employ Raul and find a way of paying him. On our return, the university allowed me to pay him with the private fees I was earning from overseas patients being referred for tubal surgery. Thenceforth, every Wednesday morning Raul and I used to make a trip to the private Avenue Clinic in St John's Wood.[12] I would take the patient's out-patient records with me, while Raul would arrive with a huge 'doggy bag' – a massive hold-all containing our latest home-made surgical instruments and a massive micro-scope head which he carried every week from our laboratory. There, over the ensuing years, we worked together building up microsurgery and reconstructed hundreds of damaged fallopian tubes. Raul soon became highly skilled and is now recognized everywhere as a world expert in fertility surgery. Eventually the University of London acknowledged his unique skills and took him on the academic payroll, employing him as a consultant and senior lecturer.

There was no reason why Raul should have agreed that, for all those years, all his private income should be ploughed back into our research, but over the next decade or more several millions were raised which ensured that – until the rules of the NHS were changed in the mid-1990s, forcing us to charge health authorities for all treatments – nearly all IVF was done on an NHS basis without any charge. Experimental treatments such as pre-implantation diagnosis and the necessary research on which they were founded were developed through Raul's colossal generosity.

Pursuing DNA mutations: cystic fibrosis

The problem with our work so far was that the diagnosis of the sex of the embryo was useful only for some two to three hundred disorders – those that are sex-linked. A much bigger problem was the diagnosis of disorders that are not carried on the X-chromosome. There are around six thousand single-gene diseases sited on the other chromosomes (the so-called autosomes), of which the most common is cystic fibrosis. Cystic fibrosis affects the lungs and digestive system. Roughly one in twenty of the population carry a genetic mutation that causes this disease. Carrying the mutation does not matter to the carrier, but because these mutations are so common, the probability of a female carrier meeting a male carrier and marrying him is high. As you would expect from the mathematics, in about one in four hundred couples in the UK population both partners carry one of these gene mutations, meaning that their children are at risk. Cystic fibrosis is a recessive disorder, so for the child to be affected it must inherit a mutation in the gene from both parents. If it inherits the mutation from only one parent, it will be a carrier. So for any couple who are both carriers, the chances are that one in four of their children will be affected; one in four will not carry a mutation from either parent; and two in four will inherit a mutation either from the mother or the father and be carriers. With both partners in one in four hundred couples being carriers, roughly one in sixteen hundred births in the UK produces a child with cystic fibrosis.

In recent years great progress has been made in keeping children with cystic fibrosis healthy, but they tend to become increasingly disabled as they get older. They cannot produce digestive enzymes from the pancreas, so babies tend to have diarrhoea, be malnourished and fail to thrive until the disease is diagnosed. When it is, their food can be supplemented with enzymes. Even then, they are likely to have serious bowel problems, such as twists of a dilated intestine, which may require emergency surgery. They also can get liver disease with gallstones, and diabetes. But the effects of the disease on the lungs are more serious. Cystic fibrosis sufferers are prone to pneumonia. Repeated lung infections cause more and more scarring and increased breathlessness, and eventually they cannot get sufficient oxygen and become blue. The lung condition is eventually fatal. Gene therapy is being tried, and Charles Coutelle now heads a team at Imperial College London trying to find an effective genetic treatment. But the only therapy that really works is the massive undertaking of a lung transplant – which means waiting for another young person to die in a traffic accident. And to survive the transplant the patient must take immuno-suppressant drugs for life – with all the risks that that entails, including an increased likelihood of suffering from certain cancers. In the past, most children with cystic fibrosis did not survive beyond their teens. Nowadays, they usually live longer, and may survive into their twenties or even thirties. But cystic fibrosis, like most genetic diseases, is well worth avoiding.

One reason for wanting to start our attempts at

pre-implantation diagnosis with cystic fibrosis was because it is so common. Also, our collaboration with Bob Williamson and Charles Coutelle made it likely that working together would increase the chance of finding a method of PCR diagnosis. Initially, too, the Cystic Fibrosis Trust was quite encouraging and kept asking us to try to find a method for identifying the responsible gene in embryos. Also, there were hints from them that they were likely to support the research financially. As it happens this support never materialized, which was irritating, not least because we were spending considerable income trying to find a way of treating the patients they kept sending us.

Progress was slow; but we had a stroke of good fortune. Alan had been lecturing in the USA and had come into contact with Dr Mark Hughes from Texas, who had attended one of Alan's talks. Our work had been widely publicized internationally, and American colleagues were getting extremely interested in our approach. Mark was a medical doctor and biochemist from Baylor College in Houston – and a gifted communicator. Of the DNA alphabet, he is quoted as saying:

If I could break open one human cell and stretch out the DNA, and then type in 12-point font (a standard type size) the four letters of the genetic alphabet contained in that DNA, it would take up a 300-volume set of Encyclopedia Britannica. We can think of the chromosome as one of the volumes of that encyclopedia set. If you have too many books, you have problems like Down's syndrome or worse. If you have too few, you have severe birth defects.[13]

His analogy was powerful but incomplete. What Mark did not say, of course, was that a single letter misprinted makes a massive difference to whether or not a gene 'reads' normally. One misprinted letter in an entry in the *Encyclopaedia Britannica*, say that on 'Liturgical Drama in the Middle Ages', doesn't necessarily prevent the reader from understanding that 'Daniel in the Lion's Ben' was one of the most popular biblical plays of that time – and working out the misspelling.

Over a period of some months during 1991 and 1992, after meeting Alan Handyside, Mark flew backwards and forwards from Houston to London. One afternoon he arrived from Heathrow at short notice, having telephoned from the airport. At his request and though they were like gold-dust, I had managed to procure him two tickets for *Phantom of the Opera*. He entered my office, took off his coat and lounged back in my easy chair, supporting his embroidered Mexican kicker boots on the copy of *Christopher Marlowe's Collected Works* on my desk. Up to that moment he never seemed to have much respect for me or my science – but when I handed him the tickets, he fell off the chair.

Soon, Mark cleverly devised a novel and highly accurate method for detecting most common mutations seen in sufferers from cystic fibrosis in single cells. After some months, we were in a position to speed the process up sufficiently to attempt to use this method to diagnose the disease in embryos. Thereafter we treated a number of patients successfully at Hammersmith Hospital, with six or seven unaffected babies being born to families who had previously lost a child or who had a very sick infant. We published this work in 1992 in the *New England Journal of Medicine*. By this time Mark had also turned his attention to Lesch-Nyhan syndrome – we were still anxious to try to help Janet. Eventually, Mark and Alan devised a method for looking for the actual mutation that caused Lesch-Nyhan disease in her case. This was no easy task. Each time this disorder occurs, there is what is sometimes called a private mutation: that is, the precise configuration of the DNA in the gene on the X-chromosome is different in each family. Consequently, PCR diagnosis has to be tailored to the individual carrier.

The tailoring was done by taking DNA from some of Janet's aborted fetuses, as well as from Brian. Eventually, having pinpointed the precise DNA letters that were causing the problem, Mark devised his PCR primers and validated their effectiveness on white blood cells taken from Janet. Then Janet came for a further IVF attempt – without her husband, but using his sperm with his approval. This time she conceived, and finally delivered a healthy boy free of Lesch-Nyhan syndrome. This was a real triumph – an example of how diligent basic laboratory science could be used to treat one specific patient. Never again, in all its future generations, would this disease plague this family.

Following this effort, we turned our attention to a number of different genetic disorders. By now there was widespread interest in the technology, and a number of units around the world were starting to use a similar process. Some of them sent colleagues to train with us, or to undertake research projects. Pre-implantation genetic diagnosis has subsequently been adopted in virtually every country treating patients by IVF.

Staining chromosomes: FISHing for a diagnosis

Sadly for Richard Penketh, in 1992 he had to leave our unit to carry on training for his medical career. By this time new techniques were being developed for staining individual chromosomes in single cells; Richard had been in the field just a bit too soon. These promising methods – which could be done quite rapidly – involved gluing the tiny cell onto a glass slide. A selected length of DNA with the appropriate sequence of letters, matching those on the chromosome of interest, would be prepared. This probe would have a fluorescent dye incorporated in it. It would then be applied to the cell and if all went well would stick to the corresponding sequence of DNA on that individual chromosome. When the cell was placed in a microscope under fluorescent light of the right wavelength, the DNA sequence would glow very brightly.

Various improvements were being rapidly developed, and we teamed up with Dr Joy Delhanty at University College, London, a leading expert in the field of cytogenetics. Eventually we had a very effective modus operandi. Alan, Kate, or one of their assistants would remove cells from the embryo to be diagnosed first thing in the morning. Our team would stick them to glass slides and then the cells would be rushed across London to UCL in a black cab, where Joy and her team would start the staining process. Once this had been done, Joy would place the slides under a microscope in fluorescent light. I say the probes glow brightly – in fact this brightness was relative, and to take good photographs an image intensifier was needed, not unlike those used by astronomers to view stars in deep space.

If things went well, we would have a diagnosis by as early as three or four in the afternoon. Joy would phone me with the results, we would telephone the patient to ask her to return to hospital, and generally we could transfer embryos before the evening. Sometimes it took a bit longer – the number of West End theatres into which I would creep after the first act had started was growing.

The technique was very accurate for sexing embryos, because we could identify both X- and Y-chromosomes, using two different coloured fluorescent dyes. But soon there were DNA probes for a number of other chromosomes, and this technique became highly useful for assessing some common chromosome abnormalities in human embryonic cells. It was not long before we could diagnose Down's syndrome, with its three copies of chromosome 21, and other chromosomal conditions contributing to the death of babies before birth. We also had a few women who repeatedly miscarried a pregnancy because either they or their husbands had chromosome translocations. A translocation is where the chromosomes are mixed up – for example, part of chromosome 1 being attached to part of chromosome 13. As the manufacture of probes got better we were sometimes able to detect the embryos that showed these kinds of abnormality and remove them before embryo transfer.[14]

The chromosome staining technique was rather winningly called FISH (fluorescent in situ hybridization) and has gone on to be a very important and useful tool in embryology. Eventually Joy's team expanded their range of probes and showed in a very elegant way how frequently chromosomal abnormalities occur in humans. FISH was gradually expanded by colleagues at UCL and overseas to the point where many different chromosomes could be examined simultaneously, and this led to the technique of aneuploidy screening (see page 371).

Diagnosing single-gene diseases of late onset

Pre-implantation diagnosis was developed to identify gene defects that primarily cause diseases in babies or small children. But there are a few single-gene diseases that manifest themselves in later life. One of the most notorious is Huntington's Chorea, a horrible progressive wasting disease. It causes dementia and is invariably fatal. There are perhaps five thousand people with it in Britain – and if they know they carry the gene, they also know that eventually the disease will definitely strike them. While it can affect very young people (particularly, for some as yet unexplained reason, if the gene is inherited from their father rather than their mother), it usually starts between the thirties and the fifties. Apart from steady mental decline and symptoms of psychiatric disorder, people afflicted with it suffer from seizures, uncontrolled writhing movement,[15] and rigidity; death occurs through pneumonia. Many sufferers commit suicide because they cannot face the illness. Huntington's is a particular problem for pre-implantation diagnosis because the configuration of the relevant section of DNA makes reliable PCR very difficult. It is only very recently that doctors in Belgium and Adelaide have managed to screen embryos for it successfully.[16]

Then there are the gene defects that can cause cancer. Most are rare, but there are a few that are rather more common: the breast cancer genes, BRCA-1 and BRCA-2, bowel cancer (*polyposis coli*) genes and those for a form of brain cancer, retinoblastoma. The issue here is complicated by the fact that carriers do not always get the cancer, the likelihood varying between 50 and 85 per cent depending on the mutation. Some people consider that embryo screening for these diseases is wrong, because carriers live full, normal lives until they get the disease and sometimes they can take steps to avoid it. Women with BRCA-1, for example, can have their breasts amputated and their ovaries removed to prevent the cancer risk. But retinoblastoma can kill children, and the gene for *polyposis coli* is a dominant one – meaning that 50 per cent of the members of any affected family are likely to get it. And very often, when these cancers do occur – as, for example, with breast cancer – they affect young people. They tend to spread more rapidly, are more difficult to treat and are more likely to be fatal.

In late 1992 I discussed the possibility of using pre-implantation diagnosis with PCR to detect cancer genes with Sir Walter Bodmer, then the head of the Imperial Cancer Research Fund, the largest cancer charity in Britain. We did some preliminary work on one breast cancer gene, BRCA-1, but at the time did not have a patient who was interested in going through treatment. One of the problems is that women generally do not find out that they have the gene until they develop cancer; and any pregnancy (especially with the stimulation needed for IVF) in a woman who has recently had breast cancer might cause recurrence of the disease.

However, we were aware of one young woman in Newcastle whose family had suffered gravely from bowel cancer: five members of her family had the disease, or pre-cancerous bowel polyps, and another three had died of it. Joy Delhanty was keen to try to make the diagnosis at UCL, so we collaborated by giving the young woman two cycles of IVF treatment in 1993. Although we paid the patient's fare to come down with her husband from Newcastle, she found the procedure difficult to bear, and eventually she gave up. We published this work some years later, in 1998. I mention this because, to do the pre-implantation diagnosis in her case, we voluntarily approached the HFEA for approval. It was given in due course without difficulty. So I assume the authority must have had a record of this and also would have known about our subsequent publication.[17] Nonetheless, over ten years later, on 10 May 2006, this disorganized government body – which obviously is a bit muddled – announced proudly that now, having consulted the public, it was going to agree to allow the screening of embryos for gene defects causing cancer in adults.

Many readers may find this quite difficult to understand. They may possibly also find it difficult to understand why 'public consultation' was really necessary.[18] Doctors are allowed by law to conduct abortions in a very wide variety of such cases – embryonic diagnosis is surely more acceptable morally. Many of the other diseases we screen for may not kill somebody till quite late in life – cystic fibrosis is usually not fatal until at least the age of twenty or thirty. What business is it of the HFEA, or for that matter a small section of 'the public', to involve themselves in a very private decision which in no way affects 'the public interest' or for that matter threatens the moral fabric of our society?

The limitations of pre-implantation diagnosis

It is a matter of constant surprise to me that so many members of my own profession, who ought to know better, do not seem remotely to appreciate the limits of this treatment. Dr Perry Phillips, an obstetrician and gynaecologist, is a director of IVF Canada. He has said of pre-implantation diagnosis: 'This is the beginning of the end of genetic disease . . . That's the dream of medicine. It's our dream. This should have the same impact [that] antibiotics did to bacterial disease.'[19] Over-statements like this, often made in good faith, are not helpful. It is, of course, nonsense to suggest that pre-implantation diagnosis can eradicate genetic disease, yet this impression is so frequently given. For one thing, in order to consider pre-implantation diagnosis, a family has to know that one or more members carry a gene defect, and in practice almost invariably the first inkling they have of this will be when they have a damaged child. But even if people did know they carried a gene defect before having children, pre-implantation diagnosis could not eradicate these illnesses, for very many of them occur as new mutations. For example, at least one-third of children born with Duchenne muscular dystrophy are born to parents with no history or evidence of a mutation which will cause this illness. Moreover, pre-implantation genetic diagnosis is useful only for single-gene defects, like cystic fibrosis; it is not a weapon against genetic predisposition to diabetes, heart disease, obesity or gallstones. Statements like those of Dr Phillips result in mis-understandings of what this treatment can achieve, and accusations of eugenics.

Pre-implantation diagnosis has many other limitations. One of the most significant is that, even though there are now a great number of refinements, PCR does not always amplify the one copy of the DNA to a million copies totally reliably. And it is a procedure that carries serious risks of contamination. It is always tricky getting just a single cell – and only a single cell – into the PCR tube, ready for analysis. Stray DNA from people in the laboratory, or from a patient treated the previous week, can easily be amplified in error and cause a misdiagnosis. A third problem is that most diseases caused by single-gene defects can have a wide variety of different mutations. Cystic fibrosis has one common mutation – so-called ∂F508 – that occurs in about 75 per cent of families. But there are well over one hundred other mutations that may cause the condition, and each may require different probes and slightly different conditions for optimal diagnosis. Some common genetic disorders – like Duchenne muscular dystrophy – have a huge number of variations in the abnormal dystrophin gene which is responsible for the disease.

Of the six thousand separate diseases that are known to be caused by defects in a single gene, the majority have not been fully sequenced or have a completely unknown abnormality of the DNA; so for these it is not possible to make a satisfactory primer for PCR. This is severely frustrating for patients, who understandably find it hard to accept that pinpointing their specific DNA configuration can take months or years and requires resources that seem quite beyond the capacity of most health services. And even once the configuration is known, working out the best way to do single-cell PCR can be a full-time project for one scientist for several months. While it is true that sequencing the DNA is rapidly getting more simple, these are still massively difficult problems. In some ways the work that was done in the early 1990s was far ahead of its time.

But for me, the biggest problem regarding the issue of pre-implantation diagnosis is concern about embryo biopsy. While it was originally thought that all the cells in an eight-cell embryo are identical, we now know that that is not necessarily true. The work that Kate Hardy initially did on our unit, since confirmed by other scientists, shows that the majority of human embryos have some abnormal cells, while others are quite normal. This is a problem, because quite frequently two or more of the cells may have abnormal chromosomes, while the rest of the embryo may be normal and indeed viable; so removal and analysis of abnormal cells will give a false result.

Admittedly, a false result is likely to result in a diagnosis of abnormality, which will always militate in favour of the embryo being discarded. But it could make gene diagnosis[20] as well as chromosome diagnosis difficult – so, at least for the foreseeable future, pre-implantation biopsy is likely to be an inefficient technique.

It also worries me that pre-implantation diagnosis has become so highly commercial. With a few exceptions it is regarded in many units, particularly in the USA, as a valuable business proposition, with fees totalling up to $30,000 being charged for a single treatment cycle. This may be cheap compared with the cost of bringing up a child with a gene defect (it has been estimated that it costs at least $30,000 a year to care for a child with cystic fibrosis), but the intensely commercial nature of so much IVF means that money is being diverted away from improving the procedure into the reaping of profits. This has undoubtedly had a serious effect on research. It is deeply disappointing to see now how little is being published in this field from the UK, for example, as so many bright people are being drawn into private practice.

We need to think much more constructively about new approaches. Improvements in DNA diagnosis are, of course, going on all the time. But one possibility is to look at the substances being produced by the embryo: that is, to look not at the spelling of the individual genes, but at what the genes are making. This would have value because it could make pre-implantation diagnosis less invasive. Instead of taking cells away, a laborious and time-consuming approach, we could analyse the fluids surrounding the embryo for the proteins or amino-acids that the embryo is making. It might also be possible to look at the physical properties of the embryo. For example, how it refracts light of a particular wavelength may give a clue as to what proteins are being produced inside it – without removing a cell to see.

The ethics of pre-implantation diagnosis

It is quite difficult to understand why such a meal is made of the ethics of pre-implantation diagnosis. Screening an embryo for defects is what nature does herself all the time. Although so many human embryos are abnormal, very few abnormal babies are actually born, because most of these embryos perish at the earliest stage of development. Take the example of Turner's syndrome. While most Turner's embryos are lost naturally at a very early stage, the defect is compatible with life. So some babies are born with it. Turner's syndrome is a disease that affects girls – they are missing one of their X-chromosomes. They grow up to be short in height, they sometimes have heart defects, they occasionally have mental problems and they do not menstruate, having no ovaries.

Similarly, most embryos with Down's syndrome are destroyed naturally. Down's causes mental deficiency and heart defects. Of course, some children born with Down's can live happy lives, though experience shows that serious problems often arise as they grow older and need more care. Some highly moral members of our society make a huge fuss about couples feeling that they cannot face the prospect of bringing up a child with Down's and opting to terminate a pregnancy when the fetus has three copies of chromosome 21. Yet routine chromosome examination of miscarriages clearly shows that far more Down's syndrome pregnancies miscarry naturally, or are lost at the embryonic stage, than are actually born. Most women who opt for pre-implantation diagnosis for Down's syndrome – going through the long, expensive and tedious process of IVF – do this because they have a moral objection to termination. However, they see nothing wrong in randomly choosing any embryo that is merely free of that defect and discarding the others. Given that we are allowed to do abortions on Down's syndrome pregnancies, I wonder why it took a year for the HFEA to allow us to have a licence to do pre-implantation diagnosis for patients prone to have three copies of chromosome 21.

Also highly disturbing were comments from some colleagues, who perhaps should have known better, about the ethics of what we were doing. Occasionally I wondered if there was the slightest tinge of irritation that we had been so successful with our researches. I remember a lecture given by invitation by Dr Robert Edwards, the father of IVF, at Hammersmith Hospital in the early 1990s. By this time, our unit was well established in pre-implantation diagnosis and we were actively researching new genetic diseases to screen. During the lecture, attended by colleagues from all over Europe, Dr Edwards pointedly criticized those units doing embryo research. In his view, he said, all 'spare' embryos should be frozen, and subsequently used only for IVF treatments. I was very disappointed. Only a few years earlier he had written: 'I believe that such studies must be carried out. Their advantages could be too great to be neglected or ignored. The ethical issues involved in establishing and studying early embryos in vitro should be acceptable in view of the potential advantages of the work.'[21]

Some years ago pre-implantation diagnosis was mentioned in the course of a debate in the House of Lords. Lord Alton, a man whom I admire in many respects for his moral views, repeatedly said of our work on pre-implantation diagnosis that it 'reduced human dignity'. It is deeply depressing to hear these things said about treatments that are done with great difficulty, for the best of motives, to help families who are suffering. I often wonder if people who make these aggressive comments recognize the profound upset they cause. When I heard these remarks in the Lords' debate, I felt greatly disheartened and almost found myself wondering whether I was wasting my time on work that seemed to be so little valued and so readily denigrated. But coincidentally, just two days after that debate, a patient called Norma came to my clinic.[22] She suffered with a disease called Conradi's syndrome,[23] which, in her case, was linked to the X-chromosome. Her whole skeleton was poorly formed – she was just over 4 feet high, and her pelvis so misshapen that she could hardly walk. She also had painful lumps at the end of her bones. Her face was somewhat deformed, too, with a prominent forehead and a slightly flattened nose. Her skin was coarse and scaly, and, worst of all (given her difficulty in walking straight), she was going blind.

Norma asked me if I would do pre-implantation diagnosis on her embryos, assuming she was fit enough to get pregnant at all. I pointed out that the DNA sequence causing her disease was not known, so PCR could not work. She replied that she knew that well enough. What she wanted, she said, was for us to sex her embryos. She recognized that if she became pregnant with a girl, there was no way of telling whether the child would be seriously disabled like her. So she was asking if we could transfer only male embryos to her uterus. But, I said, if we put a male embryo back and it carried the damaged, rather than the normal, X-chromosome, it will die at birth – almost the worst injury a mother can face. Norma said she understood that, but if it inherited the normal X-chromosome from her it would be normal and her boy's children would be normal. She wanted, she said, 'my children to have more dignity than I have had'. I felt like weeping. Partly for Norma, but, if I admit it, partly selfishly, in the relief of vindication.

'Designer' babies

One of the most troubling tendencies in comment on pre-implantation genetic diagnosis is the use of the label 'designer' babies or 'perfect' babies. With one rare exception (to which I shall turn in a moment) they are, of course, nothing of the kind. Most of the time we are simply identifying one, two or three letters in the DNA alphabet of that person – which runs to a total of well over three billion. This is hardly 'design'. Nor are the embryos that are chosen for transfer 'perfect'. All we know is that they do not have the mutation that causes one specific disease. That apart, they are taken entirely randomly. We are screening for a single rare defect – not for any other characteristic. We are not able to screen for the myriad diseases that are carried in all our genes and from which, from time to time, when the dice of life are loaded against us, we suffer.

Nor could pre-implantation diagnosis be used to ensure 'desirable' characteristics, in any case. Human traits and characteristics like musicality, intelligence, beauty, gentleness, strength, aggression are complex and produced by the interaction of many different genes with one another and with the individual's environment. It is foolish to believe that we could select embryos for those traits with the tools at our disposal.

There is one special – and rare – exception to the repudiation of the 'designer baby' argument. A very few couples in Britain and in the USA have requested pre-implantation genetic diagnosis to select embryos of a particular tissue type. Their reason for this has been that they already have an older child with a potentially fatal disorder (not necessarily genetic) who needs a tissue transplant. For example, this child may have leukaemia or some other disease and be helped to survive if compatible cells, stem cells, could be taken from a new baby. Usually the request has been for cord blood (from the placenta) which would otherwise be thrown away and which might be used to repopulate the elder sibling's bone marrow.

Such requests sound innocuous, but I believe they are fraught with risk. First, there is concern about bringing any child into the world for a specific purpose. If – for whatever reason – that child cannot fulfil that purpose, its very *raison d'être* is challenged. Supposing it turns out not to be of the 'right' tissue type or blood group to provide the desired cord blood cells. Will the child grow up under a shadow of failure? And suppose the tissue match with the sibling is good, but the cord blood contains no usable stem cells. Is this child, when it is old enough, expected to give bone marrow as a donor? Supposing the child, asked to consent, refuses for whatever reason. How does it remain integrated with its family? And what if a bone marrow graft is given but does not take, and possibly a kidney transplant is then required? What future can such a child anticipate if it wants to refuse? For all these reasons, I have been quite uncomfortable with the idea of producing this kind of 'designer' baby. From the very beginning of its life, its autonomy is compromised.

So I was quite surprised when the HFEA gave licences in this country for these procedures to be carried out. In addition to my reservations in principle, explained above, this kind of pre-implantation diagnosis is extremely unlikely to work and raises massive hopes from parents that will in most cases not be fulfilled. It is also colossally expensive. It was reported in one newspaper that one couple had spent over £30,000 in unsuccessful pre-implantation diagnosis to help save their existing child. How much better if that money had been spent searching for a compatible, unrelated donor.

The HFEA, I have no doubt, was acting out of compassion in giving these licences. That surely is a step in the right direction. But assuming there should be licences at all, and given that the HFEA makes a huge fuss about ensuring the 'welfare of the unborn child', many people were left worrying about their capability of making sound judgements.[24]

I have to admit, however, that recently I have come to feel less critical of the HFEA on this point, and to feel that I may have been wrong to be quite so negative about this treatment. While I still have misgivings, I am impressed by the accuracy of diagnosis achieved by some Italian colleagues, which at least shows that the treatment is feasible. In 2005 Dr Fiorentino and colleagues in Rome used a new method of PCR to identify the histocompatibility genes (those that determine the compatibility of tissues between different individuals) in forty-five couples.[25] They then managed to get matched genes in 15 per cent of embryos, and to deliver five live babies. They collected cord blood from these children, and three siblings requiring a stem-cell transplant benefited from the subsequent transfusion. It is very difficult to carp about a treatment that has saved lives so dramatically.

Aneuploidy screening

Aneuploidy screening is an unusual form of pre-implantation diagnosis, carried out not to prevent disease, but in an attempt to improve the results of IVF. Aneuploidy screening involves staining chromosomes, using FISH, to assess embryos. The idea is to try to exclude those that might have defects, so that we can transfer embryos that show no abnormalities. As around 75 per cent of human embryos obtained by IVF show some chromosomal abnormality, it seems to make theoretical sense to screen out those with defects.

One problem is that it has not been possible so far to stain all the chromosomes simultaneously. Screening started with just three colours (three different chromosome probes) and then moved to five. It is now possible to screen nine or ten chromosomes with different colours (multicolour FISH), and some progress has been made at examining all twenty-four chromosomes using a rather different technique. But only multicolour FISH has been used to treat patients.

Aneuploidy screening has been recommended for three kinds of patient: first, older women who have a less good implantation rate at IVF, because it is presumed that more of their embryos have chromosome abnormalities; second, those younger women who unaccountably have failed IVF repeatedly and may be producing abnormal embryos; and third, those patients who do not have any chromosomal abnormality (such as a translocation: see page 358) but repeatedly lose pregnancies through miscarriage. Aneuploidy screening is now done in many IVF centres, almost always privately.

But there are problems. It is relevant that IVF with pre-implantation diagnosis, which is almost identical to aneuploidy screening, has a lower success rate than routine IVF. This is in spite of the fact that pre-implantation diagnosis is mostly done in completely fertile patients who would not otherwise need IVF. This should sound warning bells. Aneuploidy screening may confirm that the biopsied cells are abnormal, when actually the rest of the embryo is normal. Sometimes the biopsied cells may be normal, but part or all of the rest of the embryo may have abnormal chromosomes. In consequence, some perfectly good, chromosomally normal embryos may be discarded, and some chromosomally abnormal embryos may be transferred. It is also an expensive procedure (which may, of course, be one of its attractions) – though the actual cost varies widely. In addition to the cost of IVF, in the USA it typically costs over £2,000, in London around £1,000, and in Adelaide and Melbourne about £300. As the procedure and the materials are identical in all three countries, we may feel a little confused – particularly as the units in Melbourne and Adelaide are among the most advanced and successful in the world.

Two recent good studies have been published. One has been carried out by the excellent Paul Devroey and his colleagues in Brussels.[26] They found no evidence that aneuploidy screening helped women with unexplained repeated miscarriages, irrespective of their age. The other is the important Cochrane Collaboration study, published in 2006.[27] This concluded that, as yet, there were insufficient data to determine whether aneuploidy screening is an effective intervention. Only two random-ized trials had been done, and neither had shown an improvement in birth rate or pregnancy rate. They con-cluded that until properly conducted trials are done aneuploidy screening 'should not be used in routine patient care'.

Let Richard Sherbahn, MD of the Advanced Fertility Center of Chicago, have the last word:

> I believe that there is a lot of confusion created by the aggressive 'marketing' of PGD [pre-implantation genetic diagnosis] that is done by some fertility centers and/or PGD labs. Marketing brochures, 'patient education' videos, etc. are telling consumers of infertility services that PGD can improve pregnancy rates, reduce multiple birth rates, reduce the risk of miscarriage, and reduce the risk of having a baby with a chromosomal abnormality. Patients are understandably confused when the theoretical benefits of PGD are crafted into slick marketing brochures at centers that make a substantial profit on PGD for chromosomal testing. I believe that the business of PGD has pushed ahead of science and good medical practice at some centers. In my opinion, PGD for aneuploidy screening has been implemented (by some IVF centers) before proof of (sufficient) benefit for the infertile couples.

I agree.

Cloning: A Feat of Clay

Throughout the Middle Ages, Jewish communities in Christian Europe were plagued by a particularly malevolent accusation, a version of what we might now term an urban myth. The myth, known as the blood libel, held that Jews sacrificed Christian children in order to obtain their blood to make the unleavened bread for the Passover festival. Anyone who took a serious look at the scriptures would have known that the Jewish religion has both a profound respect for life and a deep aversion to blood, and that the idea of consuming it would be high in the catalogue of sins. The reason why Jews salt all meat is to avoid the least drop of blood remaining in food. As the Jewess Fegele maintained to her torturers in the blood-libel case in Lublin in 1636, 'The use of Christian blood, children or adult, even animal blood, is forbidden to us'; and she maintained this assertion in spite of repeated torment, including the threat of red-hot irons. But throughout Europe and Russia the suspicions continued, fanned in some cases by religious and secular authorities. Behind the libel lay a vile parade of riots, arson attacks and mass killings throughout the Christian world. Incredibly, the blood libel is still stated as fact on some websites belonging to far-right extremists.[1]

It was against this tense background in medieval times that a series of fables emerged in the Jewish world detailing the physical protection and spiritual revenge of these persecuted communities. One famous story concerns the learned Chief Rabbi Yehuda Loew ben Bezalel of Prague, who lived from 1510 to 1609.[2] Around this time, the blood libel was causing untold misery to the large Jewish community. Paranoia and suspicion ran deep and it was sometimes suspected that Christians were actually murdering children themselves, then planting their remains in Jewish houses to provoke unrest. Rabbi Loew, who it is said was skilled in the mystic arts of the Kabbalah, decided that his community needed a very special sort of watchman. Like some medieval fertility specialist, the rabbi mastered the technique whereby God had created Adam, the first man, out of clay. He fashioned an enormous man, many feet tall, and gave him life by writing the most secret name of God upon his forehead. The Golem,[3] as he was known, possessed incredible strength, but, lacking a soul, had no power either to speak or to reason. He existed to do the rabbi's bidding, from fetching water to warning the community when they might be under attack.

There are various golem legends in Jewish literature – and they may have their roots in the supernatural beliefs of the ancient Near East. The Egyptians made small clay statues called *ushabti*, meant to be servants for the dead in the afterlife. Like the later golems, the clay was brought to life by an act of naming. There are various comments on the idea of a golem in the Talmud, which dates from before the fifth century CE. In medieval times, golem legends served as a source of reassurance, and also as pure entertainment. Some of them are extremely light-hearted, based on the almost sitcom-like idiocies that occur when the unthinking golem misinterprets his master's commands. But they all possess a common theme. In creating the golem, the rabbi unleashes a power more terrifying than he ever imagined. It grows until it is beyond his command – destroying the very communities it is meant to protect. In despair, the rabbi has to deactivate it, rubbing out one of the mystic letters on his creation's forehead to render it mere clay again.

This theme, of created life running amok, arrested the imaginations of generations of later writers – perhaps because, from the sixteenth century onwards, our increasing mastery of science also created a deep-seated unease. From Frankenstein's monster to twentieth-century films like *The Stepford Wives* and *The Boys from Brazil*, our culture has expressed considerable misgivings about man's ambition to control and manipulate nature. And the golem legends live and breathe today, in the highly charged and often cloudy debate surrounding cloning and techniques using stem cells. For in vitro fertilization did not just give us the means to mix eggs and sperm; it gave us the means to master life itself, to fashion human beings not from clay but from our own material, to select and grow body parts for repair and renewal – even, possibly, to extend our allotted time on the planet. It gave us, in some senses, the powers we traditionally ascribe to God.

The building blocks of the body

Unlike most of the new techniques and treatments discussed in this book, the story of cloning – and also stem cells – goes back to questions posed by a nineteenth-century university professor named August Weismann. Weismann, who taught at the University of Freiburg, was not a specialist in reproduction. He was interested in the cells, the building blocks of the human body – and in what became of them as the body develops. As we now know, when an embryo is just a clump of two, or four, or eight cells, each one of these cells has the ability to create another embryo. But as the embryo develops, it loses this potential. Its cells undergo increasing specialization, becoming brain, lung, throat, skin, hair, teeth and muscle cells.

A useful metaphor might be the effects of a standard British education on a child. Many children start out with an equal enthusiasm for languages, creative arts and sciences. By the time they are eighteen, they will have been pigeon-holed by the curriculum into one of two seemingly incompatible disciplines – they are 'artists' or 'scientists', and it will be extremely difficult for them to reverse this compartmentalization in later life. Difficult, but not impossible. And that is why it is an appropriate metaphor for embryology, which concerns itself with the same basic question. Can we reverse the specialization of human cells, turning these dedicated little units back into the raw stuff of existence itself?

Weismann thought not. His view was that as cells begin to specialize – right after the first division – they lose that vital bit of programming that once enabled them to become anything. If you split a two-celled embryo in half, he thought, you'd end up with two half-embryos. But this theory was swiftly knocked out by a brilliant biologist called Hans Spemann. Spemann used a hair from his baby son's head to split a two-celled salamander embryo in half. He got two whole embryos – suggesting that, at least at this early point in development, the cells remained able to develop into any tissue. But this then posed other burning questions. When, exactly, did cells lose this property? Did they always partly retain this ability – or did it stay buried inside their blueprint, awaiting the time when scientists could tease it out?

Spemann, who received a Nobel Prize in 1935 for his contribution to understanding of specialized tissues in newt embryos, had some years earlier devised what he called a 'fantastic experiment'. His plan was to take an older, specialized cell from an embryo, and place it inside an egg that had had its nucleus removed. What remains inside the egg after the nucleus is extracted is a jelly-like substance called the *cytoplasm* – and Spemann thought that this could somehow rejuvenate the older cell, turning it back into a young, totipotent one. He tested his theory with his salamanders and was able to create an embryo from the cell of another one – perhaps the first example of laboratory cloning. He longed for the day when he could take a cell from a human adult and do the same thing. But, armed only with his baby hairs and salamanders, Spemann was trapped in the realms of theory by a basic lack of technology. It was 1952 by the time another pair of researchers in Philadelphia, Robert Briggs and Thomas King, were able to advance Spemann's experiments.

Briggs and King addressed themselves to the business of removing the nuclei from eggs, and popping replacement cells inside them – designing their own instruments to accomplish this incredibly complex task. They experimented with frogs' eggs, producing twenty-seven tadpoles by so-called *nuclear transfer*. It was an advance – but the tadpoles were unable to develop further. Later, in the 1960s, the distinguished Oxford scientist John Gurdon was able, with great difficulty and variable rates of success, to create live frogs from embryonic cells.[4]

But even he was unable to fulfil Spemann's dream of turning a fully specialized, adult cell into an embryo.

Cloning mammals: the improbable in pursuit of the impossible?

One of the curious aspects of cloning is that this field has attracted some very strange characters. One such was Karl Illmensee from the University of Geneva, who claimed to have cloned three mice in 1979. This announcement shocked the science community and caused a real flurry in the press. The journalistic interest was all the greater because the popular film *The Boys from Brazil*, based on the novel by Ira Levin, had recently been released; but scientists viewed Illmensee's claims askance because repeated attempts at cloning were starting to persuade many that it was highly unlikely that this procedure would be possible in a mammal.

Illmensee claimed he had cloned his mice by removing the nucleus from one of the cells of a four-day-old mouse embryo, using a pipette so fine its tip could not be seen with the naked eye. The nucleus was then apparently injected into a fertilized mouse egg whose original genetic material was sucked away using the pipette. But Illmensee seemed to be rather reclusive and secretive about his work. He had the habit, it was said, of working alone in the laboratory late at night or at weekends when no-one else was around to see what he was doing. It seems he never was able to show his colleagues what precisely he had done or how he had achieved it. And he even discouraged other researchers in his lab from doing the experiment. One of his colleagues in the same lab seems nevertheless to have attempted to repeat Illmensee's experiment, but failed totally. And, to add to everybody's suspicions, it is reported that one of the students in the laboratory noticed that the micromanipulator that Illmensee had mentioned he had used one weekend was broken at the time. There were various reports that other students had examined the glassware that Illmensee claimed to contain mouse eggs, but could not find any evidence of their presence.

A commission was established to investigate Illmensee and to probe his claims. Illmensee denied any fraud, but the investigators decided that the experimental results were scientifically worthless and therefore that the experiments should be repeated. They never were; later there was a report in *Nature* that Dr Illmensee,

> accused of falsifying nuclear manipulation experiments in mice, has lost his National Institutes of Health (NIH) grant. Neither NIH nor an international commission of inquiry was able to draw any firm conclusions about fabrication, but the commission found errors and discrepancies in Illmensee's records that cast 'grave doubt' on the validity of his experiments. Further grants from NIH must await validation by Illmensee of the disputed results.[5]

This did not prevent Dr Illmensee from continuing his research, and although he never published the full details of the cloning experiment, he continued to bring out a stream of interesting papers on other aspects of embryology. And many years afterwards he would team up with a certain Dr Zavos, of whom more later.[6]

Another very enigmatic man was the mysterious Dane Steen Willadsen, clearly a deeply talented and brilliant scientist, but one who also gave the impression of being a bit of a maverick. I heard him speak only once, and my impression was of a very clever person with a perpetual grin, full of the joy of doing the impossible or, at the very least, the controversial. It seems that his *raison d'être* was summed up by his phrase: 'The role of the scientist is to break the laws of nature, rather than to establish, let alone accept them.' Willadsen worked for a while in collaboration with that phenomenal group in the University of Cambridge, mostly in an agricultural research centre. There he certainly tried the improbable. He took some early embryonic cells from a sheep and mixed them with those of an embryonic goat. A photograph of the resulting bedraggled-looking animal, its odd tufts of goat hair surrounded by patches of white wool, subsequently appeared on the cover of *Nature*. It was reported as bleating like a sheep – but, sadly, smelling like a goat.[7] This so-called Geep was not quite like the goat chimaera of ancient times – half-goat, half-man. But it raised the question whether scientists had over-stepped the mark. Now that IVF was possible and human embryos available, supposing some evil person did mix in some animal cells?[8] It was said that Steen invited some friends round one evening and barbecued one of these animals at a party. He was reported as saying: 'It was not very tasty, but that was because the sheep part of it was too old.'

Steen Willadsen is also important in this story because, more than a decade before Keith Campbell and Ian Wilmut cloned Dolly, he cloned lambs from embryo cells that had already begun to specialize. But he did not publish the news of his feat immediately.[9] What he did was first to remove the nucleus from a sheep's egg; then he took an eight-cell embryo from another sheep and removed one of its cells. The 'Frankenstein' bit was giving these two cells the spark of life. Essentially, he sent a jolt of electricity across the two cells, fusing them together.[10] Thus the egg was 'tricked' into believing that it had been fertilized – and the result was life created from life. He repeated this process to obtain a number of cloned embryos which he then coated with agar, the jelly-like polymer extracted from seaweed and algae, to protect them. These embryos were placed in a sheep's fallopian tube for about a week to mature before finally being transferred to the uterus of a surrogate mother sheep. Two lambs died at birth – and another survived, to be the first mammal cloned by the nuclear transfer method. Willadsen's feat was all the more remarkable given that, only a few weeks before the lamb was born, various distinguished biologists were saying that cloning a mammal was clearly impossible.

In a nutshell, this is how cloning takes place – in the sober and sterile atmosphere of a laboratory, rather than at the stroke of midnight to the accompaniment of thunderstorm and dry ice. The word comes from the Greek *klōn*, meaning a twig, and indeed cloning is how some species of trees, insects and reptiles reproduce. Humans have had mastery of this technique for centuries – whenever we take cuttings from a plant and nurture them to life, we are cloning. The word is pregnant with scary associations, but most of them are undeserved.

Multiplication and moral panic

It is easy to understand why an unusual person and lateral thinker like Steen Willadsen could provoke massive public comment – though ironically his cloning experiment went largely unnoticed because he did not publish any details very quickly. But it is much more difficult to explain the uproar caused by the – admittedly unwise – mention by my reticent, gentle friend Bob Stillman that he had 'cloned' a human embryo at George Washington University. The media coverage was extraordinary and, given how innocuous his experiment was, it should have given a clear foretaste of things to come.

In 1993 Bob and his colleague Jerry Hall announced that they had attempted to 'multiply' seventeen human embryos. Their paper was read at the annual meeting of the American Fertility Society in Montreal, in a low-key presentation which nobody in the audience (I was there) thought was in the least bit remarkable – beyond being a rather interesting piece of work. Their experiment was based on the idea that transferring a single embryo to the uterus had a very low success rate (around 18 per cent at that time). So Stillman and his junior colleague dissolved the outer coat, the zona pellucida, of seventeen embryos two to three days old, at which point each comprised two to eight cells, and then divided them into their individual cells. Each cell was then given an artificial zona by coating it in agar. Thereafter, the researchers simply observed these embryos in culture for two to five days to see if they would develop further. They were encouraged to see that no fewer than forty-eight new embryos were formed, though none were likely to be viable. Apart from any other reason, they had now, of course, lost their protective zona pellucida. Nevertheless, what the researchers had done was to mimic what happens in nature when identical twins are formed. The ultimate objective was eventually to consider trying to give women a better chance of conception by transferring two embryos if only one was initially available. Needless to say, the events that followed their announcement made further work unthinkable.

Bob Stillman is a cautious, self-effacing character, a caring doctor with high ethical standards, and he must have been deeply upset by what followed when the press got hold of the story. When it broke a day or so later, the front page of the *New York Times* carried the headline 'Scientist Clones Human Embryos, And Creates An Ethical Challenge'. Within hours the world's media had jammed the switchboard at George Washington University, and some very unpleasant comments were being made. 'This is the dawn of the eugenics era,' declared Jeremy Rifkin, founder of the Foundation on Economic Trends, a biotechnology 'watchdog' group in Washington. Organizing protests outside the university, he said the scientists had created 'standardized human beings produced in whatever quantity you want, in an assembly-line procedure'. Republican politicians in Washington pointed to the deficiency in bioethical leadership as Rifkin's protesters took to the streets, calling for an immediate ban on human embryo cloning.

Then began one of the most vicious debates about medical ethics since Louise Brown's birth. *Brave New World*, *The Boys from Brazil* and human eugenics had finally arrived, it seemed. 'The people doing this ought to contemplate splitting themselves in half and see how they like it,' said Germain Grisez, a professor of Christian ethics at Mount Saint Mary's College in Emmitsburg, Maryland. The reaction from around the world was, in many ways, even more sharply hostile. A spokesman for the Japan Medical Association found the experiment 'unthinkable'. French President François Mitterrand pronounced himself 'horrified'. The Vatican's *L'Osservatore Romano* carried a front page shouting that such procedures could lead humans into a 'tunnel of madness'. 'This is not research,' snarled Dr Jean-François Mattei of Marseilles. 'It's aberrant, showing a lack of a sense of reality and respect for people.' In Germany, Professor Hans-Bernhard Wuermeling, the medical ethicist from the

University of Erlangen, called producing carbon copies like this 'a modern form of slavery'. It is interesting how very ready ethicists are to attitudinize when called on to give ethical opinions about matters of public interest – sometimes, it seems, forgetting the ethics of what they are saying. They proclaimed nightmare visions of baby farming, of cannibalized babies bred for transplants and spare parts.

And some of my scientific colleagues did not acquit themselves very well either. I imagine they spoke without full possession of the facts, unaware that these were not viable embryos. 'Hall and Stillman haven't done science or medicine any favours,' acidly said Dr Marilyn Monk, a research scientist at the Institute of Child Health – who herself had worked with non-viable embryos in the past. And Dr Leeanda Wilton, director of embryology at the Monash IVF Centre, normally the sweetest-natured person (and also a scientist who has worked on many spare embryos in her time), said there were hundreds of scientists who could have split an embryo in half, just the way Hall and Stillman did: but 'they haven't done so because it opens a can of worms'.

In the face of this outcry, Bob Stillman kept maintaining that the embryos they had chosen for their experiment were those that could not have been transferred even if they had wanted to do so. They had applied for ethical approval and obtained it simply to work on embryos that had been fertilized by more than one sperm and therefore were in no way viable. And they certainly did not go beyond the limits proposed by their ethics committee. This experiment was not genetic engineering, it was not tampering with the DNA, they were not selling the embryos, it did not risk damaging any human child that might be born and it certainly did not unravel the moral fabric of American society. Admittedly, it wasn't the best science of the century, but the researchers had made some perfectly valid original observations about human embryonic development that were reasonably relevant. And, ironically, they had used one of the techniques that Steen Willadsen had advocated in cattle a decade earlier and which was being used occasionally to improve fertility in domestic animals without any fuss at all.

When *Time*'s correspondent Ann Blackman asked Hall if he was worried about the effect his research might have, he is reported as shaking with emotion, saying: 'I revere human life, I respect people's concerns and feelings. But we have not created human life or destroyed it in this experiment.'

Dolly, the cloned sheep created from an udder cell of an adult ewe, has had volumes devoted to her already. She was, by all accounts, a very charming sheep – brimming with charisma in a species not exactly renowned for it. What concerns us here is how Dolly was created, and what her creation meant. She too was brought into being by the process of nuclear transfer. But in her case, the workers at the Roslin Institute near Edinburgh did something novel. Previous experiments had involved the transfer of a nucleus from an embryonic or primitive cell, before its full programming as a differentiated cell was completed. Dolly was made by transplanting a mature, adult cell nucleus into the empty egg cell. The implications were, of course, that if the experiment could be reproduced in humans you could make a carbon copy of any person, be they fetus, child or adult.

So Dolly's arrival, as well as spawning a crop of cringe-making tabloid headlines ('Can Ewe Believe It', 'And We Like Sheep' among others), was scientifically important. It provided the final answer to Weismann's questions. The successful cloning of a mammal proved that, even as cells became specialized, they retained the ability to be 'switched back'. This raised other potent possibilities, because if we could master the means of switching, we could theoretically turn cells into whatever we chose – hearts, livers, kidneys, brains, skin, bone marrow. Cells, we realized, were nature's universal coinage; and, once we understood them, we could use them for an endless range of biological 'purchases'. We could, people theorized, perhaps reverse Time itself, become immortal.

These observations have not led us to immortality, as yet. But they have given us understanding that cell specialization is a process that might be reversed and manipulated. They have also led us to understand that the cells inside an embryo are the richest, most potent type of cells available. And you may, even if you have never fully come to grips with the science involved, have read of these 'adult' and 'embryonic' stem cells in various newspaper reports, and gained a flavour of the contentious debates surrounding these two types of matter. Most people will also be well aware of the strong, possibly misguided, position taken by the US government with regard to research on embryonic tissue, contrasted with the advances being made elsewhere. Sadly, very often there is a shudder of distaste at the very use of the word 'embryonic' – with its connotations, perhaps, of babies are being dissected in laboratories. Nothing, of course, could be further from the truth. Is there the merest hint here of a modern 'blood libel', this time against scientists?

The cloning cult

The arguments against embryo research would carry less weight if it were not for the unfortunate presence of a mischievous group who specialize in commanding media attention – media attention which then informs the way politicians and even Johns Hopkins University professors make up their minds. On Boxing Day 2002 a company called Clonaid announced that it had successfully cloned a human being. It followed this announcement with the news that, after 'Baby Eve', five more clones had been produced. The world scratched its head. The former science editor of a major US broadcaster offered to test Clonaid's claims. To date, it has provided no proof.

The opinion of most experts is that Clonaid have not produced a child – even though it is remotely possible that it possesses the means to do so. The company is an off-shoot of a New Age sect created by a former racing driver, who claimed that aliens enlightened him while he was walking the foothills of the Auvergne. Over the years this sect, the Raelians, have very successfully created publicity for themselves with a variety of carefully planned stunts. In 1992, during Operation Condom, vans decorated with paintings of contraceptives, swastikas and flying saucers pulled up outside Catholic high schools during the lunch hour. The bemused students of Quebec watched as white-gowned, bearded Raelians emerged from the vans to hand out armfuls of condoms. In 1993 Raelians gatecrashed the Montreal Jazz Festival by publicizing and then staging a 'Masturbation Conference'. They seem to delight in the bewilderment and outrage that they provoke, and I can't, frankly, see much harm in them. The cloning revelations were probably carefully devised, precisely because the very word 'clone' conjures up such fearful Golem/ Frankenstein/Stepford Wives associations in the public imagination. They knew which buttons to press – and by doing so, they also dramatically increased their member-ship figures.

Oddly, the Raelians' publicity stunt on cloning was actually taken seriously by a few people, and it is said that it influenced the US government's thinking when it was considering the ethics of embryonic stem-cell research. One area of stem-cell research is the use of nuclear transfer to obtain cells of a specific type. So the stem-cell debate got tainted with the perceived threat of cloning humans. But I think the Raelians were always a sideshow. Much more serious was the outrageous activity of Dr Severino Antinori and Dr Panos Zavos, which certainly produced ripples, if not waves, in Washington.

A perilous ambition

Dr Antinori had a clinic in Rome, ironically just a few yards from the Vatican. I visited this man a few years ago with a film crew to interview him about his practice of establishing pregnancy in women over sixty years old. Although he has been repeatedly referred to as a 'scientist' in the press, his credentials for meriting such a serious appellation are limited: a search through the peer-reviewed literature on the main international database reveals a handful of desultory publications, few showing much originality or great scientific merit. One of these, published in a normally highly serious scientific journal, has a title more suitable for a magazine like *Woman's Own*: 'A child is a joy at any age'.[11]

Some time around 1989 Antinori seems to have attempted his first transfer of a donor egg into a woman well past the natural menopause, and the press showed typical outrage. In 1994, however, he made massive head-lines internationally because he had done this trick with Rosana della Cortes, who at sixty-three was then the oldest known woman (since Sarah, the wife of Abraham) to give birth.

When I went to interview Antinori, the experience was not reassuring. He appeared bombastic, and was highly aggressive when challenged on camera. At first he spoke fluent English, but after a while, when it was clear even to him that he had been caught on record making damaging statements, he asserted he did not understand the language. We caught him spouting racist remarks about our black sound-recordist, and I felt sure there was some anti-Semitism in the way he pointedly referred to my being Jewish. He was certainly not the cuddly, warm doctor that is his preferred media image. At the end of the interview, having recognized, I think, that he had made a total fool of himself on record, he became quite violent, attempting to throw one of our very valuable film cameras down a flight of stairs in his clinic.

I suppose it was inevitable that he would seek even more publicity, and the massive attention given to cloning in the media gave him the opportunity. In due course he met up with Panos Zavos, and in April 2002 he claimed one of his patients was pregnant with a cloned human embryo. Shortly after this, he announced a series of other patients with cloned babies 'on the way'; but four years later, no baby or cloned embryo has materialized. Dr Antinori flouts professional and public opinion, while clearly enjoying notoriety. In a way, his partnership with Zavos seemed almost a perfect meeting of minds. While Antinori says he is motivated by a genuine wish to help childless couples, I think that both he and Zavos are essentially driven by the desire to achieve 'fame' by being the first to clone a human being. The really sad thing is that there are people out there who see this as an answer to their mortality. But even if he or Zavos succeeded, the methods of cloning currently available are very likely to produce many abortions, or, worse still, deformed and abnormal children with crippling fatal disorders. Yet amazingly, it seems there are enough gullible people out there sufficiently impressed by the smooth talk of these two men to ignore how serious these risks actually are.

To my mind, Zavos seems more dangerous than Antinori. I have met Zavos only once, when I debated cloning with him at the Oxford Union. He seems marginally more intelligent and even thicker-skinned than Antinori; and he too seems to have rather limited scientific credentials to take on the complex task of cloning a human being safely. Most of his publications are in the field of sperm quality and male fertility, rather than in the difficult area of the potential damage to gene expression in attempts to perform nuclear transfer. But now Zavos and Antinori seemed to have parted company, and Zavos has teamed up with Karl Illmensee – who, although his original claims to have cloned a mammal were almost certainly fraudulent, certainly would have the scientific experience and intelligence to manage a serious attempt at making a human clone.

Antinori and Zavos have undoubtedly had an influence on the regulators. In Italy, the parliament brought in draconian laws that not only forbade cloning, but made genetic testing illegal. Moreover, the Italian government, seemingly influenced by the adverse publicity surrounding Antinori, produced other legislation that makes even much routine IVF practice virtually impossible. In consequence, many Italians are now seeking treatment outside their country. Perhaps even more worryingly, Zavos has had a hugely adverse effect in the USA. Once he started strongly publicizing his claims about cloning, it became increasingly obvious that the US government was highly unlikely to fund embryonic stem-cell research, effectively banning it in all publicly funded institutions. This, together, with President Bush's personal doubts about the nature of the human embryo, has had a highly inhibitory effect.

Dr Zavos's website offers for sale a home gender selection kit – for a little under $1,000. I have no idea whether it works at all, but I imagine the Advertising Standards Agency would have a field day with its claims in the UK. Once you have clicked past the pictures of cloned pulsating sperm with which it opens, the site leads with a description of Zavos, couched in appropriately modest terms:

Dr. Panos Michael Zavos, Ed.S., Ph.D. is a Professor Emeritus of Reproductive Physiology-Andrology at the University of Kentucky, Honorary Professor at the China Academy of Science, Founder, Director and Chief Andrologist of the Andrology Institute of America, and President and CEO of ZDL, Inc., a private corporation that markets infertility products and technologies, worldwide. He received his B.S. in Biology, his M.S. in Biology-Physiology and Ed.S. from Emporia State University in Emporia, Kansas. Dr. Zavos received his Ph.D. in Reproductive Physiology in 1978 from the University of Minnesota. Dr. Zavos has a long career as a reproductive specialist and is the chief scientist in the development of several new and innovative technologies in the animal and human reproductive areas with worldwide implications. He has authored or coauthored more than 400 peer-review publications, along with a number of solicited reviews, book chapters and popular press releases.

I regret that the search engines I used – PubMed and Medline: the gold standard for medical papers – were not sufficiently sensitive to find more than a small fraction of the 400 peer-review papers that Dr Zavos claims to have been involved in writing.

Both Antinori and Zavos bring our subject – a subject that is crucial to the well-being and happiness of many, many people – into disrepute. It is very unpleasant to see the field of endeavour that one's colleagues and oneself have followed and tried to improve over so many years being hijacked in this way. But while I deprecate the activities of both, I am more scared of Zavos, because he persuades journalists that he really has done (or could do) what he claims. And now that he seems to have teamed up with a scientist who has genuine skills, I think there is an undoubted chance Zavos might even make a more serious attempt at human cloning.

The dangers of cloning

At present, cloning is not remotely close to being a safe technology for humans. Most of the original sheep embryos that were cloned by Wilmut and Campbell did not survive. Even now, nearly ten years after Dolly's birth, well over 90 per cent of cloned embryos are not viable for long. When Dolly was born, most of the other sheep produced in the same way had massive enlargement of various organs, abnormalities of the placenta or huge derangements of growth. Some were born dead; the others died for no apparent reason soon after birth. It is probably true that Dolly was a normal sheep; although she did die somewhat prematurely, with advanced arthritis, her death seems to have been due to a lung infection not uncommon in sheep, and arthritis is a fairly frequent disease in this species.

Since that time, most other cloned animals have had

abnormalities. In nearly every other species that has been cloned, whether the resulting animals be mice, dogs, monkeys, pigs or primates, there has been a high rate of malformation as a result of this tampering. Cloned cats have frequently suffered from abdominal malformations, respiratory disease and septicaemia.[12] Cloned cows have been born with shortened, deformed legs and multiple abnormalities of their internal organs. Pigs have been produced with abnormal testicles and deformities of the bowel. Goats that have been cloned have died suddenly and without any obvious cause immediately after birth.[13] And evidence is constantly leaking out that cloned animals do not always express genes normally.[14]

To create Dolly the sheep, 277 eggs were collected from approximately 50 animals that had been treated with FSH. This number of eggs produced 29 embryos that seemed sufficiently normal to transfer. No fewer than 13 ewes were used as host mothers, and although this is a highly fertile species where most embryos would normally produce a baby, the birth of only a single lamb resulted.[15] This is a success rate of about 3.4 per cent of transferred embryos and 0.4 per cent of reconstructed eggs. Since that time the efficiency of the procedure has improved a little, but in most species where it is successful at all, the efficiency is still only about 2–3 per cent. Humans, being much less fertile naturally, would yield a much lower rate of success.

So, in the unlikely event that a woman did get pregnant with a clone, the risk of producing an abnormal human is high.[16] Eventually a seriously ill baby is almost certain to be born. What is so disturbing about Zavos's claims is that he repeatedly asserts that any abnormalities caused by cloning can be screened before birth, so that there is no risk of an abnormal child being produced. This was a boast he made at the Oxford Union debate, with such confidence that a highly intelligent audience was largely taken in. He clearly does not understand the limitations of the technology of pre-implantation diagnosis, discussed above (see page 361). It may be used to hunt down an abnormal chromosome or a specific gene defect, but abnormalities of the kind produced by cloning cannot be detected reliably by any known method. And given what we do not know about why apparently normal animals have unpredictably died after birth, it is absurd to consider risking humans in this way.

What precisely is going wrong? There are still many puzzling issues about cloning. One unresolved issue is that of ageing. The nucleus that is transferred is from an adult cell, taken from a fully developed organism. In Dolly's case that was an udder cell, highly specialized to yield milk. All the specialized cells of this kind in the body are called *somatic* cells. The other kind of cell is a *germ* cell – the egg or sperm, both of which are effectively just a means to an end. Nearly all somatic cells are programmed to renew themselves by dividing and multiplying.[17] Rather like an electric battery, each somatic cell can do this only a certain number of times before it dies. This makes them very distinct from germ cells, which do not divide, but lie waiting to do the vital job of starting new life. We can even measure the shelf-life of a somatic cell by reading its DNA code. To put it simply, each cell contains a chunk of DNA coding that doesn't do the same job as its other molecular colleagues. The *telomere* is a short stub of DNA at the tip of the chromosomes that shortens every time the somatic cells renew themselves. When it runs out, the cells are dead.

It is not yet clear how cloning affects this. Some experiments have shown that cloned animals have shortened telomeres, suggesting that even though they are born immature, some ageing has already taken place in their bodies. But in another study, researchers at a leading biotechnology company were delighted when they produced twenty-four healthy calves from adult cells. Even though their progenitors were of advanced age, the telomeres of the newborn cloned calves were no shorter than those of normally produced calves. Quite what these findings imply is unclear, but it is possible that cloning truly could turn back the clock. Could cloning slow, rather than hasten, the ageing process? What seems more certain is that cloning, in most animals, does risk altering the way genes work. As we have already seen in chapter 7 on pre-implantation diagnosis, genes produce proteins depending on the sequence (or 'spelling') of the DNA in the gene; if there is misspelling, a gene may produce an abnormal protein which leads to a disease. But we now know that it is also possible for genes to have their function altered without any change to the actual DNA spelling. Under certain conditions, chemical groups can be added to a gene that switch it on or off. If it is switched off, then it won't spell out the message to give instructions to make the protein for which it is responsible. The effect of switching genes on or off like this is called 'epigenetic'.

Epigenetic effects are quite complicated, and we are still trying to understand them. As it happens, there are different chemical additions that can be made to the DNA – some of them act on genes that are on chromosomes inherited from the father, others on genes from the mother. Genes that act preferentially on one parental chromosome rather than another are called 'imprinted genes'. 'Imprinted genes' do not obey the traditional laws of genetics. With most genes, both parental copies are equally likely to contribute to the outcome. But the way an imprinted gene will work (or 'express') depends on which parent it was inherited from. With many imprinted genes, a cell will use only the copy from the mother to make protein, and with other imprinted genes, only that from the father. And it turns out that imprinted genes are particularly important during early development; so if something goes wrong with imprinting, the fetus is likely to have abnormalities.

So, given that a cloned animal does not receive paired chromosomes, one from each parent, it may seem unsurprising that, somewhere along the process, imprinted genes do not always express normally. This may explain why certain serious abnormalities, including those of the placenta, occur during early development. But it seems that cloned animals frequently show other epigenetic effects too – with genes, though normally spelt, expressing abnormally. As it turns out, there is still a considerable lack of solid research in this whole area. Until that work is done and these changes are better understood, any human cloning is fraught with danger.

Even if people did begin to clone themselves or their loved ones, a major drawback would become apparent as soon as the clones grew up. No clone is a complete replica of its progenitor, because it carries some genes from the mitochondria in the cytoplasm of the egg into which the cell nucleus was placed. Admittedly, this is only a tiny amount of the total DNA (about 16,500 DNA letters in the mitochondria, compared with over three billion in the nucleus), but these few genes, inherited only from a mother, are important in how we handle energy (see page 261). Identical twins, because they come from the same egg, are closer to each other genetically than are clones because they share the same mitochondrial, as well as nuclear, DNA.

On top of this, there is a much greater influence – that

of nurture and the environment in which we find our-selves. If I gave a cell of mine to be cloned into an egg tomorrow, and the resulting embryo grew into a baby, that baby – born sixty-five years later, from a different uterus, subjected to different nutritional and hormonal influences before birth and delivered in a different environment – would be very different from me. Once born, it would have different mother's milk, would experience different influences at different times during development, would eat a different breakfast, go to a different nursery, see television programmes that I never saw as a child, and go to a different school in a totally different society. That child might resemble me in some ways, but it would be a quite dissimilar person. Ira Levin's thriller *The Boys from Brazil* tells the fable of the exiled Nazi scientist Josef Mengele hiding himself in the wilds of Paraguay, and cloning a number of boys from Hitler's bodily remains. But no matter how those children were brought up, none of them would be a replica of the Führer. Indeed, it is an interesting thought to consider that an Adolf Hitler clone born a few decades after his death, breast-fed lovingly, cared for with gentle affection, given self-esteem and educated at a decent school, might end up as quite a pleasant, if strict, supermarket manager, or an affa-ble and somewhat disorganized stationmaster in Vienna's main railway terminus.

Why clone?

So what is the purpose of making cloned animals and humans? At the time Dolly the sheep was born, a number of possibilities were suggested. Some people thought cattle that provided good meat, or gave good milk yields, might usefully be cloned to improve livestock. Another suggestion was that cloned animals, after genetic modification, might be useful to provide specific proteins or medicines. For example, a cloned cow that has also been genetically modified could be used to produce blood-clotting factors in its milk for haemophiliacs, or various protein-based pharmaceuticals. Another argument for cloning was that it might be possible to breed animals whose organs were suitable for human transplantation. The pig, having many organs of similar size and shape to those in humans, might be an ideal candidate. Given that there is a huge shortage of organs for donation – perhaps only one-fifth of the people who are ill and need organs that would save their lives receive them – using a suitable animal for the purpose would seem to be a worthwhile goal. So researchers have tried to modify the genes of cloned pigs in the hope that, by altering certain proteins the pig produces, they might avoid its liver, heart or kidneys being rejected by the human immune system. It has also been suggested that the study of animal clones might also give better animal models of diseases. If a series of identical animals are given a standard or varying treatment, then the statistical analysis of their response to treatment may give insights into improving that therapy. Finally, if perhaps paradoxically, cloning could actually be used to enhance biodiversity. A number of rare species (the Tasmanian Tiger and the White Rhino have been cited, among others) are currently endangered or likely to die out, and cloning might ensure their survival.

Each of these justifications for cloning is flawed. Rather than improving herds of livestock, cloning might be disastrous. It certainly would be a risky technology for these purposes, because flocks or herds that are not genetically diverse would be much more prone to being wiped out by a single catastrophic disease or infection. Nor am I convinced that using cloned animals to produce particular pharmaceuticals or proteins is a sound strategy. Using other forms of genetic modification, for example transgenic animals (see page 476), would be easier, once we can simplify the technology to produce large transgenic mammals or birds.

I remain profoundly unconvinced, too, about making cloned animals as organ donors. If cloning does produce abnormalities of gene expression or premature ageing, then transplanting an organ from such an animal might be catastrophic for the recipient. The organ would have the genes of its donor. Genes that did not function properly or that caused the failure of an organ after transplantation would spell medical chaos for the recipient. Nor are clones good models for human disease. The use of identical mice as models can already be achieved by careful inbreeding, and it is doubtful that clones would give much additional advantage.

The many claims that cloning would be useful to protect endangered species are equally unconvincing. Seemingly advanced by scientists trying to find some justification for their work, they merely give the impression of a sop offered to environmentalists who are, quite reasonably, querulous about humans manipulating animal genetics. Could cloning really protect the rhino? Is it seriously likely we could recreate a woolly mammoth from frozen skin cells stored in the icy tundra somewhere on a remote hillside? If a cloned embryo were produced, what animal could be the surrogate mother? It seems unlikely that such an embryo would 'take' inside the uterus of a modern African or Indian elephant. The preservation of endangered species is much more likely to result from sensible and responsible use of this planet's resources and more recognition of the importance of maintaining different species' native habitats. If technological solutions are to be used at all, then carefully applied 'low' technology – better understanding of breeding, perhaps the use of properly run zoos, controlled and sensible release of species into the wild – is likely to be more valuable. If 'high' technology is needed, then there may be a case for embryo transfer, embryo freezing and other sophisticated techniques. Cloning will almost certainly remain in the domain of fictional works such as Michael Crichton's *Jurassic Park*.

Is there any place for human cloning?

It has been claimed that no fewer than 8 per cent of Americans would like to be cloned if the technology were available. Of course, human cloning should be beyond consideration at present, given the risks – outlined above – of producing a damaged fetus. So it is interesting to consider the boasts of Antinori or Zavos. If they cloned a human now (so called reproductive cloning) they would risk the strong probability of a human child with serious abnormalities. Leaving aside the moral argument entirely for a moment, imagine the consequences for them. They would inevitably be sued, and it is almost certain that they would lose their case in a court of law – even if an adult parent had given 'informed consent', signed waivers, or provided any other protective instrument. Which expert witnesses would come to their defence when the full weight of serious scientific opinion is absolutely condemnatory of them?

These advocates of cloning might consider one other matter. Between 2 and 3 per cent of human babies are born with abnormalities; perhaps another 25 per cent of pregnancies miscarry. Were they to be sued by one of their disgruntled patients because of a miscarriage or fetal abnormality, they would almost certainly lose in a court of law because so few experts would be prepared to say that such an outcome was definitely not the result of the cloning 'experiment'. Given their massive financial liability, it seems unlikely that any underwriter would be prepared to offer them indemnity insurance beforehand.

Suppose cloning was safe, however, and there was no evidence that the risk was any greater than with any other reproductive technology. Under what circumstances might cloning be useful?

Dr Antinori argues that he wants to undertake human cloning to allow infertile people to have children. It is true that there might be a very limited place for cloning when a man or woman cannot produce germ cells, and therefore cannot make an embryo with their own genetic background. So a man not producing sperm or a woman with no eggs might justifiably consider cloning. Theoretically, this could be marginally better than accepting genetic material from a sperm donor, and it may be that in the fullness of time our society comes to see this as an acceptable treatment. The other arguments for cloning – for example, to replace a lost child who has died prematurely or a deceased partner – seem very dubious. Bereavement cannot be resolved in this way, and a clone may turn out to be a severe disappointment. As such a clone developed, he or she would inevitably acquire divergent tastes, interests and abilities. Imagine the predicament of a grieving parent who had cloned a dead child in order to assuage his or her loss. Then consider the bitter disappointment when Johnny version II had no interest in football, or preferred his stamp collection to a girlfriend –

turning out, in fact, to be a horrifying apparition of the Johnny version I so sorely missed: physically his double, but mentally someone else.

Almost the only other motivation for cloning would be to reproduce oneself. I suppose people who are troubled by their own mortality might be pleased to think of their clone living on after them, but it seems unlikely that this procedure would really alleviate their sense of personal inadequacy – or bolster their feelings of personal greatness. Imagine an heirless tycoon who clones himself to create a son – who in turn shows more interest in saving the planet than playing the stock market, or isn't even particularly bright.

The unfair pressure on these clones, brought to life in order to recreate someone else, living in someone else's image, in their perpetual shadow, could be considerable. Ethically, such a practice would be indefensible because of the possible risks of psychological harm to any child. Opponents of cloning suggest that any child born after such a procedure might feel that they had only limited personal autonomy and were in thrall to the person from whom they had been cloned. Many people who are deeply opposed to cloning further argue that cloning might encourage parents to value children as a kind of commodity, instead of loving them for their own sake.

Another objection to cloning is that it could lead to a form of eugenics. Thus some totalitarian state might decide to clone aggressive, strong, violent people to provide the invincible army of the future. Currently, though, some aggressive army units do a pretty good job with the form of engineering already available and applied to the eighteen-year-old recruits who are attracted to the armed forces. Military training is almost certainly much more efficient than cloning will ever be, because it does not take eighteen to twenty years to establish the army a fascist government might need. It might also be useful to consider where a government would manage to recruit all those women it would need to act as the surrogate mothers for such a fighting force.

A fairly commonly expressed argument against all forms of reproductive manipulation is that their use is a misapplication of extremely scarce resources. It is often reiterated that cloning and like technologies divert limited resources from more pressing social and medical needs. Medical procedures should be used, it is argued, for projects or health care that are most likely to improve the health of the public. This is clearly true; but I do not think that cloning practitioners would be likely to offer their services in the public sector.

Nevertheless, I find myself genuinely doubtful about nearly all the arguments against reproductive cloning. In Britain there are around 25,000 human clones living quite happily already. They are not disadvantaged, they do not find their autonomy threatened, and they certainly do not lack human dignity simply because they have precisely the same genes as an identical twin brother or sister. They would surely argue that their being a twin does not raise any of the fears that are associated with a golem.

But for the foreseeable future there is an excellent reason not to contemplate human reproductive cloning – that very real risk of producing a seriously damaged infant. It is likely to remain an objection for a long time to come; and, as we shall see in the next chapter, it may have an important impact on the way we manipulate human embryonic stem cells.

Stem Cells: Nature's Magic

In 1666 a giant shark was caught off the coast of Livorno, in Italy, and the fishermen – scared witless by this thrashing monster from the deep and worried about the small size of the deck of their boat – bludgeoned it to death. They landed the shark, and it so happened that Nicolas Steno, a Danish scientist living in northern Italy at the time, heard of the catch. After a short piece of bartering, he obtained its severed head. Steno was an expert anatomist, best known for dissections and anatomical studies of fish. He subsequently demonstrated his surgical prowess by a highly smelly dissection of this enormous but unappetizing fish-head in front of the assembled courtiers in Florence. Once this display was concluded (with considerable fortitude being shown not only by Steno, but by the bystanders too) he presented a publication of his study to the Grand Duke Ferdinando for the potentate's amusement.

So began the first case of aggressive competition between scientists studying reproduction – and it has echoes in some of the conduct of science today. The support of Duke Ferdinando led to Steno being appointed as the court scientist for other dissections. Steno soon noted that the females of not only fish, but animals that had live-born offspring, also had 'testicles'. He decided correctly that there are real structures, eggs, which originate in the female 'testicle', and that they pass from there into the uterus. Steno had a Dutch friend, Jan Swammerdam (1637–80), who studied insects and frogs. His meticulous research on frogs included boiling their eggs to make them easier to dissect, and making careful drawings of the egg. He showed an egg dividing into two separate cells after fertilization – the fundamental process at the start of vertebrate life.

Swammerdam was convinced he had already seen human eggs before he heard of Steno's theoretical proposition. In the middle of graduation week at Leiden, in 1667, Swammerdam had been asked by his tutor, Professor Johannes van Horne, to come to help him with a private dissection in his own house of a human female corpse. Van Horne partly needed his pupil's dissection and drawing skills, and partly wanted him to join him in publishing the findings. During the dissection both men became convinced that the ovaries they saw looked similar to those in other animals. Moreover, they thought they could identify eggs inside them, or at least vesicles that contained eggs. After the dissection, and once the graduation ceremonies were over, Swammerdam heard nothing more. Van Horne, in spite of his promises, had not prepared a manuscript and publication was not imminent. Then, suddenly, some nine months later, Swammerdam had an urgent note from van Horne asking him again to come and help him with a dissection, this time of 'a delirious virgin who has drowned herself'.[1] Then came the bombshell. During a royal visit to Leiden Swammerdam learnt that, while he and van Horne had been dithering about producing a manuscript, Steno had published his account of the reproductive system of the shark, recognizing the importance of its ovaries.

Swammerdam was furious with van Horne and critical of his tardiness in publishing. Now Steno would get the credit for all the effort that the pair of them had made. But Steno was all sweetness and light. Writing to Swammerdam, he said:

> Dearest Friend, I await with great expectation the observation of yourself and the most distinguished Doctor Van Horne on the testicle, and I am far from aggrieved that I have been preceded in this matter by my Teacher and friend, and will solemnly swear that had I known it, I would not only have mentioned him by name but would have made public his observations ...

Meanwhile, another friend, former fellow pupil and most ambitious colleague of Swammerdam's, Regnier de Graaf (1641–73), a physician from Delft, had been doing extensive research in between seeing his sick patients. De Graaf had heard all about van Horne's unpublished dissections – indeed, Swammerdam had told him. But de Graaf had himself been working on dissections of the reproductive systems of female rabbits, hares, cows, sheep and pigs as well as humans, and felt certain he had established that embryos came from eggs. So, to pre-empt his former professor, de Graaf published a 1,500-word précis of the book he was to publish, but which was not yet ready to be printed. Rather in the way I might apply for a patent to get priority on an idea, de Graaf sent this précis to another former teacher, Dr Franciscus Sylvius, who was also at Leiden University. A few weeks later, van Horne read the précis as it came off the university presses. He was horrified: as Matthew Cobb says in his book, he was about to be upstaged by a previous pupil for the second time in just a few weeks. He tried to protect his intellectual property by getting depositions from his collaborators, but the damage was done.

Accusations of plagiarism, lying and stealing other people's discoveries followed. The Royal Society in London reviewed the evidence and said Steno had published first. There was huge bad feeling. Johannes van Horne never did publish his pinnacle achievement. In 1670 he suddenly died following an acute contagious illness that affected many people in Leiden. He perished before he could publish. Meanwhile de Graaf, aggressively accusing 'spiteful fellows' that might 'snatch from me my glory', published *The Generative Organs of Men* in 1668. He followed this with his 1672 work *The Generative Organs of Women*, which was of critical importance at the time. In it, he described follicles, the fluid-filled blisters surrounded by cells, on the surface of the ovary. These, he noted, grow from microscopic size at the start of the menstrual cycle. However, with the primitive microscope at his disposal he was unable actually to see a human egg, so remained unaware that each follicle contains just one egg, to be released during ovulation.

De Graaf was competitive, aggressive and hot-tempered. Sadly, these attributes are often still an advantage in the testosterone-driven world that is modern science. The story is interesting because it pinpoints the moment when science became uncompromisingly cut-throat. Though there was generally little financial reward for being first to make a scientific discovery, there was – and still is – immense prestige at stake; and of course, then as now, with fame came funding. Here were three friends and their revered teacher, who at one point had studied together, all of whom had postulated the most revolutionary idea in biology – the idea that mammals had eggs – and each of whom felt certain he had been the first to observe it. Nicolas Steno, at least, was gracious enough to back down on de Graaf's publication. His other Dutch contemporary was not so magnanimous. Swammerdam persistently claimed to have identified ovarian follicles before de Graaf. Nevertheless, that ephemeral structure, a fluid-filled cyst 2 centimetres across, seen at full size for only a few hours before ovulation, is now called a Graafian follicle. Bitter feelings between de Graaf and Swammerdam continued to simmer. De Graaf published a rebuttal in his defence, but went on brooding over what he was convinced were unfair allegations – and in so doing allegedly contributed to his own premature death at the age of thirty-two.

Stem cells: matter of infinite possibility

With the discovery of the egg came renewed interest in Aristotle's notion of pre-formation. The proposition was that each egg had a miniature organism inside it with all its organs developed; conception just kick-started its growth. Moreover, the tiny creature – if female – would have its own eggs already inside it. Inside those eggs would be more minuscule organisms, and so on.

The advent of microscopes had convinced many of the best scientists of the age of the truth of pre-formation. Swammerdam had reported seeing pre-formed structures inside insect eggs, while Marcello Malpighi (1628–94) had reluctantly supported pre-formationism after observing what looked like embryos in unfertilized eggs. The microscope had another impact: it made it possible to imagine minute structures. Now there was no reason to suppose that organisms could not get infinitely small. The theory was appealing for religious reasons – it meant that God could have created all future generations of humans at the same time. This seems a good biological metaphor for stem cells, which, as we shall see, have become the focus of controversy and competitiveness in today's scientific race to be first with the big discovery.

Stem cells are the precursors, or parents, of other, more specialized cells. The most obvious precursor cells are the undifferentiated cells in an embryo at the very beginning of development, a few days after fertilization. Eventually these cells, which lack any specific function of their own, give rise to around 210 different cell types,[2] forming skin, fat, neurons, red blood cells and so on as the fetus develops. For this reason, embryos are considered a potentially truly magical resource. The ability of these early embryonic cells to generate any kind of specialized cell, and therefore any kind of tissue, means that they hold unique promise. If we could harvest these powers by working out how nature performs this truly magical trick, we could grow all kinds of replacement tissues to treat many serious diseases. So, over the past decade or so, scientists have focused increasingly on trying to find the factors that command stem cells to grow into particular cell types. Some progress has been made and now, in the laboratory at least, it is possible to persuade embryonic stem cells to grow into a range of different cell types.

Embryonic stem cells are derived from the inner cell mass of the blastocyst. The inner cell mass is that part of the embryo which becomes the fetus (the outer cell mass becomes the placenta and membranes). To obtain these cells, the embryo is treated chemically and the inner cell mass cells separated. The cells are then grown in culture in contact with 'feeder' cells and dosed with certain proteins that stimulate development of stem cells. True embryonic stem cells have the unique property of being 'immortal'; once a stem-cell culture (or 'line') of this kind is obtained it will divide and grow indefinitely. These cells can then be treated with different compounds so that they can be enticed to grow, for example, into nerve cells, bone or beating heart muscle.

Growing stem cells from the embryos of any mammal species is difficult, and growing human stem cells in this way is a really complex task. Often, cells that look like stem cells are taken from embryos and grow in the culture dish but, when they are carefully examined, clearly do not meet the true criteria to prove they really are stem cells. So a number of tests must be done to confirm that stem cells are really present. First, they must be shown to be 'immortal' – dividing continuously. They can also be tested to see if they are expressing the genes – making the proteins – that are specific to stem cells. Another test is to inject some of these cultured cells into an immune-deficient mouse (which, being immune-deficient, will not reject the cells) to see what tissues are produced. These injections may be made into a mouse embryo or into an adult mouse. Injected into an embryo, they should form a chimaera which contains the whole range of different cell types; injected into an adult, true stem cells form a growth that contains a tumour with many different cell types in it.

The use of embryos as a convenience – simply as a factory for human spare parts – is controversial, and so a vigorous search for alternative sources of stem cells has continued. It turns out that there are many possibilities. For many years, it has been well recognized that the bone marrow contains stem cells. All blood cells – red cells, white cells and platelets – have a short life. Red blood cells, for example, last only about three weeks before they die and are removed by the liver. So the red cells have to be constantly replenished in their millions; and this is accomplished by the stem cells in the bone marrow, which can develop into all the different kinds of cell that are present in the bloodstream.

This remarkable potential of bone marrow has been exploited for many years. It saves lives. The treatment of childhood leukaemia often depends on it. Leukaemia is a disease where some of the white blood cells – perhaps because of a mutation in a particular kind of blood stem cell – grow in abundance and swamp the other blood cell types. Killing these cancerous cells with chemotherapy alone inevitably kills other blood cell types as well. So one strategy is to destroy all the patient's bone marrow cells completely, and then replace them. Usually this is done by chemotherapy, sometimes in combination with radiation. The patient may be exposed to a very high dose of X-rays – quite like being fairly close to a nuclear explosion but without the bang. This destroys all the bone marrow and the stem cells, which are more sensitive to X-rays than other cells. Left untreated at this stage, the patient would rapidly die from anaemia, through loss of red cells, and infection, through loss of the white cells which maintain the immune system. So a transfusion of bone marrow, containing new stem cells, is given from a compatible donor. This is usually a close relative to ensure as good a genetic match as possible and so avoid rejection reactions. The new blood stem cells repopulate the hollow bones and start to form new red and white blood cells and platelets. These new stem cells will continue to replenish the blood system for the rest of the life of the leukaemia victim.

There are other tissues that are rich in stem cells. We constantly lose the most superficial layer of skin, and these cells need to be continuously replaced to keep the skin in good health. This repair is carried out by the stem cells in deeper layers of the dermis. Stem cells have also been identified in relatively large numbers in muscles, the liver tissue and the brain, as well as in several other organs around the body.

Usually, the stem cells that grow in adult tissues like the marrow or skin are programmed to differentiate only into that specific kind of tissue. But there is intriguing recent evidence that many of these 'adult' stem cells have wider potential. For example, Professor Nick Wright, researching at my own hospital in 2000, found that, after bone marrow transplantation, some of the transplanted bone marrow stem cells had grown into tissues not connected with the marrow. In a fascinating study, he and colleagues examined the livers of nine female patients who had received bone marrow from male donors.[3] What the researchers found was that a number of the hepatocytes – liver cells – had a Y-chromosome, strongly suggesting that the parent cells must have been bone marrow stem cells from the male donor. Since then other studies have found evidence that, after transplants, bone marrow cells can be found in the brain, the heart and the walls of blood vessels. Moreover, after experimental damage in mice, research workers in Milan transplanted bone marrow cells into the bloodstream. They migrated to the site of injury and helped its repair.[4] In the 1990s, another study took bone marrow cells and injected them into mice that had been exposed to radiation and had suffered damage to organs and tissues. Like some microscopic emergency team, these cells appeared to migrate to the places where they were most required and repair them – becoming skin, lung, intestine, kidney, liver, pancreas, muscle, blood vessels, heart and brain. A team at Sweden's Karolinska

Institute reported blood cells turning into lung cells; at Yale, bone marrow cells apparently became lung cells; at Iowa, skin stem cells apparently became bone marrow.

So it is now clear that adult stem cells do have considerable potential to grow into types of tissue other than that from which those stem cells were derived. The burning scientific question is whether some adult stem cells have a similar potential to embryonic stem cells, which are thought to be much more 'plastic'. This 'plasticity' is highly important, as it seems that most adult stem cells are limited in the range of cell types into which they can turn. So, for example, *multipotent* stem cells have the ability to become a variety of cells, but for the most part within a particular organ. We find them not in every part of the body, but mainly within the bone marrow, skin, retina, brain and muscles, and inside the teeth. *Pluripotent* stem cells can develop into virtually any tissue, and at present there are only two known sources of these – embryos and germ cells. *Totipotent stem cells*, however, can form not only all the cells found anywhere in the body, but also cells in the extra-embryonic tissues – the placenta, and the membranes that surround the fetus in the uterus. Totipotent cells may be derived from very early embryos within a few days of fertilization.

Sources and uses of stem cells

Stem cells are not quite the modern-day equivalent of Rabbi Loew's magic golem-generating formula, or a scientist's Holy Grail. There are limitations in how they develop, and these are still being discovered. It seems that the 'younger' the stem cell, the more potential it has. So an embryo is the most promising source. Recently there has been increasing interest in stem cells found in the baby's cord blood at birth. Stem cells can definitely be found in the cord blood, and as the umbilical cord and placenta are normally just thrown away, incinerated or put on rose bushes as fertilizer, their use would be un-controversial. This is why a number of women are arranging to have their babies' cord blood frozen and stored in liquid nitrogen, in case treatment of an illness in later years may be helped by the use of their own stem cells.[5]

Another source of stem cells is the germinal ridges in the fetus. During early development, germ cells, which will eventually become the sperm or eggs, stream into the fetus from outside the embryo through the umbilical cord. These cells settle in two tiny structures known as the germinal ridges; these structures will eventually give rise to the ovary or testis after further development. Brigid Hogan, working in Tennessee with mice, has demon-strated that these germ cells are pluripotent stem cells. The application of this particular discovery in health care is complex and fraught with ethical difficulty. A huge number of fetuses are destroyed by abortion each year in the UK and many more are lost by miscarriage. Altogether, probably around four hundred thousand fetuses are discarded; and, although the fact may not be easy for some people to come to terms with, most of these are thrown away and destroyed by incineration. Some of these fetuses – perhaps one-sixth – could yield sufficiently fresh tissue at the right stage of development for the germinal ridges to be 'harvested'. Admittedly, identifi-cation and isolation of the germinal ridge cells is difficult. But this represents a huge amount of material which, if treated appropriately, could give rise to banks of stem cells that could be used to treat an extensive range of crippling diseases. I personally do not see why we should not use discarded human material in this way, provided proper consent is given by the parents. But the suggestion that this tissue might be exploited causes outrage among many people who see in such use a threat to our moral values.

The properties of embryonic stem cells are truly remarkable and seem unique. There is growing evidence that they can be directed towards growing into cell types that might be tailored to meet an individual patient's needs. There is also increasing evidence that, with various forms of genetic manipulation, they could be grown into tissues that would not be rejected by the recipient's immune system. The promise is that they could be injected to repair the heart after a heart attack has killed muscle, used to produce insulin-secreting cells to cure diabetes or skin grafts for burns patients, employed to prevent people dying of liver failure or injected into the brain to combat some serious neurological disorders.

One of the earliest uses of embryonic stem cells was in work done in Sweden in the 1980s, when researchers injected nerve cells taken from a fetus at about six to eight weeks' gestation into the brains of patients with advanced Parkinson's disease. Parkinson's is a horrible neurological disorder, causing increasing paralysis, tremor, limb stiffness, loss of balance and difficulty in walking. As the disease advances it becomes less responsive to drug treatment. Speech, swallowing, writing and, often, even thinking become difficult. In the advanced stages many sufferers are completely frozen stiff, hardly able to move and confined to their beds. It is a very common disorder, affecting about one in two hundred elderly people; in the UK, it is estimated that there may be as many as sixty thousand new cases each year. The disease is caused by a loss of a particular kind of nerve cell, the dopamine-producing nerve cells in one part of the brain, the substantia nigra. In most cases its cause is unknown. Occasionally, it seems, it can be produced by brain injury – as is probably true in the case of Mohammed Ali, the world heavyweight champion boxer, who, like all boxers – no matter how good or skilled – received repeated concussive blows to the head.

In advanced cases of Parkinson's, drugs and surgical treatment to the brain become increasingly ineffective. The Swedish researchers hit on the idea of transplanting immature cells dissected from the brains of aborted fetuses. It was hoped that these might mature to produce dopamine-producing cells in the brain in the area of the substantia nigra. Amazingly, the transplant did seem to give relief to some sufferers. A number of these men and women were investigated at Hammersmith Hospital some years after the fetal cell transplant, and the use of the positron emission (PET) scanner showed that in some of them there was an increase in dopamine secretion in the cells in the substantia nigra. This was associated with a noticeable improvement in their symptoms. But the effect was patchy. In some cases, this remarkable relief was short-lived; others got no better, and some became even worse. So there has been considerable doubt about the effectiveness of this kind of fetal cell transplant. Eventually, further trials, comparing fetal cell injection with an injection of sterile water (a so-called sham injury), suggested that any needling of the brain would have been as beneficial. However, it remains the case that some patients did significantly better than others with the fetal cell therapy; so it may be that the precise part of the brain from which these cells were derived, or how they were purified, may have made a big difference. Nor do we know how many stem cells were present in the grafts. One very difficult problem with these fetal transplants was that material from at least seven, possibly more, aborted human fetuses was needed to get sufficient cells to treat just one human patient. This heightened the concerns that many people had about harvesting cells from aborted fetuses cells for such treatments.

Following this very early experimental work, it was natural that much scientific attention came to focus on embryonic stem cells as a means of generating dopaminergic nerve cells. The American actor Michael J. Fox, himself suffering from Parkinson's, became a passionate advocate of embryo research and was highly supportive of the work done by biologists at Harvard. In one series of experiments there, researchers injected embryonic stem cells into rats whose dopamine cells had been destroyed. Of the twenty-five injected, fourteen began to produce dopamine and showed an improvement in their symptoms. But at that stage it became apparent that injection of these embryonic stem cells carried a major drawback. Placed inside the body, they had a tendency to form tumours – not exactly a plus for people also suffering with a brain disease. More recently, Dr Ron McKay and his colleagues in Bethesda in the USA have grown and purified dopaminergic neurons from embryonic stem cells in the laboratory, and then transplanted them into the brains of rats with a condition resembling Parkinson's disease.[6] Before the transplant, these rats had difficulty in walking; afterwards, they showed marked improvement, and the transplanted cells were shown to be producing the neurotransmitter dopamine in large quantities. Dr McKay did not report any evidence of tumour formation. More recently still, Dr Yasushi Takagi, working in Kyoto, has made a similar advance in monkeys, and the prospects of generating neurons from embryonic stem cells to treat this debilitating condition grow increasingly promising.[7]

As I write this, the world is agog with the news that a laboratory based in North Carolina has successfully grown pieces of human bladder from adult stem cells and transplanted them successfully into seven patients aged between seven and nineteen.[8] Some of these transplants have now lasted five years. In his laboratory at Research Triangle Park, Anthony Atala has been able to grow uteri, vaginas, blood vessels, kidneys and even an apparently fully functioning rabbit's penis – culturing cells to grow around a scaffold made from nylon or collagen. His privately funded work uses a mixture of adult and embryonic stem-cell tissue and, it is claimed, opens up the prospect that, one day, the miseries of the transplant waiting list may be over. The main source of organ donors – healthy young men killed in car and motorbike crashes – has been dwindling in the west for a long time. So most patients in need of a heart, lung or liver transplant have to wait for somebody else to die. In addition to that, those who receive a transplant will have to take debilitating immuno-suppressant drugs for the remainder of their lives. For many people, Atala's technology raises the prospect that people could receive organs cultured from their own cells.[9] This means that they would have to wait only as long as it took to grow the organ, with no risk of their body rejecting it.

But a strong word of caution is needed. Growing any organ in culture is very much at the limit of what is currently feasible. It may just be possible to replace a complete bladder, but the problem of the organ's blood supply remains, and at present that seems likely to pose continuing difficulty. A large organ transplanted without an adequate supply of oxygen will die. Also, growing big organs in culture will take a considerable amount of time – many weeks or months, at least. A seriously ill patient on a list for a heart transplant, for example, may not be able to wait so long with his or her own organ failing. Moreover, a bladder is a relatively simple organ; it is most likely to be much more difficult to grow, say, a kidney. Apart from its highly complex blood supply, the kidney has a delicate and complex set of cells of many different types, and a multiplicity of exquisitely organized collecting tubes for purifying the blood and for producing urine. Growing an organ that complex in culture, while keeping it oxygenated and healthy, will continue to be a tall order.

The *Superman* star Christopher Reeve was another passionate advocate of embryonic stem-cell research – not least because he was paralysed in a horse-riding accident and hoped that, one day, this branch of science could lead to a cure. Sadly, Reeve died before his hope could be fulfilled. But it is possible that embryonic stem cells may indeed offer hope to people with spinal damage. A team at the University of California, Irvine, cultured embryonic stem cells and directed them to grow into *oligodendrocytes* – the neural cells that form a substance called *myelin*. Myelin is the fatty substance that covers the brain, enabling the cells to communicate with one another through electrochemical charges. These cells, when injected into rats that had had their spinal cords severed, helped them walk again. Timing was crucial in this experiment – rats treated up to seven days after the injury improved, but those treated two months later showed no improvement. Even so, it seems that this experiment may indicate exciting possibilities.

But once again a serious caveat is needed. Rats with

severed nerves – even those with severed spinal cords –
hold surprises. They can make an extraordinary recovery
without any treatment at all. This Californian experiment
may seem convincing, but the results need very careful
analysis. Moreover, it would be essential to examine the
effect of similar injury in other animals before coming to
firm conclusions. A key problem is that the spinal cord
carries many different kinds of nerve signals – some
convey painful messages to the brain, others sensations of
heat, cold, vibration or position in space. Other nerves
carry messages from the brain to initiate movement.
Yet others carry automatic responses – instructions to
start digestion, to sweat or to continue breathing. After
spinal cord injury there is considerable formation of scar
tissue, and it seems very unlikely that the injection of any
number of foreign stem cells will improve that healing –
and get all the nerves of all the different types in the spinal
cord joining up to each other correctly. In this respect one
may think of the spinal cord as a gigantic telephone communications cable: if it is severed, it won't work (and
different parts of the brain cannot receive and send
messages) unless the wires are reconnected in the same
configuration as before the injury. For this reason, some
researchers in the UK are trying a quite different tack.
Geoffrey Reisman, of the National Institute of Medical
Research and University College London, is attempting
what he thinks is likely to be a more successful approach.
He hopes to encourage the stem cells already in the
nervous system to recognize nerves with a particular
function and to get them to proliferate and repair an area
of damage after spinal cord injuries.

Why use embryos at all?

The discovery that stem cells taken from adult tissues can grow into different cell types has led many people, erroneously, to argue that we do not need to experiment with embryonic stem cells. At present, embryonic stem cells seem the most likely to be therapeutically useful. Moreover, they can be produced in considerable numbers and grown into colonies or banks of cells. These collections of cells provide an effective test-bed for assessing the effects of different chemicals, each of which, when added to the colonies, helps them differentiate in a particular way. One problem with adult stem cells is that they are located in inaccessible places, such as the brain and bone marrow; another is that they are in limited supply. Moreover, because they come from adults, they have suffered the attrition process of ageing, and may have a shorter lifespan than their immortal embryonic counterparts. Those who argue that adult stem cells make the use of embryonic cells unnecessary fail to recognize the information that the study of embryonic stem cells continues to give us.

There is also considerable scientific dispute over some of the most optimistic reports of adult stem-cell activity. Some participants in this often heated and politically charged debate point out that there are a number of reasons why adult stem cells may only *appear* to be as potent as embryonic ones. Many cells look similar; but to know for certain whether a stem cell has truly switched roles, it has to be observed under a microscope to see if it is creating the right proteins. Even then the results vary. Furthermore, few of the studies of adult stem-cell activity determine whether the transplanted cells actually repair damaged tissue effectively. Moreover, some studies report that bone-marrow cells that became neural cells had double the number of chromosomes. This may be because they are just fusing with other cells, rather than becoming new cell lines. If they have an abnormal number of chromosomes, these cells are most unlikely to function properly and problems will arise when they divide. A team at the University of Minnesota has discovered a type of adult cell called an MAPC (multipotent adult progenitor cell), which may just be the medical equivalent of Superman's kryptonite. These cells appear to be capable of turning into a wide variety of cells, such as neurons for the brain or endothelium, the cells that line blood vessels. But these magic cells seem extremely rare – some estimates suggest that fewer than two thousand MAPCs exist in an adult mouse. It may be that a tiny quantity of very elusive cells is actually doing the work for which a much wider category of adult stem cells is being credited. If they are (and some scientists even doubt their existence), it may be extremely hard to isolate them.

The arguments roll on, and they have become highly contentious because of the ethically fraught question of using embryos. In a country like America, where government policy and funding have long been coloured by religious views, there are strong incentives for researchers to prove that adult stem cells are just as useful as those from embryos. In turn, some proponents of embryonic stem cells have been too quick to focus on the failures of the opposing camp's work.

Any work with human embryonic stem cells does present some very difficult technical problems. For a start, these cells are not very stable: work done with human embryos in my own lab by Kate Hardy and others shows how frequently embryonic cells develop strange chromosomes, or divide abnormally. This could be a very serious problem when generating stem cells. Another problem is that embryonic stem cells divide very slowly – perhaps one cell division every twenty-four to forty-eight hours. So it may take a long time to generate sufficient stem cells to treat an individual patient. And this slow turnover implies another potential difficulty. In any culture of embryonic stem cells, there may be some cells that are dividing rather more quickly than average. Although it might be thought advantageous to encourage these cells, their faster pattern of propagation may mean that they are genetically abnormal. But because they are growing faster than the 'normal' stem cells, selection pressure may result in these fast cells outgrowing those dividing less frequently. This in turn could result in many embryonic stem-cell lines having abnormalities that might affect a patient's health if they are eventually transplanted.

It would be nice to think that embryonic stem cells could be injected directly into a patient, where (as some mouse experiments showed with the use of marrow cells) they would migrate to the site where they are needed and develop into the right specialized cell type. But experimental work has shown that embryonic stem cells injected in their undifferentiated state form tumours, and therefore this would be a highly dangerous therapeutic strategy.

So researchers are under pressure to find ways of ensuring that these stem cells are fully differentiated before they are transplanted to human patients. But even then there are risks. An injection into the brain sufficient to alleviate or cure Parkinson's disease will require more than a hundred thousand cells. No matter how carefully these cells are tested, purified, and re-tested, it would be hard to guarantee that there were no 'rogue' cells in the stem-cell culture; and it would presumably take only a single undifferentiated cell to multiply and cause a cancer for the whole treatment to go disastrously wrong.

Then there is the knotty problem of 'de-differentiation'. We may be completely successful in obtaining stem cells and directing them to differentiate into a particular kind of specialized cell type. But if there are abnormalities in the way these cells continue to divide, there is the possibility that they may become undifferentiated again, reverting to their embryonic state. We also have to confirm that differentiated stem cells – which have been artificially treated in culture to maintain their differentiated state – continue to function normally after transfer to the patient. There would be no point in going through the difficult and potentially danger-ous process of transplanting dopaminergic neurons into the brain, only to find that after a few weeks or months these cells just stop working or do not continue to produce dopamine in sufficient quantities to be useful.

In spite of all these concerns, it still seems that adult stem cells are not likely to be as useful as embryonic stem cells. The apparent plasticity of embryonic cells remains of key value. This means that, if research is to continue and we are to develop treatments, we will continue to need human embryos. At present, there is an obvious source for them: IVF treatment, because it entails the creation of some embryos that will not be transferred. Some of these unused embryos may be frozen for later use, but many are discarded. It therefore makes good sense to use them to make stem cells, enabling the development and application of treatments for hundreds of painful, debilitating and sometimes fatal conditions.

Generating stem cells by nuclear transfer

There is potentially, however, another way of obtaining

embryos to create stem cells. It is possible to collect eggs and undertake somatic cell nuclear transfer (SCNT) – the process that was used in the creation of Dolly the sheep. An egg from which the nucleus has been extracted is implanted with the nucleus taken from a suitable adult cell, tricked into believing it has become fertilized, and – with considerable hard work – we could eventually have an embryo which would, with further manipulation, yield useful stem cells. This process has been referred to as 'therapeutic cloning' – as opposed to 'reproductive cloning', which allows the embryo to develop normally inside the uterus, eventually giving rise to cloned children.

'Therapeutic cloning' like this would have certain advantages, the main one being that embryos created by nuclear transfer carry the genes of the person from whose body the nucleus used to start the process was taken. This means that any tissues derived from stem cells created from that embryo would not be rejected by that person's immune system. In effect, this process would allow the treatment of the person giving the nucleus.

However, there are horrendous difficulties to overcome with SCNT. First, in order to produce the nuclear-transferred embryo, a human egg is needed. Human eggs are in very short supply. IVF patients are naturally very reluctant to give up eggs because they feel they need all their own to ensure that they have sufficient embryos to give themselves the best chance of fertility treatment succeeding. And, as we have seen, getting egg donations from other sources is very difficult because the process of ovarian stimulation and egg collection is so demanding for most women. Payment might just be one way around the problem, but because of the risk of exploiting poorer people, many countries have banned or put strict limits on payment.

But the problems are by no means limited to the supply of eggs. We have seen how low the success rate is with any mammalian cloning procedure. Several hundred attempts at creating embryos were made before Dolly the sheep was produced. Although the process has become a little easier since then, the success rates in most species even now are well under 10 per cent; and in humans, the rate is certainly going to be lower still. Experiments with various monkey species, our nearest animal relatives, demonstrate how difficult it is to clone any primate.

And achieving the cloned embryo is only the first difficult step. Next comes assurance that it is functioning normally and producing cells with normal chromosomes in which the genes are working normally. As we have already seen, most clones have abnormalities of gene expression, problems with imprinted genes or other epigenetic changes that are likely to be deleterious. But even assuming everything is working smoothly, creating stem cells from a cloned embryo presents a whole new suite of problems. Embryonic stem cells are difficult to manufacture in any mammal, and in humans the process seems especially difficult. Many units around the world have been attempting to produce embryonic stem cells. They have been pooling all the spare embryos they obtain and still, after months of work, have not produced a single line of embryonic stem cells. Most of the successful units have taken a year or more to produce just one or two lines of stem cells. Even when these stem cells are successfully grown, and dividing happily in culture, further tests have shown that they are not completely normal: often, some genes are not expressing properly, or are over-expressing some proteins. So the combined achievement of producing a cloned embryo that is 'normal' in every respect and then deriving stem cells from it that work normally is still a very tall order.

An example of the difficulties involved in making embryonic stem cells – even without the added problems of cloning by nuclear transfer – is offered by the UK Stem Cell Bank run by the Medical Research Council. At the time of writing this cell bank has been operating for four years. By regulation, all units given approval to make embryonic stem cells in the UK (and there are around fifteen of them) are required to submit samples of all their cultured stem-cell lines so that they can be grown up for general use by any interested researcher. At present, even with all those units submitting samples from all the stem-cell cultures they have obtained, the bank has not been able to release any stem cells for general research, because none have been verified as normal.

All this means that the therapeutic applications of SCNT – though it sounds an immensely promising technique – may be very limited. Nevertheless, competition to break through these numerous difficulties to generate cells by these methods has been intense.

The pit of fame

South Korea is a country approximately one-third the size of the UK with a roughly equivalent-sized population for its total area. It is becoming increasingly wealthy – currently the average per capita income is rising fast, and GDP is about two-thirds that in the UK. This astonishing economic growth has occurred over the past thirty years as a result of the country's position at the centre of a technological boom, particularly in the communications and information technology industry. Korean science is rapidly catching up with the best research in the world, and the whole nation is very much geared to the commercial exploitation of science and technology. In the summer of 2005 it seemed as if a really major scientific breakthrough had taken place at Korea's leading university – Seoul National University. For once, Korea seemed to be leading the world in one of the most high-profile areas of science.

Professor Hwang Woo-Suk looks more like an insurance broker, or a mild-mannered solicitor, than a national hero. But national hero he had certainly become. There are few countries in the world that treat scientists the way we idolize Madonna or David Beckham, but in South Korea Dr Hwang was very much a celebrity. His origins are very humble indeed. He was born in the impoverished, hilly part of central Korea to a farming family who lived just at subsistence level. His father died when he was five years old, leaving six children and an impecunious widow who was so poor she had to borrow money to pay for the family's only cow – the centre of its very existence. The young Hwang grew very fond of the cow, and vowed when he was older to become a vet. After doing well at school, he went to the prestigious Seoul National University, qualified as a veterinary surgeon and decided to do a PhD. Over the following years he became expert in various animal reproductive techniques. He researched domestic animal infections, studied the causes of abortion in cattle, developed ICSI in pigs and examined parthenogenesis in that species, and did research on various aspects of animal embryo freezing. Then, in 1999, he hit the headlines in Korea with the announcement that he had managed to clone a cow – one of the first scientists in the world to accomplish this difficult feat, so soon after Dolly's birth. Further laudatory publicity followed when he explained how he was working towards producing a BSE-resistant cow. And by 2002 he said he had cloned other animal species, including a pig.

Soon Dr Hwang had built up a very sizeable research group and rapidly published a string of papers aiming to improve animal husbandry and various animal IVF techniques. By 2003 he was a national icon – and then came his most amazing announcement to date. In February 2004 he stated that he had managed another first. He had done what every other researcher had found impossible. Having experimented with over two hundred human eggs, he had managed to clone a human embryo and derive stem cells from it. The leading periodical *Science* published the details and Hwang was launched into orbit. It is true that a few months later there were some slightly disquieting reports in *Nature*, in which it was alleged that two of his female laboratory colleagues had donated eggs for these experiments – but Hwang denied this, pointing out that he had already stated in *Science* that all his eggs came from IVF patients, who had donated with informed consent. In any event, nobody took these rumours very seriously, for Dr Hwang's stock was rising so high that he was now fêted everywhere. Foreign scientists treated him with huge respect and he was greeted with obsequious deference at international meetings. The noted biologist Dr Gerald Schatten, from Pittsburgh, added an important feather to Hwang's cap by starting a collaboration with him. It seems that Schatten has a penchant for numerous collaborations all around the world – in the past three years or so, he has appeared as senior author not only on publications from his own university in Pittsburgh, but also on others from Oregon, Washington, Providence, and Emory University, and on yet more overseas, from Sendai in Japan, Birmingham in the UK, Santiago in Chile and Buenos Aires. Now his name would start appearing on

Hwang's high-profile papers. The Korean government and national media, of course, were delighted that local work done in Seoul was truly competitive in world terms.

Then, in May 2005, Hwang made an announcement of truly Nobel Prize status.[10] He reported in *Science* that he had created eleven separate lines of human embryonic stem cells by nuclear transfer. Remarkably, he had used only 185 eggs,[11] so in just one year he had improved his success rate in getting stem cells fourteen-fold – at a time when most European workers were happy to get a single cell line in a year's work and were not even deriving embryos by cloning. Almost more remarkable still was that the announcement that each of the nuclei that had kick-started these eggs into development had come from a real patient who needed stem cells. One patient suffered with a genetic immunodeficiency disease, others had spinal cord injury, some had juvenile diabetes. So Hwang was proposing that, in due course, these three diseases might be treated by transplantation of stem cells. Because the nucleus that had initiated the growth of these embryonic stem cells had come from a patient, all the cells derived from this process would be genetically matched to that particular patient, and would not be rejected by his or her immune system.

Hardly surprisingly, perhaps, Hwang was treated like a film star. Women supporters formed a fan club in his honour, devoting a website to him. Television stations wanted him on chat shows. The Korean post office issued a commemorative postage stamp in his honour, and in June 2005 the South Korean government named Hwang the nation's first 'top scientist', conferring on him a grant of 3 billion won (about £1.7 million) annually for five years, accompanied by an offer to fund two more laboratories at a cost of around £30 million.

Then, just over a month later, the hardworking Hwang (who claimed that he got to the lab every morning at 6.00 a.m., left every night at midnight and only saw his wife when she was asleep) rushed in with another scientific first. He announced the delivery of Snuppy: an Afghan hound of a beautiful brown and gold colour, and the first dog in the world to be cloned.[12] Gerald Schatten, his collaborator, was pleased to observe that Snuppy was just 'a frisky, healthy, normal, rambunctious puppy'.[13] And why clone a dog? According to Hwang, 'Dogs share physiological characteristics with humans. A lot of diseases that occur in dogs can be directly transferred to humans.' I do not think Hwang was referring to rabies; what I think he was surely trying to imply was that the dog might be a good model for studying human disease.[14] And why call him Snuppy? Easy: Snuppy is short for Seoul National University puppy.

In October 2005 Hwang – at the peak of his success – announced that he was setting up an international consortium on stem-cell research. As part of this project he opened the World Stem Cell Hub, which was intended to act rather like the MRC's Stem Cell Bank in the UK, as an international repository. He also announced plans to open an international network with his first local branches in the United States and the United Kingdom.

But by this time there were increasingly disquieting rumours that just would not go away. The *Korean Times* alleged that some of the eggs that had been obtained for Hwang's research had been bought – in spite of his clear statements to the contrary in his scientific publications. Some people said that certain women had been paid $1,400. Selling human tissues is illegal in Korea. Soon, there were even worse rumours – it seemed that some of the younger women in the lab might have been aggressively coerced into submitting to ovarian stimulation and donating their eggs, and one or two of them had clearly felt that their further promotion might depend on their co-operation. Before long, certain associates of Hwang were sensing that the ship was sinking. In November 2005 Gerry Schatten pulled out of a partnership with Hwang, citing his reason as questionable ethical practices in obtaining the donor eggs. Gerry seems to have felt so let down by his friend that he confessed to a *Washington Post* reporter, 'I now have information that leads me to believe he had misled me. My trust has been shaken. I am sick at heart. I am not going to be able to collaborate with Woo-Suk.' In the same month Hwang apologized for telling untruths about his research workers' eggs and said that he did not realize that his lab had paid other women for their eggs. In view of the growing concerns that were gathering around him, he seems to have decided at this stage to resign as head of the World Stem Cell Hub. But in spite of the growing disquiet, his fan club stayed loyal, with over five hundred women offering to donate their eggs to his lab.

One month later, Schatten asked *Science* to remove him as the senior author of Hwang's May 2005 report, suggesting that some elements of the paper might have been fabricated.[15] Then one of the co-authors of the paper,[16] Dr Roh Sung-Il, a hospital administrator and specialist in fertility studies, maintained that nine of the reported eleven cell lines had been faked and that some of the photographs that had been taken, and published, were doctored.

From then on it was downhill all the way. Seoul University launched an investigation, as did the two key journals, *Science* and *Nature*. Hwang then admitted he had had only eight cell lines when the 2005 article had been submitted and asked *Science* to withdraw the paper. He also suggested that someone unknown to him in the lab had committed the fraud and that some of the embryonic stem cells had been switched round. Soon, Seoul National University found evidence suggesting that at least nine of the eleven cell lines were bogus. By Christmas, Hwang was saying he would resign his professorship. It later turned out that all the cell lines were false.

On 12 January 2006 Dr Hwang appeared at a televised news conference and apologized abjectly. Behind him stood twenty members of his research team, all looking deeply uncomfortable. Bowing low before the cameras, he said: 'I feel so crushed and humiliated that I hardly have the energy to say I am sorry.' Crying, he said, 'I seek your forgiveness.'

But the awful slide into the pit did not end there. Almost exactly a year after his dramatic publication in *Science*, Hwang was indicted for allegedly accepting $2.1 million in private donations based on the outcome of the falsified research. He was also accused of embezzling about $831,000 in private and government research funds, and of buying human eggs for research, a violation of the country's bioethics law. His defence rang rather hollow: 'It was clearly my mistake. I was not involved in the process, I just received results. I fully trusted [my colleagues] without doubts, but it was clearly my mistake to approve the results.'

I have told this sad story in some detail because it gives a flavour of the aggressively competitive, testosterone-driven atmosphere that surrounds so much research in all universities – particularly reproductive research. The stakes are high and the rewards are great. The researchers needed to impress. International scientists sought justification for being involved in a controversial area, and these events suggest that an international collaborator may have wanted to extend his influence. A leading journal seems largely to have seen a remarkable success and not to have checked its facts adequately or made completely sure of its peer-review process. Government and university saw the advantages of commercial application of a new technology, and national pride inhibited serious analysis of what had been done. This is not just a Korean problem, but I suppose that any country which emphasizes the commercial value of its science to so great a degree may be especially prone to this sort of terrible incident.

We should make no mistake. In many ways Dr Hwang is a tragic figure and he is probably now suffering terribly. But the biggest losers in all of this are, of course, patients. Perhaps one of the worst aspects of the whole case is the harm it has done to science – and more particularly to the ill people who might benefit from the responsible application of that science. Research into stem-cell biology has certainly been damaged, and the public will henceforth be even more sceptical about the good faith of scientists. And, sadly, the image of IVF technology will be a little more tainted – making it even more difficult for many infertile patients to seek the sound, honest treatment that is available from official health-care bodies.

Would better regulation have helped to prevent this scandal? I doubt it. Korea already had legislation in place and sound mechanisms for ethical approval. Ultimately, aggression, the ambition to be first and dishonesty may always combine to draw a scientist off the straight and narrow. Probity cannot be enforced. What we really need to do is to change the culture in which science is increasingly pursued in many parts of the world; but I think that so radical a change would be difficult now.

Habits are ingrained; it is long time since de Graaf published.

Religious attitudes to stem-cell research

For some people, generating stem cells by nuclear transfer entails the creation of life as a means to an end, and is therefore a grave sin. But it seems difficult to view the embryo as a human person before it reaches a stage where we can discern any evidence of a nascent nervous system – which occurs at around fourteen days. Until that stage, the embryo is surely no more than a clump of cells. And perhaps there is a moral imperative to use these cells to save lives if we can.

Surely, the issues around embryonic stem cells pose no greater problems than the issues around IVF itself. This view is effectively that of the British government. Since January 2001 the cloning of embryos to obtain stem cells – therapeutic cloning – has been permitted in Britain, although reproductive cloning of humans remains a criminal offence. The picture elsewhere is less permissive. In the USA, embryo research is often confused with the issues raised by the political hot potato of abortion. Since 1994 there has been no US government funding for embryo research. This withholding of support is probably in opposition to a majority of the American people, who according to recent polls seem to want the research to go ahead, tend to see its benefits and are unmoved by any sentiments about an early embryo being a human. Researchers can, nevertheless, exploit a legal loophole that permits access to stem cells that were not created via government funding. Since 2001 President Bush's administration has also permitted research using the sixty or so stem-cell populations created from embryos before the ban; but of these sixty, only around a dozen are considered to be even potentially usable. America remains mute and frozen in its attitude to this vital field of research, leaving people like Gerry Schatten with an alternative: to become a little like the soldiers of fortune of old – scientists offering their skills abroad.

Just as conservative America's stance on abortion hinders stem-cell research, so it is the Jewish attitude to the same subject that sanctions it. Judaism permits abortion when the unborn child poses a threat to its mother's life. Even when it does not, it considers the fetus to have a lesser status than a child. Judaism generally does not object to the use of embryos left over from IVF treatment – although most authorities are undecided on whether they may be created specifically to be a source of stem cells. There is a chasm of difference, legally speaking, between finding a beneficial use for embryos that are the by-products of a permitted procedure, and creating them as a means to an end. But at the same time, there is a strong principle favouring anything that confers health on humankind. One American strongly in favour of stem-cell research is Rabbi Moshe Tendler, who sits on the US government's National Bioethics Advisory Commission and has written movingly on the moral imperative to conduct research. A particularly Jewish concept, to which he refers in the statement below, is that of making 'fences' – called *gezerot* in Hebrew – to protect the spirit of the laws set out in the Torah. Fences exist to rule on matters that are not mentioned in the Torah, sometimes because the circumstances were irrelevant in ancient times. Nearly two thousand years ago, the rabbis defined *gezerot* that forbade polygamy and opening someone else's mail, matters not mentioned in the Torah, in order to protect the wider notions of propriety and respect that quite clearly are part of Torah law.

> A good deal of rabbinic law consists of erecting fences to protect biblical law. But a fence that prevents the cure of fatal diseases must not be erected, for then the loss is greater than the benefit. In the Judaeo-biblical legislative tradition, a fence that causes pain and suffering is dismantled. Even biblical law is superseded by the duty to save lives, except for the three cardinal sins of adultery, idolatry and murder. Life-saving abortion is a categorical imperative... Mastery of nature for the benefit of those suffering from vital organ failure is an obligation. Human embryonic stem cell research holds that promise.[17]

We find a similar degree of tolerance in Islam – for which forty days after conception is considered the point at which an embryo assumes fuller human status. Also, since in most cases Islam prohibits the adoption of excess embryos left over from IVF, some scholars argue that it thereby creates a positive obligation to use them for research purposes. However, unlike Roman Catholicism, Islam has no Vatican to promulgate centralized rulings on the subject. In the main, stem-cell research is viewed favourably in Islamic states, but there are different rulings as to the legality of using cloned embryos, or those discarded as a by-product of IVF.

Christian attitudes also vary considerably. St Thomas Aquinas seems to have thought that an embryo had no soul at the moment of conception, acquiring it slowly thereafter. But the Vatican takes a more rigid view, arguing that human life begins at the moment of conception. In the *Donum Vitae* (Gift of Life) of 1987, the papacy declared that, 'from the time the ovum is fertilized . . . [it] demands the unconditional respect that is morally due to the human being in his bodily and spiritual totality . . . [and] his rights as a person must be recognized, among which in the first place is the inviolable right of every human being to life'. Accordingly, it regards embryo research and cloning as 'a gravely immoral act, and consequently gravely illicit'.

This strong position may not reflect the views of most Americans, though. In an American Harris poll conducted in August 2004, 73 per cent of US Catholics voted in favour of embryonic stem-cell research and only 11 per cent were against it. Even among the deeply religious evangelical Protestants, only 20 per cent were against the use of embryos. For many Christians, including some moderate Catholics, the potential benefits of embryo research outweigh the negatives. Michael Mendiola, professor of Christian ethics in the University of California at Berkeley, argues that it is a question of tolerating a lesser evil to gain a greater good. Researchers, he suggests, should continually try to move beyond the use of human embryos, but to do this, it may be necessary to employ them in the short term. Protestant theologians like Gaymon Bennett, also based at Berkeley, point out that we have never been able to define the moment at which an embryo becomes a person, while we know only too well how much grave suffering there is in the world. This marked contrast makes it reasonable to use embryos to alleviate some of that human misery.

Could we perhaps really create life?

President Bush may be reluctant to fund stem-cell research, but make no mistake, whatever the protestations of my American colleagues, they are still leading the way with some very exciting developments in this field. In

2003 Karin Hubner and her colleagues at the University of Pennsylvania in Philadelphia published the record of an extraordinary advance which really alters our ethical understanding about the beginning of life.[18] They were able to persuade mouse embryonic stem cells to differentiate into eggs.

Their approach was simple enough, although it still requires considerable skill in growing embryonic stem cells. It seems that under routine culture conditions some of these cells just formed egg cells on their own. The need was to identify which cells were taking on the characteristics of eggs and to nurture them. Moreover, these cells, when separated and isolated, had a tendency to form follicle-like structures, with the usual clumps of feeder cells around them. What is more, they appeared to start to undergo meiosis – the process whereby germ cells (and only germ cells) throw off one set of chromosomes as a preparation for fertilization. After approximately one month in culture, some of the egg cells separated from the feeder cells in a fashion strongly reminiscent of what happens at ovulation. With further growth it seems that some of these 'egg' cells formed bodies that resembled embryos – even though there had been no attempt to fertilize them. They produced the proteins usually produced by mammalian embryos after a few days' growth and generally gave the impression that they were attempting parthenogenesis (see page 37).

It is not yet clear whether these mouse 'eggs' will be capable of fertilization and further development, but if they are, this is revolutionary and a technique that could presumably also be applied to humans. If human embryonic stem cells have similar powers, these cells could generate an abundance of eggs for therapeutic cloning and also, of course, alleviate the current shortage of eggs for donation. A year after the Philadelphia study, a suggestion that human embryonic stem cells might have the ability to form egg-like cells was published by Amanda Clark and colleagues from San Francisco.[19] Although Dr Clark's cells do not show all the properties of eggs – and are still a long way from being fertilized – this is clearly an area that will develop fast.

Notwithstanding all these advances, the properties of all stem cells remain a huge enigma. Karim Nayernia from Göttingen, in Germany, has just published two remarkable papers which challenge many preconceptions about the potential of stem cells.[20] He and his colleagues have been working not with embryos (which is illegal in Germany) but with bone marrow cells. Using marrow cells taken from mice, they dosed them with the active principle in vitamin A (chosen because it is known to have an important effect on the testis during development and is needed to produce sperm). Eventually their 'adult' stem cells turned into early male germ cells, the precursors of sperm. They produced many of the proteins that early male germ cells should produce and, when transplanted back into the testis of a mouse, divided like sperm-cell precursors. But they did not undergo meiosis. This very interesting experiment is important because it demonstrates how much we do *not* know about stem cells. Although embryonic stem cells look likely to be more useful to make new cells and tissues, Dr Nayernia's work demonstrates that we have much to learn. And, remarkably, just two months or so before publication, he demonstrated that some of these cells can actually fertilize eggs and produce live offspring.[21]

This is a very rapidly developing field. Just a few days after writing the above section of this chapter, I had an extraordinary experience in Adelaide. In recent years, reproductive science in Australia has been increasingly at the cutting edge, and the meeting I was attending (where I gave a very low-grade paper) was a showcase for the excellent research currently being produced in that country. Adelaide itself is very much a leading centre for reproductive research, but it would be almost unthinkable to have a good academic meeting on reproduction in Australia without some contribution from Melbourne by Alan Trounson.

During a lecture on stem cells, Alan suddenly put up the slide of a photograph of what looked very like a normal human egg. The impression I first got was that this had been produced from a human embryonic cell line. Alan has been working on generating germ cells in mice; he has not been able to reproduce Dr Hubner's work completely, but he too has shown that it is possible to grow ovary-like structures in culture from mouse embryonic cells, and to do so in a very controlled fashion. Alan was convinced that if embryonic stem-cell cultures were dosed appropriately, they could be persuaded to grow eggs or sperm preferentially, rather than any other cells. He proved this by exposing his stem-cell cultures to growth factors collected from the testes of newborn mice. Treated with the substances that testicular tissue produces, they grew very effectively into ovary-like structures, and the eggs that were formed were surrounded by two layers of feeder cells, just as in normal ovarian tissue. No zona pellucida was formed, so these are in fact not normal eggs; however, it seems likely that, with further research, cells like this will be capable of undergoing fertilization – raising some curiously challenging questions about our parentage and our origins.

But what I found so intriguing is that the eggs Alan showed were produced not from female embryos, but from stem cells that had been derived from males. Alan's extraordinary photograph was arresting because, using his technique, it now would be possible for a male to produce eggs.

Sperm stem cells: an under-researched resource

Hitherto in this chapter I have avoided any mention of my own research interests in stem-cell biology. For the past decade I have been working with Carol Readhead, a molecular biologist at the California Institute of Technology in Pasadena, near Los Angeles. Carol is an expert in making transgenic animals, which is why we originally collaborated (see chapter 10). But she has long had the idea and the goal of trying to generate stem cells from a novel adult source.

One very under-researched area is the production of stem cells in the testis. Unlike the ovary, which is invested with all its eggs before birth, the testis continues to produce new sperm throughout most of its reproductive life. These germ cells are derived from stem cells that lie in the lining of the little tubules in the testis. They are constantly dividing to produce spermatogonia, the precursors of the cells that will eventually become spermatozoa. Theoretically, there are a number of reasons why these stem cells may be particularly useful biologically. First, the testis is unduly rich in them, and, of all the 'adult' stem cells available in various tissues, they are among the most easily accessible. It is a relatively easy matter to take a biopsy from the testis – indeed, frozen pieces of testis are very frequently stored by fertility clinics, so there would be little difficulty in getting patient consent for this research or to use cells as treatment. Second, we are able to grow them in culture and, as we shall see, this offers interesting prospects for modifying them, like other stem cells, for therapeutic use. And, because they are germ cells, they are very different from all other adult stem cells, which are essentially geared to produce the tissue or tissues of one organ. Although these stem cells produce only sperm, those sperm do lead to the most primitive stages of early human development, so it is very likely that sperm stem cells may be pluripotent.

One potential use for sperm stem cells that could be applied almost immediately would be to preserve male fertility when it is threatened by cancer treatments. A number of potentially fatal conditions in young people can be cured, but only by treatment that may cause sterility. These diseases include leukaemia, Hodgkin's disease, and one or two other cancers that occasionally affect boys or young men. Often the only cure will be by chemotherapy; sometimes radiation is needed as well. If the radiation is in the pelvic region (possibly because that is where the cancer is mainly situated) the effect may be to sterilize the boy. As mentioned earlier, stem cells are highly sensitive to radiation – and very sensitive to most forms of chemotherapy as well. If the stem cells are destroyed, the ability to make sperm will be permanently damaged, once recovery from the cancer has been achieved.

A potential strategy is to store as many sperm samples as possible in liquid nitrogen and then to conduct artificial insemination years later, after the cancer cure. But obviously younger boys will not produce semen, and therefore in their case routine storage before chemotherapy or radiation is not a practical proposition. However, their testes do contain sperm stem cells waiting to be activated by the hormone system. So Carol and I, together with our friend Outi Hovatta from Finland, considered taking biopsies from the testicle, isolating the spermatogonial stem cells and then freezing them in liquid nitrogen. This has been successfully achieved in mice. The difficulty is in putting the stem cells back into the testis, once the cancer treatment has been accomplished. In an undamaged testis (one not affected by radiation, for example) it has been possible – using delicate microsurgery and exquisitely fine glass pipettes – to transfer sperm stem cells back into the tubules of the testis. Under ideal circumstances, they seem to be able to migrate to the lining of the tubules, where they can link up with the feeder cells that nurture them. Once this happens, they seem able to recommence cell division – giving rise to new sperm, thus restoring fertility. Some progress has been made in animal research already. The main difficulty is getting the stem cells back into the testis, but the problems are largely technical and are likely to be overcome fairly soon.

It is intriguing to consider that it might be possible to use cells from the testis as a resource to treat male patients who need their own stem cells, for whatever reason. Admittedly this approach would not be suitable for females; if they needed a stem cell transplant, any cells derived from the testis would need to be modified so that they would not be rejected by the woman's immune system.

One of our difficulties in the UK has been in trying to get approval to do the appropriate experiments to examine the properties of spermatogonial stem cells. Carol and I have only just got a licence to compare sperm stem cells with embryonic stem cells. There is clearly an important comparison to be made, but for a while it was surprisingly difficult to persuade the regulatory authority that this research required us to grow human embryonic stem cells for comparison. It should, I think, have been obvious; but we were initially told that we should just use sperm stem cells. The authority seemed unduly troubled by the thought that serious researchers might want to conduct unnecessary or frivolous experiments, and for some time it looked as though our application might be turned down, even though the government's own local ethics committee had given us its full approval.

I am very hopeful that sperm stem cells may have an extremely valuable therapeutic application. They are much more plentiful than embryos, and potentially much more easy to obtain. A key advantage is that their use raises none of the ethical problems usually associated with embryonic manipulation. Moreover, the ability to grow such cells in culture means that we should be able to manipulate them genetically. If we can modify their genes successfully, then both men and women could be beneficiaries of this technology. As we shall see in the next chapter, genetic modification is both the great hope and the great danger in the whole field of reproductive medicine.

The Trojan Horse

Genesis, which is where I started, is not the only ancient text to show how having children changes our feelings about mortality. The *Epic of Gilgamesh*, perhaps the oldest extant work of written fiction, tells of a real monarch, the King of Uruk in Sumer in Mesopotamia. One version is probably more than five thousand years old, written on twelve large clay tablets. In the seventh century BCE it was inscribed in cuneiform and placed in the archives of King Ashurbanipal.[1] The epic treats fertility as mortal man's chance of living on after death. At the end, the tyrant Gilgamesh, two-thirds god and one-third man[2] (presumably it was the man part that did all the womanizing the poem describes), has failed in his attempts to become immortal. He reflects that 'For all the people, whoever they may be, funerary statues are made for future days, and set aside in the temples of the gods. Their names, once uttered, do not sink into oblivion. Aruru [the earth goddess], the older sister of Enlil [the god of wind], provides them with offspring for that purpose.' Through his offspring, Gilgamesh ensures he will not be forgotten.

The drive to survive and the quest for reproductive control

The drive to reproduce is an extension of the drive for survival. Both are almost certainly hard-wired into our brain, aspects of the same genetically determined instinct that has evolved through all animal species. Just as humans have always tried to ensure their own survival, so we have also invariably attempted to use whatever means we have to encourage reproduction, to ensure the next generation. Though we think of reproductive manipulation – assisted reproductive technologies, cloning and genetic modification – as modern phenomena, humans have always tried to meddle with reproduction.

Having children was as life-altering to cave man as it is now to commuter man. Surprising as it may seem in view of the population explosion this planet has seen in the past fifty years, humans are naturally one of the least fertile species. Gestation is lengthy and the infant has a very large brain compared with its body size. So, for a safe birth through the tight female pelvis, delivery occurs when the newborn is very immature. As Stephen Jay Gould once observed, humans are born in an embryonic state.

Our defenceless infant is a burden. It requires regular feeding – in hunter–gatherer societies, breast-feeding continues for at least two years, sometimes much longer. The growing child is almost two years old before it can take more than a few steps and needs constant care for the first years of life. Even at this age, full maturity for most mammals, a small child would have been difficult to protect if a family of early humans came under attack by predators. So it should not be surprising that humans have tried whenever possible to take control of their own reproduction.

Apart from breast-feeding, the natural way of inhibiting female fertility, there were always other ways of manipulating reproduction. Prehistoric peoples almost certainly used infanticide and possibly used abortion. Experts agree that infanticide was common, particularly if a baby was deformed or if the mother had another infant. We can learn something from modern hunter–gatherers. One 1960s study of Australian Aborigines estimated infanticide rates of between 15 and 50 per cent among them. This seems an extraordinarily high estimate but it is salutary to consider that, at the present time, there are a number of villages in India where males are preferred and no girls have survived birth for some years.

Primitive birth control was probably practised. There is some archaeological evidence, backed by modern anthropological observations, that contraceptive herbs were used forty thousand years ago. Archaeologists found piles of husks of starflower (borage), which is useless as a foodstuff, in the Doura caves in Syria. Was it valued as a medicine? After all, today, many women take it to relieve premenstrual tension.[3] There are several modern tribal groups that use plant derivatives as contraceptives. The Deni Indians in the Amazon use a plant similar to curare. Laboratory tests suggest that this plant may make men infertile for up to six months – or longer if they don't survive the contraceptive.

In 1908 the archaeologist Josef Szombathy was excavating in the hills along the River Danube, close to the insignificant village of Willendorf, when he found a strangely shaped lump of limestone just 11 centimetres long. After cleaning the object, he discovered a carving of a grotesquely obese woman that was to become one of the icons of prehistoric art. The figurine, now in the Natural History Museum in Vienna, has massive breasts and a big belly. Her corpulent thighs do not hide her very obvious vulva, while the sculptor has reduced her arms to a merest hint over the top of her breasts. What seems particularly odd is that the figure's head is featureless, her face covered with a spiral of tight curls. Does this statuette, now thought to be some thirty thousand years old, represent one of our earliest records of human attempts to influence fertility?

Szombathy called her the Venus of Willendorf, following the Marquis Paul de Vibraye, who in 1864 had found an ivory statuette in the Dordogne, now in the Musée de l'Homme in Paris, consisting of legs, a torso and a prominent vulva. He called it the *Vénus impudique* (immodest Venus), playing on the term *Venus pudica* (modest Venus) used to describe Roman statues where the goddess attempts to cover her breasts and pubis. The comparison was unflattering, contrasting the supposedly rough sexuality of prehistoric art with the decorous sensuality of Roman statuary.

Around two hundred similar figurines have been found scattered throughout Europe and as far east as Siberia, most of them around 23,000–25,000 years old.[4] The Galgenburg Venus, now in the local Weinstadt Museum in Stratzing, Austria, is one of the oldest finds. Only about 7 centimetres high and carved from shiny amphibolite slate, it was dug up in 1988 at a site where Paleolithic hunters lived. Sadly, it broke into several pieces during the dig. This female statuette has a curiously provocative pose, the trunk turned to one side, the left arm raised into the air and the right resting enticingly on the thigh. Her apparent dance makes it tempting to see evidence of a fertility rite. Like the Venus of Willendorf, she has a prominent vulva and is labelled in the museum as 'Fanny'. This may, of course, be typically subtle Austrian humour.

Many Venuses suggest some ritual connection. They often taper to a point with the legs almost fused – like a mermaid's. Perhaps these figures were stuck into the ground where they could be admired, adored, perhaps worshipped? Analysis of wear patterns can also be intriguing: a 28,000-year-old pendant from the Grimaldi caves in Italy shows a woman on each side, one apparently pregnant and one not. Patterns of wear on the pregnant side suggest that the belly has been rubbed; perhaps it was a charm.

Masculine symbols have been uncovered, too. The anthropologist Nicholas Conard of Eberhard-Karls University of Tübingen and his team found fourteen fragments of highly polished siltstone in the Hohle Fels Cave near Ulm. Their careful excavation of these separate pieces, all around 28,000 years old, took place over five years.[5] The reconstructed piece of stone closely resembles a human penis, with a series of etched concentric rings, one close to the tip giving the appearance of the edge of the glans. Moreover, it is life-size: for those wishing to know more about prehistoric man, it is just under eight inches in length. As pointed out by a number of breathless journalists, its surface is so highly polished that it could have been used as a dildo. It would not, however, fulfil that function now – at least, not without discomfort: for, some time after it was made, it was used as a very different kind of tool. Parts of its surface have become very roughened where it seems to have been used as a hammer to shape flints.

This object is not unique. Among other examples, there is even an intriguing one consisting of two phallic protrusions at an angle to each other. Serious archaeologists often identify this as a spear-straightener or other innocuous object, but the shape is so phallic that it is easy to see the sexual possibilities. We know that ritual deflowering is carried out in some tribal groups, and this was suggested as one potential use. Perhaps they were just for pleasure. The notoriously sexually active bonobos – the apes that are our closest genetic relatives – have been seen using similar-shaped objects in just this way.

Earliest written records confirm humans' determination to control their fertility. In Mesopotamia, the region between the Tigris and Euphrates rivers, some thirty thousand baked clay tablets have been found, inscribed in Sumerian or Akkadian. The biggest single source is the seventh-century BCE library of King Ashurbanipal at Nineveh. These records make it clear that women held many of the same legal rights as men. Monogamy was normal, although a man was entitled to marry again if his wife was infertile. Divorce was accepted; adultery was illegal, and adulterous women could be punished with death. Adopted children were given the same rights as genetic children.[6]

Over one thousand tablets deal with medical matters. Medicine was mostly sorcery – about which we should not feel too superior: as we have seen in this book, the practice of IVF in some supposedly sophisticated clinics still is. A different god or goddess protected each part of the body, and illnesses arose from the actions of deities or malignant spirits. For example, the 'Treatise of Prognosis' says of the male patient that 'if his penis and his testicles are inflamed, the hand of the goddess Dilbat [equivalent to Venus] has reached him in his bed'.[7]

The tablets also tell us something about the Mesopotamian understanding of fertility. Pregnant women were vulnerable to harm from Lamashtu, a female chimaera with the head of a lioness, the ears of a donkey and the feet of a bird. There were also risks from unfulfilled ghosts such as those of women who died during labour. Protection came from amulets depicting the head of the demon Pazuzu, as well as prayers to the goddess of fertility, called Inanna by the Sumerians and Ishtar by the Akkadians. The worship of Inanna/Ishtar was widespread, and Babylon, Nineveh and Uruk all had temples to her. One tablet refers to her as having 'full power of judgement and decision and the control of the law of heaven and earth'. Archaeologists have found numerous statues of the goddess offering her breasts as a sign of fertility.

In ancient Egypt, too, fertility was a matter of religion as well as science. The theme of fertility runs through much of Egyptian myth. The goddess of human fertility, Isis, was widely worshipped. Set, god of chaos and evil, became infatuated with her and pursued her in the form of a bull. Apparently Set lived on a diet of lettuce, regarded as an aphrodisiac probably because its milky white fluid evoked comparisons. Isis was quicker than Set. Frustrated, he ejaculated on the ground. Isis, outraged, showed her disgust verbally. But the bull's sperm was fertile and seeded the desert, giving rise to various plants.[8]

Her husband Osiris, god of the underworld, was dismembered by Set.[9] Isis eventually pieced Osiris back together but could not find his penis – Set was digesting it. She substituted a wooden phallus, and her temples therefore contained phalluses made from stone or wood. Phallic imagery was also central to the worship of Min, the god of animal and crop fertility, who was depicted with an erection. Prudish Victorian Egyptologists pictured him from the waist upwards only.[10] The use of phallic objects in a religious context was common in this period. The excavation of Harappa in Pakistan, one of the major centres of the enigmatic Indus civilization, yielded numerous stone phalluses, possibly precursors of the *lingas* representing the creative power of the Hindu god Shiva.

Funerary stela – inscribed stone or wooden slabs – show infertile Egyptian couples praying for children in temples. Small figures of naked women with explicit genitalia, sometimes also with babies, have been found in tombs, and have been interpreted as charms to enhance a woman's fertility. Women in childbirth were protected by various goddesses, including a pregnant hippopotamus called Tawaret. She was considered malevolent in pre-dynastic Egypt, but was rehabilitated due to the ferocity with which female hippos defended their young. Archaeologists have found her image on long, flat ivory objects known as apotropaic (protective) wands, some of which are inscribed with the names of a mother and child. They may have been laid on the abdomens of pregnant women to protect their babies.

The Egyptians recognized a class of physicians who were neither sorcerers or priests. Herodotus reported in the fifth century BCE that an Egyptian physician might specialize in treating one disease: oft-quoted job descriptions include the 'Physician of the Eyes' and 'Shepherd of the Anus'. The Ebers papyrus, written around 1550 BCE, describes the treatment of diseased breasts with a prayer to Isis and a mixture of calamine, cow's brain and wasp's dung.[11] It suggests ways to encourage milk to develop, and how to tell whether breast milk is good by its smell – good milk smells like grain, while bad milk stinks of fish. The papyrus suggests that women with spots before their eyes are infertile, perhaps referring to diabetes which, untreated, results in infertility in men and women.

The other key document in our understanding of

Egyptian attitudes to fertility comes courtesy of the remarkable William Matthew Flinders Petrie. Born in 1853 in Kent, he was self-taught and without formal qualifications. Yet by the end of his life in 1942 he had published over one hundred books and nearly nine hundred articles in learned journals. As a boy, Petrie was excited by mathematics and geometry; by the age of thirteen he had already convinced himself that he should go to Egypt to see the pyramids.

One of Petrie's most significant finds was the Kahun papyrus, discovered in 1889 at the Fayum site of Lahun. The papyrus, one of the very oldest in existence, is dated with a note on the back to 'the 29th year of the reign of Amenenhat III' (c.1825 BCE). It is a treatise on gynae-cology, divided into thirty-four paragraphs, and can be seen today in University College London. Unfortunately its condition is such that much of it is undecipherable. A paragraph on the diagnosis of pregnancy, for example, is tantalizingly fragmented: 'you will [examine] her: if you find the muscles of her breast khasha, you will say this is a woman giving birth; if you find them kenken [soft?], you say she will give birth late; but if you find them like the colour [you will say she will not ever give birth]'.[12]

Egyptians believed the female reproductive organs were connected to the rest of the body, and relied for their health upon the free flow of bodily materials. This could be tested by fumigation. The Kahun papyrus lists a number of pregnancy tests. One involved placing cut onion into the woman's vagina and then smelling her breath. Another was for the patient to sit on a mixture of date flour and beer: if she vomited, it was a good sign, indicating that her passages were open.

The Ebers papyrus recommends contraception by inserting linen soaked in a mixture of honey, acacia, dates and colocynth (a small yellow fruit also known as bitter apple) into the vagina. Acacia, if fermented and mixed with water, produces lactic acid, a spermicide still used in contraceptive jellies. Tests on acacia suggest that it might encourage abortion even without fermentation. Colocynth, another ancient drug, is still used as an abortifacient by some Arab women.

One memorable contraceptive involved placing crocodile dung 'near the entrance to the womb'. This possibly had a mystical significance – the crocodile was a symbol of Set, associated with miscarriage and abortion. But more practically it might have acted as a barrier, making it harder for sperm to reach the uterus. It might also have been highly efficacious by causing the lady's partner to experience total and immediate detumescence.

By the time of the ancient Greeks there is plentiful evidence of reproductive manipulation. Near the end of Act III of Aristophanes' comedy *Peace* (421 BCE), Hermes invites Trigaios to enjoy a lady friend, Opora. Trigaios worries what will happen if he 'over-indulges himself'. 'Opora' means 'fruit', and Aristophanes is playing on words. 'Not if you add a dose of pennyroyal,' advises Hermes. Pennyroyal remained popular as an abortifacient for centuries. It has the desired effect mainly because it is highly toxic.

The sixty texts associated with Hippocrates list several abortifacients which might be applied to the genitals, among them a pessary of an appetizing mixture of crushed beetles, herbs, cuttlefish eggs and wine. According to *Diseases of Women*, best of all was the squirting cucumber. Aristotle, on the other hand, suggested that women should 'anoint that part of the womb on which the seed falls with oil of cedar, ointment of lead, or frankincense commingled with olive oil'. Modern gynaecologists might flinch at the toxicity of lead, but olive oil would have reduced sperm mobility.

Aristotle (384–322 BCE) drew a fundamental distinction between matter and form: matter was the raw material, but it could be changed into different forms. The female role was to provide matter in the form of menstrual fluid, which was then acted upon by the more concentrated male semen. Aristotle urged his readers to 'Compare the coagulation of milk. Here, the milk is the body, and the fig-juice or the rennet contains the principle which causes it to set.' Male fertility could be tested by putting a sample of semen in water: if it floated then it lacked the necessary consistency to form a fetus.

Aristotle relied upon knowledge gained from studying chickens' eggs. He cracked open fertilized eggs at regular intervals throughout incubation, observing the changes as chicks developed. He believed wrongly that the embryo got its entire nourishment from the white of the egg, and noticed that the yolk grows larger as it liquefies during development. He even questioned whether an unfertilized egg is alive. Aristotle's emphasis on the importance of menstrual blood may have come from observation of the blood spots sometimes found in the birds' eggs. In spite of these errors, Aristotle trod the path that eventually led to modern developmental biology.

By Roman times, manipulation of reproduction had become more sophisticated. Soranus of Ephesus was the pre-eminent gynaecologist of the ancient world, working during the reigns of Trajan (98–117 CE) and Hadrian (117–38 CE). A pioneer, he recommended the use of a vaginal speculum centuries before it was accepted practice. He distrusted pregnancy tests relying upon the supposed passage of odours between the genitals and rest of the body. He commented that pregnancy was sometimes harmful for women, advising sexual abstinence for medical reasons.

For contraception, Soranus advised readers not to have sex during the most fertile period of a woman's cycle, which he thought was during and immediately after the menstrual period.[13] This was not very helpful, but this idea ran right through to the eighteenth century. To encourage conception he recommended certain positions, particularly sex with the man behind. He suggested the woman should lie down afterwards to retain the semen, useless advice still given in many clinics to this day. Soranus also prescribed contraceptives:

> Conception is prevented by smearing the orifice of the uterus all over with old olive oil, honey, cedar resin or the juice of the balsam tree...or put a lock of fine wool into the orifice of the uterus; for such things as are astringent, clogging, and cooling cause the orifice of the uterus to shut before the time of coitus and do not let the seed into the womb.

A particularly sought-after contraceptive was a type of giant fennel known as silphium. Found only in North Africa, silphium was harvested to extinction in the third or fourth century BCE; and in the second century BCE the Greek writer Polybius famously lamented the de-population of his country as a result of its use. Polybius suggests that fertility is not in the hands of the gods:

> Men had fallen into such a state of pretentiousness, avarice, and indolence that they did not wish to marry, or if they married to rear the children born to them, or at most as a rule but one or two of them, so as to leave these in affluence...it was of no use at all to ask the gods to suggest a means of deliverance from such an evil. For any ordinary man will tell you that the most effectual cure had

to be men's own action, in either striving after other objects, or if not, in passing laws making it compulsory to rear children.

By the next century the Roman state, with a falling soldier population, was acting to encourage population growth.[14] Julius Caesar (c.100–44 BCE) awarded land to men with three or more children, while Augustus (63 BCE –14 CE) gave preference to candidates for public office in proportion to the number of children they had. In 17 BCE Augustus wrote: 'If we could survive without a wife, citizens of Rome, all of us would do without that nuisance; but since nature has so decreed that we cannot live comfortably with them, nor live in any way without them, we must plan for our lasting preservation rather than for our contemporary pleasure.' For the Romans, a successful marriage was one that produced children. Augustus also made it harder for the celibate to inherit property, and limited the amount that could be left to spouses in the wills of childless couples. Other laws encouraged widows and divorced women to marry again.

It is clear, then, that there were increasing attempts to manage fertility, both encouraging and preventing childbirth. But there is also increasing evidence that both the Greeks and the Romans manipulated fertility by more sinister means. There is no question that the Romans practised eugenics. A father had the absolute legal right to destroy a deformed child. Infanticide was sometimes by simple exposure: occasionally a baby might have been rescued from a hillside, but most seemingly perished. Various other methods of killing were sanctioned and widely practised, being quite openly accepted as normal behaviour. Archaeologists have found evidence of this throughout the Roman Empire, including Britain, and infanticide was also widely employed in Sparta and

Carthage. There is considerable argument about whether females were more likely to be killed; possibly male infants were more valuable. In an interesting piece of research, Dr Marina Faerman and colleagues of the Hebrew University in Jerusalem studied the bones of some one hundred neonates excavated from under a Roman bath-house in Ashkelon.[15] Testing DNA fragments obtained from nineteen of the left femurs that they identified, the researchers found that fourteen were male, five female. They suggest that the high proportion of males indicates that females may have been selected for preservation and perhaps reared to be prostitutes in this building, which may have doubled as a brothel.

Infanticide was so common in most ancient cultures that the Romans found the Jewish practice of condemning it strange. Josephus (himself a Jew, though completely Romanized), wrote in puzzlement: 'The Law orders all the offspring to be brought up, and forbids women either to cause abortion or to make away with the fetus.' So eccentric was this attitude considered that when Tacitus, in Book 5 of his *Histories*, wrote of how 'all their other customs, which are at once perverse and disgusting, owe their strength to their very badness', he included among them: 'It is a crime among them to kill any newly born infant.'

The eugenics movement

I have no patience with the hypothesis occasionally expressed, and often implied, especially in tales written to teach children to be good, that babies are born pretty much alike, and that the sole agencies in creating differences between boy and boy, and man and man, are steady application and moral effort. It is in the most unqualified manner that I object to pretensions of natural equality. The experiences of the nursery, the school, the University, and of professional careers, are a chain of proofs to the contrary.[16]

Thus wrote Sir Francis Galton in 1869. Galton was a skilled statistician, developing a number of important advances in the analysis of biological data. He was also the first person to provide scientific proof that fingerprints are unique to an individual. In 1885 this remarkable polymath, who as a six-year-old child prodigy had read the entire works of Shakespeare and had gone on to become an explorer, researcher, sage, writer and mathematician (although, in spite of his reverence for those who got really good degrees at Cambridge, he initially only scraped through his Tripos), coined the word 'eugenics'.

Galton is now regarded as a flawed individual who started a movement that did considerable social harm. A bit of a social climber, he observed that after his marriage 'we led a life that many of our social rank might envy. Among our friends were not a few notable persons, a full half of whom were known to me through my wife.' He was proud of being elected early to that rather self-satisfied club, the Athenaeum (to which his cousin Charles Darwin also belonged), and no doubt equally gratified by being appointed a Fellow of the Royal Society at the age of thirty-eight.

He observed that many prominent people came from a small number of families, which he took to indicate not that the privileges conferred on them by their environment and education gave them a head start but that their positive traits were inherited. He was particularly impressed by judges and by members of the House of Lords. On this basis he would no doubt have argued vehemently against the recent reform of that institution by

Tony Blair's government – but, having sat through a great many hours of debates on the subject, I cannot recall a single hereditary peer who used a Galtonian argument in favour of retention of a privileged Upper House,[17] or who argued that noblemen were of particularly sound or high intellectual achievement. Drawing on Darwin's theory of evolution, Galton asserted that, by assisting those with socially undesirable traits, politicians and policy-makers undermined society. He felt it was preferable to let these negative traits die out. His views, which now seem misguided and arrogant, reflected and fed a middle-class nervousness that the human species was getting weaker as those who might improve society with their hereditary ability were being held back by having to pay, through taxation, to support the less capable members. Galton himself espoused a kind of positive eugenics, coming up with schemes to encourage those with admirable traits – among which he counted health, energy, ability, manliness and courteous disposition – to have children, and recommending voluntary celibacy for those with undesirable traits – such as alcoholism, criminality or immorality, and feeble-mindedness. He seems to have been infertile himself, or at least in an infertile marriage; certainly, he never had children. Equally certainly, he was not celibate. Dan Kevles, in his wonderful book on the history of eugenics, points out that during his many travels in the Middle East Francis Galton led 'a very oriental life'.[18] In one of his letters to Boulton, a friend also travelling in the region, he writes how he had negotiated for a pretty Abyssinian slave. Boulton, soon after, writes back: 'What an unfortunate fellow you are, to get laid up in such a serious manner for, as you say, a few moments of enjoyment.' Galton seems to have felt deeply guilty about contracting venereal disease and probably thought he was responsible for his barren marriage.

Just as it is important to be careful about assessing intelligence and ability on the basis of looks, one should avoid paying too much attention to physiognomy. Having said this, there are several photographs of this enigmatic man that seem to show an extraordinary smug countenance. It was a short downward step from propounding voluntary celibacy to the less 'desirable' to suggesting that those with 'undesirable' traits should not be allowed to breed. Galton funded first a fellowship, and later a Galton Eugenics Professorship, at University College, London, for which he left an endowment of £45,000 – then a very considerable sum. Oddly, the current holder of that appointment is the eminent Steve Jones, a brilliant and engaging geneticist who, I suspect, views the origins of his post somewhat dubiously. By contrast, the first incumbent, Professor Karl Pearson, set out in the spirit of his benefactor to prove with statistics that there was an inverse relationship between fertility and social class.

Pearson was unquestionably a great statistician, but his excellent logic was confused by biased interpretations of heredity. There was also more than a tinge of racism in his views. In the *Jewish Chronicle* of 29 July 1916 there is an account of Galton's earlier visit to the Jews' Free School (then in Bell Lane), where he took composite photographs of the children to try to analyse some of their physical and hereditary traits:

> They were children of poor parents, dirty little fellows individually, but wonderfully beautiful as I think in these composites…The feature which struck me as I drove through the Jewish quarter was the cool scanning look of man, woman and child, and this was no less conspicuous amongst the schoolboys. There was no sign of

diffidence in any of their looks, nor of surprise at the unwonted intrusion. I felt, rightly or wrongly, that everyone of them was coolly appraising me at market value, without the slightest interest of any other kind. [Galton's words, quoted by the journalist]

While there is not the slightest sign of the quite prevalent anti-Semitism of the time, Galton's stereotypical assumptions about the objects of his observation is revealing; and Pearson praised this attempt at composite photography, saying, 'we all know the Jewish boy'. Later, in the 1920s, Pearson, who in practice was probably a good deal less racist than many of his contemporaries, is reported as saying that the Jewish children of the East End of London, while no less intelligent than Gentiles, 'tended to be physically inferior and somewhat dirtier'. Other eugenicists went much further in ignoring the influence of social circumstances on observable characteristics, for example looking for indications that traits such as drunkenness ran in families.

A related movement of increasing salience in the first years of the twentieth century was the campaign in favour of contraception, so vigorously promoted by Marie Stopes and Margaret Sanger. Stopes was certainly passionately interested in the manipulation of human reproduction, and although her pioneering work is now regarded as a great contribution to the freeing of women, there is no doubt that she espoused eugenics. Her colleague, Margaret Sanger, something of a feminist, felt compelled to write in 1919: 'More children from the fit, less from the unfit – that is the chief issue of birth control'; and in *The Control of Parenthood* (1920) Stopes typically argued for compulsory sterilization of 'the insane, feebleminded . . . revolutionaries . . . half-castes'. Pearson initially shunned contraception, but eventually came round to the idea. For their part, Stopes and Sanger enthusiastically embraced the notion of eugenics, actively promoting birth control as a way of preventing the perpetuation of negative traits.

Racism was probably less of an issue in Britain than in the USA, where increasingly it marched hand in hand with eugenics. By the 1920s, the American military was measuring the intelligence of its recruits in a systematic way. The psychologist Carl Brigham, who wrote up his analysis of the army data in his book *A Study of Army Intelligence*, confided to a colleague that 'we are on the right track in our contention that the germ plasma coming into the country does not carry the possibilities of that arriving earlier'. A widely held belief was that two out every five of immigrants coming into the USA in steerage were 'feeble-minded'. Brigham concluded from his investigations that the Alpine and Mediterranean races were 'intellectually inferior to the representatives of the Nordic race'; and anybody in the slightest doubt about his curious conviction that blacks were inferior to whites had only to peruse his report for the army, in which he asserted that the average black person in the USA had 'the mental age of a ten-year-old'.[19]

Doctors in the USA had been sterilizing the 'feeble-minded' since the end of the nineteenth century. Various states passed eugenic laws, and in 1927 a test case in Virginia involving a seventeen-year-old girl called Carrie Buck gave the eugenics movement a real boost. Its supporters saw the Buck family as a perfect example of the kind of people who should be eliminated. Carrie was illegitimate and from a poor family, and, after being raped by a nephew of her foster parents, had given birth to a child. At the age of just seven months, this baby was judged, along with her mother and grandmother, to be feeble-minded. A suit was brought to the Supreme Court, where no less a figure than the great Justice Oliver Wendell Holmes Jr declared it legal for the state of Virginia to compel Carrie Buck to submit to being sterilized. This victory for the eugenics movement ushered in fifty years of legally sanctioned forced sterilizations across the country. In his judgment, Wendell Holmes commented that:

> We have seen more than once that the public welfare may call upon the best citizens for their lives. It would be strange if it could not call upon those who already sap the strength of the State for these lesser sacrifices, often not felt to be such by those concerned, in order to prevent our being swamped with incompetence. It is better for all the world, if instead of waiting to execute degenerate offspring for crime, or to let them starve for their imbecility, society can prevent those who are manifestly unfit from continuing their kind. The principle that sustains compulsory vaccination is broad enough to cover cutting the Fallopian tubes... Three generations of imbeciles are enough.

The case set a precedent for the eventual sterilization of around 8,300 people in the state of Virginia alone.

There were, of course, many objections to the sterilization laws. One important voice raised in dissent was that of the prominent New York lawyer Charles Boston, who was heavily critical and poured scorn on the idea of hereditary criminals: 'If criminal tendencies were hereditary, then there would be more substantial reason for sterilization of reckless chauffeurs than rapists.' But in spite of these more sensitive and liberal attitudes, a number of states operated a similar policy of sterilization for criminals, which continued until a 1942 test case in Oklahoma, when a man called Jack Skinner was to be sterilized as a repeat offender under Oklahoma's Habitual

Criminal Sterilization Act (1935), according to which the state could consider sterilization for individuals who had been convicted three or more times of crimes 'amounting to felonies involving moral turpitude' – he had been arrested for stealing chickens in his youth and later had had two convictions for armed robbery. Liberal elements in the state were waiting to make a challenge to such a case, and the sterilization order was overturned with the ruling that there was no evidence that criminality was hereditary. In *Skinner* v. *State of Oklahoma, Ex. Rel. Williamson*, the Supreme Court ruled that compulsory sterilization could not be ordered for a crime with the intention of thereby weeding out unfit individuals.

Jack Skinner was saved from vasectomy, though, because of something of a technicality. In section 195, the 1935 Act specifically stated that 'offenses arising out of the violation of the prohibitory laws, revenue acts, embezzlement, or political offenses, shall not come or be considered within the terms of this Act'. The court unanimously held that in this provision the Act violated the Equal Protection Clause of the Fourteenth Amendment. In passing judgment, Justice William Douglas stated:

> Oklahoma makes no attempt to say that he who commits larceny by trespass or trick or fraud has biologically inheritable traits which he who commits embezzlement lacks. We have not the slightest basis for inferring that that line has any significance in eugenics, nor that the inheritability of criminal traits follows the neat legal distinctions...the power to sterilize, if exercised, may have subtle, far-reaching and devastating effects. In evil or reckless hands it can cause races or types which are inimical to the dominant group to wither and disappear. There is no redemption for the individual whom the law touches. Any experiment which the State conducts is to his irreparable injury. He is forever deprived of a basic liberty.

Sterilization of the mentally ill, on the other hand, continued in the USA until the 1970s.

Eugenics had very powerful supporters, and even on the eve of the Second World War organizations such as the Rockefeller Foundation had been giving money to German eugenics research centres. It is not my plan to record or discuss in this book the appalling record of Nazi Germany. But the history of eugenics will always be inextricably linked to the devastation caused by the racial beliefs promulgated by Adolf Hitler and other German criminals. However, while that period would finally bring eugenics into utter disrepute, and there was widespread revulsion about Nazi ethnic cleansing, we should never forget that many of these attitudes are far from dead. Such opinions about superiority and inferiority are not confined to any one nation or political group. Given the power that our increasing knowledge of genes and their possible manipulation brings, there are some salutary lessons in recent human history.

The population bomb

A continuing concern of the eugenicists was that 'inferior' groups and races should not be allowed to breed to the detriment of the 'superior'. It is not surprising, therefore, that attempts to encourage birth-control measures in developing countries were often met with suspicion. Rapid population growth seemed to make sense to some governments as they aimed for economic development; for example, in Chairman Mao's 'Great Leap Forward' of the late 1950s and early 1960s, population growth was held to be a key condition for progress.[20]

In the latter half of the twentieth century the world's

population more than doubled, rising from 2.5 billion in 1950 to 6 billion in 2000. Most of this growth took place in the world's poorest countries. In Europe and the USA, meanwhile, birth rates stabilized and started to fall. It is often thought that the use of contraception was an important reason for this, but the stabilization of population growth in the west had much more to do with better education and social infrastructure, good economic circumstances, and improved hygiene and general health care. Malthus, writing in the 1790s, had been convinced that the UK would be uninhabitable by the 1930s: the rise in population growth, he thought, would mean that our agriculture could not possibly sustain the size of the population he anticipated. Malthus had reckoned, for example, without the invention of the flushing water closet, reducing the incidence of the water-borne diseases that had been so potent a cause of infant mortality. Increased infant survival had the effect of reducing the pressure to have large families to ensure that at least some of one's children would survive to support their parents in old age. So, while medical technology played its part, in advanced countries these pressures abated as social conditions improved, so that more children tended to flourish throughout childhood and advance into adulthood. In many parts of the developing world these changes are still slow in coming, and population growth still tends to be highest where social infrastructure is most primitive.

Twentieth-century economists in the Malthusian tradition viewed the rise in third world population with alarm. Particularly prominent among them was Paul Ehrlich, who in 1968 published a highly influential book called *The Population Bomb*. Ehrlich did not mince his words, beginning his book by asserting that 'the battle to feed all of humanity is over'. He argued for better population control and more equal distribution of food, but was pessimistic about the chance of success. Describing population growth as 'a cancer', he expected large-scale famine in the developing world. He was not alone.

Faced with spiralling population growth, governments in the developing world agreed on the need to do something. In the same year as *The Population Bomb* was published, the UN Conference on Human Rights in Tehran declared family planning as a basic human right. By the early 1970s I, as a young reproductive scientist, had got involved in this movement. My work on the potential for reversible sterilization was considered important by population agencies at the time, and I received various invitations to try to help promulgate sterilization in the developing world – in countries such as Bangladesh, the Philippines, Thailand and India – on the grounds that it need not be regarded as a permanent procedure. The World Health Organization in Geneva was heavily involved in the thrust to encourage sterilization, and devoted a considerable portion of its budget to researching and promoting new, acceptable methods of contraception.

In 1976–8 I was a member of the Scientific Steering Group in Geneva that studied these issues. A turning point came in my thinking when I began to realize that in some parts of the world there were quite shocking misuses of medical resources, and possibly corruption. In India, a mass sterilization campaign was promoted which resulted in men being paid to be sterilized. It is said that some individuals had their vasectomies repeated three times, with three separate payments. Others were in their eighties when they had the operation. Soon there were stories of people being forced off trains to be sterilized. We were told in Geneva that the Indian government had met its target of sterilizing seven million people – and yet even so the population certainly did not fall. So the government led by Indira Gandhi, who believed strongly in a targeted approach, introduced measures to enforce sterilization and vasectomy on poorer people. There were calls for the Prime Minister's resignation; in response, she declared a state of emergency, suspended democracy and censored the press, while her son Sanjay oversaw a programme of compulsory sterilization aimed at all men with two or more children. There were violent clashes which even saw police firing on civilians. These draconian moves eventually contributed to Mrs Gandhi's defeat in the Indian general elections. In Geneva, several of us on the WHO Steering Group were profoundly disturbed by this brutal implementation of enforced sterilization, and quietly resigned.

In 1994, the UN International Conference on Population and Development (ICPD) in Cairo tried to encourage a marked change in the controversial history of India's attempts to bring down the rate of its population growth by coercive sterilization. But with US money (a pledge of $325 million), the Indian government signed a deal to 'reorient and revitalize' family planning services in Uttar Pradesh – India's largest state and a development black-spot. Certainly, until recently there has been huge pressure on many women to get sterilized, even though it is now accepted that this will be unlikely to have a major impact on population demographics.

The frightening scenario that the world's population might increase to around 20 billion by 2025 – a view widely promoted as recently as the mid-1990s – does not look remotely likely now. Current predictions are that fertility rates will eventually drop and the transition to stability will be completed around the world, although this is happening at different rates in different areas, depending mostly not on reproductive technology but on the speed of social reform. In Africa, the transition is still only just beginning; in much of India, too, growth remains rapid. China, on the other hand, changed its policies and is close to stabilizing its population growth. The success of the 'one child' policy is partly attributable to strong government control: couples must apply to local authorities not only before getting married but also before the wife gets pregnant. Contraception – particularly female sterilization – is easily available and widely used, while when it fails abortion is compulsory after the first birth. Of course, there are well-documented concerns about all this, ranging from the coercive nature of the policy to the occurrence of female infanticide.

The 'total fertility rate' (TFR) in China is now down to 1.8, which is below the 'replacement rate' of 2.1. The replacement rate is the TFR required for population to remain constant – which means that each couple must produce on average 2.1 children (the 0.1 allowing for deaths before the child reaches reproductive age). Even once the TFR falls below 2.1, population will continue to grow for a bit, as for some time there are proportionally more young people of reproductive age. China's population is now expected to stabilize by 2050 at around 1.6 billion.

In most western countries, a TFR of 1.85 would actually represent an increase. The population of Europe is currently falling at a rate of about seven hundred thousand people per year, and throughout the developed nations the TFR is well below replacement. The exceptions are Albania, with an above-replacement TFR, and the US, which is roughly at replacement level. Such low TFRs have in the past been associated with bubonic plague or extended wars. Governments in countries such as Japan and Germany are already concerned about having a shrinking workforce in years to come. There is no doubt that lowered fecundity, combined with longer life expectancy, is changing the demographic structure of the British nation. The oft-mentioned 'pensions crisis' exists because there are proportionally fewer young people of working age to support financially those beyond retirement age, which leaves any government with difficult decisions to make.

As this book was being written in 2006, the President of the Russian Federation, Vladimir Putin, in his state of the nation address, expressed concern that his country's population was declining by at least seven hundred thousand people a year – greater even than the decline in Europe. He outlined a ten-year national programme to address this problem, including an increase in child-care benefits to support young mothers, particularly when they have had a second child. In Britain, too, the decline in fertility is now making the headlines. How interesting (and how unrealistic) that only very recently fertility specialists at the European Congress on Human Reproduction and Embryology called for free IVF on the National Health Service as one solution to halting the decline in our population.

But it is my belief that the technology I have discussed in this book, as well as carrying modest promises for improving the lot of many humans – particularly, it must be said, in the developed world – unquestionably contains seeds that could lead to our own destruction.

Trojan horses and changing genes

What a thing was this, too, which the mighty Odysseus wrought and
endured in the carven horse, wherein all we chiefs of the Argives
were sitting, bearing to the Trojans death and fate. But come now,
change thy theme, and sing of the building of the horse of wood,
which Epeius made with Athena's help, the horse which once
Odysseus led up into the citadel as a thing of guile, when he had
filled it with the men who sacked Ilium.

Thus wrote Homer, in the *Odyssey*, of how the ten-year
siege of Troy finally ended. The Greek attackers appeared
to have withdrawn; outside the city walls, the army had
burnt its camp and abandoned the siege, leaving behind a
giant wooden structure, a holy replica of a horse, apparently as a kind of peace offering or parting gift. At least,
this is the story that Sinos, the sole Greek left behind by
his comrades, peddled to the Trojans. Despite the warnings of Cassandra and Laocoön,[21] the Trojan defenders
dragged the horse inside the city and celebrated their long-
awaited freedom and the power the divine horse of the
Greeks would bring them by getting thoroughly drunk.
Unknowingly, what they had in fact brought within their
gates was death and destruction. For Epeius, who had
built the horse, had made it hollow. It contained thirty-
nine hand-picked Greeks, warriors armed to the teeth and
led by the indomitable Odysseus. During the riotous party
inside Troy, the Greek warriors slipped out of the belly of
the horse and opened the city gates to let in the Greek
army, which had been hiding nearby. The city was put to
the sword, pillaged and burnt. Laocoön and all men were
killed, Cassandra was raped and the women and children
became Greek slaves.

The image of the Trojan Horse has become a metaphor

of how to hack into a computer system and destroy it. Similarly, the phrase 'computer virus' is established in the day-to-day dictionary of information technology. The use of the word 'virus' is apt. As it turns out, biologists are using real viruses as Trojan horses, and in doing so presenting humankind with one of its most promising opportunities and gravest threats.

When a virus infects a cell, it first attaches to the outside of the cell using a 'friendly' protein. Viruses themselves are very small particles; so small, in fact, that they are too tiny to carry all the DNA they need to make a cell for themselves or to replicate. So, once attached to the host cell, the viral particle releases into that cell its own set of genetic instructions, injecting either a small amount of DNA or its 'message', RNA. In effect, the virus 'hijacks' the cell. It recruits the cell's ability to make the proteins it needs and forces the cell's own machinery to help manufacture thousands of new viruses. The result is pillage. The cell eventually breaks down, releasing more virus particles. Some viruses – among them herpes and HIV – do not reproduce immediately they have attached themselves to the host cell, but instead go to sleep, sometimes for years, before reproducing.

Dr Rudolf Jaenisch, a brilliant German researcher at the Salk Institute in La Jolla, California, was the first to use the ability of viruses to merge with the DNA in embryonic cells in 1976. For the past twenty or so years, Dr Jaenisch has been at the Whitehead Institute for Biomedical Research and is also professor of biology at the Massachusetts Institute of Technology, where he has continued his pioneering work into mouse models of human diseases, such as cancer.

Jaenisch's work heralded a biological revolution.[22] In one series of experiments he injected a monkey virus into four-day-old mouse embryos. After these had been trans-ferred to the uterus, many healthy baby mice were born; but about 40 per cent of them now had an altered genome. The DNA of the monkey virus had incorporated itself into the genes of these mice.

This was an extraordinary step. It implied that scientists might now be able to change the genes in animals. And the promise held up. Over the next two decades, using this technology, we would learn a great deal more about how genes work, what individual genes do, and what happens when they are modified or absent. The ability to introduce new genes into mice – to make them 'transgenic' – has turned out to be truly one of the most important advances in medicine of the last century. It has given us insight into development, disease and drug actions. It has probably been one of the most valuable single steps in the understanding of cancer, arthritis, heart disease, inflammation and infection. Just one example is the work of Dr Richard Palmiter, one of the very first pioneers in the whole field, who is still demonstrating how valuable transgenics are in understanding the workings of the neurotransmitters in the brain.

In the field of transplant medicine, the possibility of making transgenic large animals offers the prospect of 'manufacturing' replacement organs – such as the heart, liver, or kidney. Transgenics are already helping to produce pharmaceuticals: for example, the FSH that is used for virtually all the treatments described in this book is now made by genetic modification technology. Products such as insulin, used to save the lives of millions of diabetics, and the blood-clotting factors used to treat haemophilia (see page 322) may be made by modifying the proteins in the milk that sheep, goats or cows produce. Transgenic animals are also helping with our understanding of gene therapy – the great hope for the treatment of many serious diseases in the twenty-first century. They are also increasingly used as a means to understand how new drugs act and assess their safety for use in humans.

And the technology has even wider applications. It is now being established in industry and, to some extent, in agriculture. In one intriguing application by a Canadian company, scientists at Nexia Biotechnologies spliced spider genes into the cells of lactating goats. The milk the goats subsequently produced contained many very fine and immensely tough strands of silk. These polymer fibres, which can be produced in large quantities by this method, have been extracted and woven into a thread (which they called Biosteel), from which a light, tough, flexible material that can be used for surgical sutures, ultra-light military armour and tennis racket strings can be manufactured. Transgenics can also be used to increase milk yield: there are now cows that produce more milk, or milk closer in composition to that needed by human babies. Modified sheep grow more wool, and cattle grow faster to provide better meat yields and less fatty beef. Rapidly growing salmon are being produced for the farmed fish industry – for example, in Canada a transgenic salmon, AquAdvantage, grows four times faster than wild salmon, reaching about six times normal size in a single year. It is also hoped that transgenic technology may help to produce disease resistance in farm animals (this would certainly seem to be a better bet than cloning).

Leaving aside for a moment the enormous ethical implications of these developments, one constant problem has been the sheer inefficiency of the process of making transgenics. In mice, the species in which there has been the most success, the chance of producing a transgenic is usually only 2–5 per cent. In larger animals, such as the pig, perhaps 0.5 per cent of attempts end with an animal that may be useful. So scientists have continued to search for ways of improving the method. In the first major experiment, conducted in 1980 at Yale University, Jon Gordon injected a DNA sequence of interest directly into an egg, within hours of fertilization.[23] This was done before the first cell division so that the DNA would be incorporated into every 'daughter' cell that was generated. He used pieces of DNA merged together from two viruses, choosing this particular viral DNA because he had tests that could detect it; thus he could demonstrate it had successfully combined with the DNA of the embryo. After injection, the developing embryos were immediately transferred to a surrogate mother mouse. From several hundred injected embryos, seventy-two animals were born, and just two of these showed evidence of having incorporated the viral DNA into their genes. Gordon's novel approach provided the basis for the standard method of making transgenic animals, though various improvements have been made over the years. One of the many problems of this technique is that many copies of the DNA sequence of interest have to be injected. In one experiment at this stage, Gordon had to introduce thirty thousand copies of the piece of DNA into one egg in order to get the desired transformation. Another problem was that the DNA was incorporated into the genome randomly, and did not always work in a way that could be predicted.

A practical breakthrough was made in 1982, when Richard Palmiter and Ralph Brinster showed the potential power of this technique to alter inherited characteristics of an animal. They injected a gene which produced increased excretion of growth hormone in the mouse.[24] The mice that were born – so-called giant mice – were around 60 per cent larger than other mice from this strain. But the technology remained very tricky and the problems that Jon Gordon described in his original paper were not easy to iron out. I remember visiting Jon in, I think, about 1984, when he was working in New York. He had a reputation for being somewhat taciturn, but he was all courtesy. I was completely bowled over by two aspects of my visit. The first was his remarkable technique of microinjection into fertilized mouse eggs, involving the extraordinarily delicate use of fine glass injection pipettes and micromanipulators. I did not think then that we would adopt a very similar approach for pre-implantation diagnosis in our own lab a few years later. Jon then took me down to the animal house to see his transgenic mice – and it was there that I realized both the power and the uncertainty of this technology. At the time, he had been trying to study a variety of transgenes, and what was really deeply upsetting at that stage of this work was the number of unexpected mutations that insertion of a transgene produced. Some of his mice had skeletal deformities and could not walk normally; some had bald patches in their fur; some were blind, with failure of development of one or both eyes. The technology has now become much more sophisticated and most of these problems have been resolved, but the visit gave me a powerful reminder about our ethical responsibilities in doing this work with any animal species. Returning upstairs, I was also bowled over by Jon's highly intelligent and beautiful assistant, Beth Talansky, who like Jon was kindness itself on my visit to their laboratory. After watching her injection procedure for a bit, we started to chat about the Jewish attitude to the morals of doing such research. Both she and Jon Gordon impressed me deeply in their highly ethical and caring approach and the recognition that this was a technology that, though unquestionably important, needed to be pursued with great caution.

There have been a remarkable number of improvements in transgenic technology since those early days, and it is beyond the scope of this book to discuss them all. Since Jaenisch, Gordon and Palmiter did their early work, scientists have produced transgenics by introducing genetically modified cells into mouse embryos (so-called chimaeras), by electrofusion (passing an electric current between the embryonic cells and the DNA), by transfer of modified stem cells, by particle bombardment with a gene gun, and by cloning technology (nuclear transfer). All have considerable drawbacks.

One of the biggest difficulties is that all of them require access to an embryo. Obtaining embryos, particularly in large animals like cows, sheep or pigs, is not easy; it is labour intensive and highly inefficient. Like the Grand National of IVF treatment, using embryos for this purpose presents a series of hurdles. First of all, there is the need for an IVF-like facility. Even in an ideal facility, many eggs that are injected or modified do not implant or, if they do, grow for a while and then stop developing. Many unmodified offspring are born that show no evidence of the DNA being present; the DNA may be present but without expressing, or, if it expresses, it may express abnormally; other genes near the site of DNA insertion may not work normally; the gene inserted may work only for a short time – and so on. Not all these difficulties are associated with the use of embryos, but it is a major disadvantage.

Some twelve or so years ago I first met Carol Readhead, then working in Cedar-Sinai Medical Center in Los Angeles. As I mentioned in the previous chapter, she is an outstanding molecular biologist, a South African who, having done her PhD in Cambridge, moved with her eminent astronomer husband to the USA. Carol was interested in the work Alan Handyside and I were doing in pre-implantation diagnosis, and had some very bright ideas about how we might improve our technique for PCR (see page 332). Eventually, I had an opportunity to invite her back to the UK for a visit. I was running a two-day international symposium called 'Controversies in Reproductive Medicine' and had invited sixteen speakers from around the world to present contributions on different aspects of the challenges in our field. On the second evening, at the end of the meeting, the invited speakers came back to my house for dinner and the wine flowed freely. At that time (it was some months before they made the announcement in *Nature*) I was fairly convinced that certain scientists were trying to clone a sheep in Scotland, so I sat the delightful speaker from the Roslin Institute between two of the most flirtatious female members of my team, giving them instructions to ply him with enough drink to find out precisely what was happening in his laboratory. At the end of a very productive evening we eventually called him a taxi and he went, ecstatically happy, back to his hotel. I was fully debriefed at our rather headachy laboratory meeting the following morning.[25]

That previous evening, once the dinner guests had left, Carol and I had sat drinking yet another bottle of burgundy. Our conversation became increasingly incoherent and eventually I fell asleep in my chair. About three months later, I received a phone call from Los Angeles. 'Robert,' said the voice, 'that idea you had. You ought to try it.' I vaguely recognized Carol's voice (she has never quite lost a very faint South African intonation that is more apparent when she's excited). 'What idea?' I said.

'The idea about the sperm,' came the little voice. 'What sperm? That must have been your idea, Carol.' We argued for about ten minutes about whose idea it wasn't. Then we agreed to team up for the attempt to make a transgenic animal by what seemed like an entirely novel method.

Knowing that the testis produces stem cells that eventually make sperm, we decided to try to introduce DNA of interest into these stem cells. If we could make modified sperm, we would get around one of the major difficulties in transgenic technology: we would no longer need embryos, because transgenics could be produced by natural mating.

We had two strategies to follow. One was to inject the DNA directly into the tubules of the testis; the other was to extract stem cells, modify them and then return them to the tubules by injection. This basic idea has led to over ten years of collaboration, a collaboration which still continues and has required us both to fly backwards and forwards between our respective laboratories. One very big step forward was the idea of the Trojan horse.

After various false starts, Carol and I started to use viruses as our Trojan horse, deriving considerable help from an idea that was being considered for gene therapy in tissues like the brain and lung. In some brilliant research, Luigi Naldini at the Salk Institute in La Jolla, California, had used certain viruses to get DNA into neurons for the possible treatment of neurological disorders.[26] What we wanted to do was to use his technique to get not into somatic cells, but into the germ-line. Using the right virus, such as HIV, we could modify it to contain the DNA we particularly wanted to be incorporated permanently in any offspring. The viral particles we injected would be Trojan horses which, to mix metaphors, would piggy-back the DNA into the precursors of the sperm. When they divided, leading to the production of sperm, the hope was that each spermatozoon might carry a copy of the DNA fragment we wanted to insert. Of course, the HIV would have to be modified so that it did not replicate. It would also need to be rendered harmless, so that when (with my habitual carelessness) I squirted solutions containing hundreds of thousands of viral particles around the laboratory before injecting the testicles of our mice, though Carol might get wet, she would not have a serious reason to be angry with me.

I shall not describe the ups and downs of this experiment: making viruses that did not work, a technician who we think may have given us some corrupted viral material, animals that did not arrive from suppliers on time, a rival scientist who seemed to want to publish the results from our researches before we did, my numerous botched surgical jobs at injecting these incredibly fine tubules – which didn't hurt the mice at all, but dented our small budget and meant more tiresome transatlantic journeys. We repeatedly got entirely healthy offspring which, when we analysed their DNA, showed no evidence at all of having incorporated our 'signalling gene'. This signalling gene, which could be detected on PCR, also produced the green fluorescent protein which, when bathed in light of the right wavelength, should have glowed to show that we had actually got the DNA incorporated into the genome of new baby mice.

After some years of this work I made yet another visit to California, where Carol and I – both of us by now being distracted by many other projects – were going to attempt a further experiment. By this time Carol had moved her laboratory to Caltech: the Californian Institute of Technology, one of the most advanced science universities in the world. Inevitably, as is always my fate, Carol's lab was in the sub-basement of one of its dark buildings; one loses any sense of time down there, but at least there is no warming from the intense Southern Californian sun. It was a Sunday; having arrived only two days earlier, I was feeling jet-lagged, and Carol was dog-tired, having worked continuously until six-thirty that evening. We both wanted to stop, but eventually I thought I ought to maximize my time by doing one more mouse experiment. Carol offered to help, but I irritably said I was quite capable of injecting mouse testicles on my own and suggested that instead she go next door to an adjoining lab where she could do something useful – run the PCR on DNA from previous mouse experiments we had done three or four months earlier. She was reluctant; feeling, as I did, very averse to the idea of getting yet another negative result, she had been putting off the evil hour for months. By this stage we were both getting dispirited. Eventually I persuaded her that I would not break anything and she could trust me on my own. So, once she had run some gels, she prepared the darkroom for a search to see if any of our DNA had been incorporated. After an hour or so, I had finished with our mouse experiment. I let the mouse gently recover from its anaesthetic in some warmed cotton wool, and once it was back in its cage, whiffling around and happily exploring its environment, I slipped into the laboratory next door. It was now nearly eight in the evening, and the echoing basement was ghostly, completely deserted. Suddenly, I heard an exceedingly strange high-pitched noise behind the darkroom door. Banging on the door, I went in. There was Carol, wearing her Cyberman helmet in the ultraviolet glow, staring fixedly at the gel and – well, I can only describe it as squeaking. On the gel, we could clearly see seventeen bands of our DNA signalling gene. Around 80 per cent of the litters that had been born in that particular experiment were transgenic.

There was no music as such, but we waltzed around the laboratory, up and down the tired benches of fading bottles and chemical reagents. This was surely the start of a good paper in *Nature*. I asked Carol what the best restaurant in Los Angeles was. She said she was certain it was L'Orangerie – but it had a three- or four-week waiting list to get in, being in an upmarket, celebrity part of town on the edge of Hollywood. I told her that of course we would get a table, and she kept on insisting it would be impossible. I rang and, using a ploy that would normally have acutely embarrassed me, announced myself in a superior English voice as 'Lord Winston'. That was sufficient. It wouldn't have worked at The Ivy in London, but Californian restaurateurs sometimes bow before kings and aristocrats. Carol offered to get the car but I insisted, wisely, on a taxi. At the end of the meal and after some very good Californian wines, we were not exactly going to be in a fit state to drive.

Since that work was done, we have extended the technique to see if we could repeat the same process in the pig. Currently, this really interesting research is being vigorously pursued by our lovely team of enthusiastic scientists, Feride Oeztuerk-Winder, Anil Chandrashekran and Ellen Poon. There is a very good reason for doing this, as pig organs – particularly the heart and kidney – are not dissimilar in size from those in the human and function in a similar manner. If we could find a way of modifying the pig's proteins, we might have a transgenic animal whose tissues would not be immediately attacked by the human immune system. There is no doubt that if we could use the pig for xenotransplantation in this way, many thousands of lives could be saved. One of the beauties of this experiment is that any transgenic pigs would be entirely normal and would not suffer. They would be kept in very good, clean environments and looked after very well – much better, indeed, I suspect, than most meat animals. In a society that farms animals for food, it would seem difficult to deny that farming animals to save human lives carries undoubted ethical advantages.

Ethical concerns and religious positions

The justification for transgenic research is obvious; but the ends do not necessarily justify the means. Certain aspects of transgenic research raise very serious issues about animal well-being. One concerns the inefficiency of the techniques. Even with the various improvements that have been made, probably only 25 per cent of transgenic embryos survive until weaning and only around 5–10 per cent survive to adulthood. The later the point at which the pregnancy fails, the graver the implications for fetus and mother. Mutated genes can also have a serious effect on animals' health. Some of the mice that have been created have a high probability of developing cancer. That makes them good models for studying the mechanism and treatment of that disease; but do we humans have the right to inflict that kind of pain on any other species? And if we do, what are the circumstances where it is ethically justified? Is human welfare the only consideration? Should we not be concerned with the welfare of other life-forms?

From time to time, there are demands for universal protocols for transgenics. For example, there is widespread agreement that it is reasonable that transgenics should be created and used for biomedical research where there is no other valid way of obtaining the information. Some people suggest that, because of the inefficiency of the technique and because of the unpredictable nature of these deliberately induced mutations, there should be stricter guidelines about what is acceptable. One of the advantages of the method that Carol and I devised is that many of the objections about inefficiency fall by the wayside. It is undoubtedly a more humane methodology: no embryos are manipulated and there is much less risk of pregnancy loss. The use of natural mating, followed by in-breeding where necessary, offers a huge advantage. Either the newborn animal is a transgenic, or the whole protocol has not worked. Of course, there are still problems about how genes may be expressed and the effect this may have on the animal's health, but most transgenic animals are, in practice, completely well and fit.

Another question is whether scientists should focus more on in vitro transgenic methods – that is to say, replacing the use of live animals with the study of cell cultures, where the genes inside cells have been altered. This is certainly desirable and is occasionally useful, but a key aspect of transgenic research is that live animals offer a unique and dynamic way of examining how genes function, how their function may change, what happens with ageing, and so on. Yet another issue is the problem of dangers to the environment. The use of transgenic fish has certainly occupied the thinking of the Royal Society – the world's oldest and most respected science academy. One of the Royal Society's concerns is the risk of transgenic fish escaping from a fish farm and mating with wild fish, thus potentially causing havoc in the marine environment, and possibly even changing the course of evolution.

An increasingly knotty question is whether it is possible to patent a transgenic animal. One immediate problem with such patents is that more animals may suffer because more experiments will need repetition. Moreover, there is growing concern that the fruits of scientific research – particularly research of this kind – should be freely available for general benefit. For this reason the pharmaceutical industry has agreed that the details of new mutants and strain characteristics should be widely published so that unnecessary duplication is avoided.

There are also a number of measures that can be taken to reduce the hardship undergone by animals. One is that any transgenic embryos that have been used to produce a particular transgenic line are kept frozen, stored in liquid nitrogen, and made freely available to other workers in the field. It is also widely agreed that more use of the transport of frozen embryos in appropriate containers should be made, as transferring these to the uterus of a surrogate mother when they arrive at their destination is clearly less stressful than transporting live adult animals across long distances by air or by road.

Schopenhauer once remarked that the denial of rights to animals is a doctrine peculiar to western civilization and reflects a barbarism that has its roots in Judaism.[27] This seems a travesty both of the Jewish Bible and of the Christian tradition that emerged from it. There is a notion in both Christianity and Judaism that we humans are responsible for all God's works, including the animals. The Jewish position is clear: humankind has a moral duty to protect animals at all times. Animals may not even work on the Sabbath, and (in respect for its instincts) the ox may not be muzzled when threshing corn. 'Should I not have pity on Nineveh . . . wherein there are more than six score thousand persons . . . and also much cattle?'

asks God of Jonah when the prophet complains that God has destroyed the gourd which sheltered him from the midday sun. Of course we use animals for food. But slaughter must be carried out in the most humane manner possible. At all times we are expected to alleviate *za-ar ba'alei chayyim* – 'the pain of living creatures'. With regard to causing pain for serious human benefit, Moses Maimonides (d. 1204), the greatest philosopher of his age, argued that it is 'permissible for the needs of man'. This is why he argues that killing animals for the trivial purpose of sport is wrong. In various opinions, the rabbinical authorities go to great lengths to decide when the infliction of pain is permissible, and (like the Home Office regulations) when it is not permitted, in cases where the pain is excessive, or the procedure is not aimed at improving human health.

There is no question that, in spite of Schopenhauer, Judaism absolutely forbids cruelty to animals. But in Jewish law, animals cannot have 'rights'. On this point, ironically, Schopenhauer is correct because, of course, an animal lacks the capacity to institute judicial proceedings to prevent acts of cruelty of which it may be a victim. This is essentially why Judaism has to take on the role of protector, acting on behalf of the animal to safeguard it. Most Jewish sources regard the use of animals in experimental research as undoubtedly justified provided there is good cause and this work is done with humanity. There is a paramount principle that human life must be protected. The use of animals, even breeding a transgenic mouse that develops a cancer that causes it pain and suffering, may be permitted so long as the work is done in such a way that suffering is minimized. Under these circumstances, animal experimentation which is of only potential value to humans is also permitted. While there is no express prohibition on manipulation of genes, there is a general principle that we humans have a responsibility for the planet and our environment. So, while we may have dominion over animals, we may not be reckless about the environmental problems we could cause. This view, which originates much earlier in Jewish and then Christian thinking, was formulated by the thirteenth century.

In my field of medicine, progress could not have been achieved without animal research. For example, it would be unthinkable to transfer a human embryo after IVF without checking first in animals that the different procedures used did not cause abnormalities in any offspring. And our ability to prevent fatal genetic defects that kill thousands of babies in their early years was gained only by ensuring that sampling the embryo did not dangerously impair development.

There are a number of erroneous perceptions about research using animals. One of the commonest is that animal testing and transgenics are unnecessary and do not give reliable information about how humans may react to a treatment or drug. This is factually incorrect. Ninety per cent of the time information derived from animal work, particularly from transgenic experiments, is utterly precious and could not be achieved otherwise without threat to human health. Second, there is a widespread view that animals suffer terribly during experiments. But this is not usually the case, and is certainly not true in, for example, the kind of work I and many of my colleagues have done. Our use of transgenic animals mainly means that they may have tissues that glow in the dark under certain conditions; they live happily and mate normally. To this day, the great majority of animal research, including that with transgenics, involves little or no suffering at all in the animals used. And where the slightest suffering is involved, in the UK there is stringent control by the Home Office. This ensures that, in the very rare cases where pain might occur, this is minimized and anaesthetics or painkillers are responsibly used by fully trained persons under supervision. If an animal is suffering in a way that cannot be immediately palliated, then the experiment must cease. It may require humane killing of the animal; but then, our society is perfectly ready to allow humans to kill animals for the purposes of fulfilling our requirement for nutritious food. If the slightest transgression in any aspect of animal research occurs, the Home Office takes over. Even if one animal more is used than was specified in the original research application, this will lead immediately to the permanent loss of that researcher's licence – and effectively the end of a career in the field.

The third erroneous perception is the most evil: that it is justifiable to offer violence to people like myself who are committed to this work. Even more wicked is the violence offered to people who, themselves uninvolved in the work, may offer services (banking, food supplies, building works) to those who do animal experiments. It is shocking that, in a free country, my family is apprehensive that I dare even to write in defence of this work, which I (and the great majority of the British population) consider totally justified.

Why not take the next step?

The ability to modify DNA in the way that I describe is one of the most powerful technologies ever invented by man. The pioneering work done by my colleagues so many years ago has now started to take a turn that will have huge implications for humanity.

For some time now, doctors and scientists have been using gene therapy in attempts to cure very ill patients. Fatal immune-deficiency in children has been cured by modifying the genes that these children carry, and some important progress is being made with other disorders. At the time of writing, the US National Library of Medicine records 374 approved trials in gene therapy for various cancers, including melanoma, or tumours of head and neck, brain, breast and prostate. There are recent reports that people dying from malignant melanoma have had their lives saved using this technology. There have also been attempts to treat rheumatoid arthritis, sepsis and HIV, and a number of genetic diseases, such as haemophilia, muscular dystrophy and cystic fibrosis. It is fair to say that so far, these attempts at gene therapy have been very fraught and successful treatments rare. The patients who are being treated are inevitably near the end of life: gene therapy has been sanctioned in most cases only for people who are desperately ill and as a last resort. Of those 374 trials currently registered, few have registered much success. However, the important point is that these treatments are all concerned with somatic cell therapy: that is, they may change the genes in some of the patient's tissues, but never those in the ovary or testis. The treatments do not alter the germ-line and therefore the effects cannot be inherited. Indeed, some time ago, scientists universally agreed that there should be a total ban on any attempts to change the human germ-line. Somatic-cell gene therapy is dangerous enough, and the idea of harming a future generation was unthinkable.

But it is very clear that one reason why germ-line gene therapy has never been contemplated is a practical one. Making transgenics in any species is highly unreliable, of poor efficiency and unpredictable. Supposing these problems could be solved? I think that experiments like those Carol and I have done open new ethical issues. If transgenic modification becomes easy, it seems almost certain it will eventually be used in humans. Some scientists – among them James Watson, joint winner of the Nobel Prize for discovering the structure of DNA – in championing germ-line therapy, have pointed out that it must surely be better to treat these diseases before they cripple people, rather than wait until they are in extremis. And, of course, modifying the germ-line could pre-empt the disease process.

And yet the enthusiasm for tampering with the human germ-line to prevent serious disease is unquestionably out of all proportion to what could be achieved. Most 'genetic' diseases – cancer, heart disease, diabetes, arthritis and so on – have a genetic *basis*. For the most part, they are the result of many genes working together in a particular way. Changing one part of the human genome would be quite unlikely to have much positive effect on these conditions. So the only medical application for germ-line therapy at present would be in combating those genetic disorders that are caused by adverse mutations in a single gene.

The trouble is that, although there some six thousand single-gene disorders – collectively a large number – individually they are rare. Cystic fibrosis, as we have seen (page 352), is the most common, but many others affect perhaps no more than two dozen families in a country the size of the UK. For these, rather than attempt to change the genome, it is simpler, cheaper and infinitely safer to use pre-implantation diagnosis to screen embryos in those families where the mutations are known to occur.

There may possibly be one or two exceptions to this. Perhaps most notable is beta-thalassaemia, the disease of haemoglobin that makes people very anaemic. It is caused by a mutation in just one or two genes, and there are clusters of sufferers in several countries around the world. Beta-thalassaemia was first described in 1925 in Italian children, but it is common in many Mediterranean countries, China, South-East Asia and parts of Africa. In all, about half a million children suffer from it. At about nine months of age, these children develop anaemia and increased activity in the bone marrow, and their spleen and liver enlarge to make more blood cells. These tend to break down rapidly and the children become jaundiced. Sometimes they have gallstones made of blood products. In an attempt to combat the anaemia, the red bone marrow increases in volume. The bones of the skull swell, and sometimes the children's faces become deformed. Weakening of bones makes the children prone to fractures, and marrow expansion can also lead to compression of the spinal cord, producing partial paralysis. These children survive only with regular blood transfusions and doses of folic acid; and that treatment produces its own side-effects. Iron overload and infections resulting from the repeated transfusions eventually cause heart disease, hormone disorders such as diabetes, impaired sexual development, liver and lung failure. The condition is invariably fatal eventually, though sufferers reach young adulthood, living to the age of twenty or twenty-five in a crippled state.

On the island of Sardinia, around one in six of the population carry the mutation. Caring for those affected takes around 60 per cent of the island's health-care budget. The only actual 'possibility' of a cure might be a bone marrow transplant – but at perhaps £150,000 a child, this is completely unrealistic. It seems highly tempting to consider gene therapy for the whole population – to prevent this killer disease in the first place. Supposing there were a simple mode of treatment – perhaps an injection into the testicle of newborn boys to prevent their carrying the disease – so that in time, their children would be free of the affliction?

The first problem is that there is no single mutation: there are about 20 common ones and another 150 or so that cause the problem in the genes that make red cells. So any germ-line gene therapy would not be as simple as, say, a vaccination; it would probably have to be tailored to the individual's genome. But there is another intriguing issue. Just possibly this mutation has been selected for in evolution; surprisingly, it may even be advantageous in some circumstances. Skulls have been dug up on Sardinia and carbon-dated to around three thousand years ago. A number of them show the facial and forehead deformity we associate with the production of extra bone marrow. These people have lived for three millennia, harbouring a fatal gene that should have been bred out. How come it has survived?

It turns out that blood disorders like thalassaemia may possibly protect against another fatal human disease – malaria. People with thalassaemia do not get malaria. The distinguished haematologist Sir David Weatherall has recently offered a very interesting, and paradoxical, explanation for this.[28] It seems that very small babies who are exposed to the malarial parasite at a very young age may actually be more susceptible than others; but, having had exposure to the malarial parasite as infants, they develop immunity which protects them if they survive into adulthood. Whatever the precise mechanism at work, it could be a catastrophe to engineer the thalassaemia gene out of this population. If global warming continues and the mosquito *Anopheles gambiae*, the principal vector for malaria (responsible for 500 million cases of malaria in Africa annually), sweeps up into the Mediterranean in future, our meddling could contribute to wiping out a complete island population.

Enhancing humans: nightmare or promise?

But perhaps the biggest threat facing us all – a true Trojan horse – is the prospect of germ-line therapy to enhance humans. We already undoubtedly tamper with many of our attributes. We use orthodontic treatment and breast augmentation to increase our physical attractiveness, sportspeople train vigorously to increase muscular strength and physical co-ordination, and vast numbers of people take hormone replacement therapy or use laser treatments to improve vision. In one particular area, human cognition, we stake a huge amount on the best education for our children at school and university. But these methods of enhancement all take effort and are of limited efficacy. Why not do the job properly and permanently by manipulating the genes?

Lee Silver, in his book *Remaking Eden*,[29] argues that it is only a matter of time before we try to manipulate the human mind and senses through genetic engineering. In his view, we shall soon eliminate alcoholism and tendencies towards extremes of anti-social behaviour. Visual and auditory acuity will be enhanced in some people, he asserts, to improve artistic potential. Next will come higher intelligence and better memory. Then he foresees a time when we shall use genes from other species to confer on ourselves the ability to see objects in complete darkness using sonar (like bats), or a much more sensitive sense of smell using the genes of dogs, or even genes that might help to expand human perception still further using such abilities as radiotelepathy! I wonder. Is it any accident that Lee Silver dedicates his book to Joseph and Ethel Silver for 'creating me in the old-fashioned way'?

Such speculation is not exactly helpful if we are to have informed and reasoned discussion about the future of reproductive technology. It elicits hostility from many quarters, not only from people with a strongly spiritual outlook. Leon Kass, biologist turned philosopher at the University of Chicago and former chairman of the President's Council on Bioethics, feels that embryo research stands at the top end of a slippery slope. 'In leading laboratories, academic and industrial, new creators are confidently amassing their powers and quietly honing their skills, while on the street their evangelists are zealously prophesying a post-human future.' Kass believes that, ever since IVF, human life has been in danger. The fertility specialist has subverted normal values by becoming the third parent. Now, if left unchallenged, scientists will genetically engineer babies, grow newborns in laboratory vats and use them as helpless factories to produce medical products and human parts. His nightmarish vision of the future is echoed by another member of the PCBE, the renowned neo-conservative spokesman and Johns Hopkins University professor Francis Fukuyama, who argues that the new procedures being developed will allow us just as easily to enhance the species as to alleviate suffering – raising once again the spectre of the eugenics attempted by Nazi scientists in their quest to create a race of muscular, athletic, 'pure' Aryans.

Clearly there are massive risks in this technology. There are all kinds of legitimate objections to genetic enhancement. First, it increases inequalities, in a human society which already has too many of them. Do we want to increase strife by creating more underclasses? And if we succeed in enhancing our children, we really do risk making them into a commodity. We may find that we do not like the result of what we have done to them, and reject them. Alternatively, we may have given them – permanently – attributes which we felt were desirable in our society and our generation, but which they feel to be a disadvantage in theirs. Nor should we forget that this technology would produce potentially permanent, irreversible changes in future generations. Frank Ruddle, the pioneer who worked with Jon Gordon, once said, with perhaps a little too much wide-eyed enthusiasm, 'One of the wonderful things about transgenesis is that we can do all at once what evolution has taken millions of years to do.' That is true – and one of the reassuring things about evolution is that it works through gradual change.

The results, too, are likely to be entirely unpredictable. We may think we know what a particular gene does when it is present, or its expression is modified. Silver's view of gene action, like that of so many neo-Darwinians, seems very simplistic. For example, it takes no account of the epigenetic effects I attempted to describe on page 396. We are eons away from understanding the interaction of genes and their delicate balance – a delicate balance that we could so easily upset by trying to grab an illusory power.

Perhaps the greatest threat concerns human values. We think we live in a society that no longer needs God, or at least no longer needs spiritual values. On the contrary, we do – and more than ever. When we say we believe that human life is sacred, whether we consider ourselves religious or not, we are asserting that there is a special value to humanity. For the religious it may be expressed as a feeling that we are made in the image of God. For the non-believer it is because we believe that humanness has a unique and supreme quality. If we end up changing the human genome, eventually we shall be changing our species. And once we create superhumans, what value can *human* life have?

Of course this technology contains dangers – but this is true of all humankind's technological advances. Ultimately, demagogues like Kass and Fukuyama do not impress with their rhetoric. I do not think we have handled these issues well in our society; strict regulation without genuine consensus, and without the co-operation of the clever people who may outwit the regulators, may offer short-term safety but is certainly not going to do our society good in the longer term. And effective international regulation seems a fantasy. In a global and commercial environment we shall need much more than senseless rules, and temporary solutions for complex problems.

What we need, above all, is to learn how to use our power with wisdom. We have dragged a Trojan horse into our citadel. But the critics are wrong to baulk at valuable technologies that open doors for others of more questionable merit. The Trojan horse may be inside the city, but we shall be able to contain the threat it carries in its belly. What we must not do is allow ourselves to get drunk on celebrating the power that it might bring us. Rather, that power must be used responsibly. Its misuse will be minimized if we always keep our own humanity in mind. In this respect, reproductive technology, with its ability to affect the happiness of individuals so profoundly, is a good model for how we should deal with the rapid advance of technological knowledge in general. Scientific knowledge is the property of all of society. We need to ensure that it is firmly embedded in a culture of openness and ethics, and taught, pursued and developed for the benefit of humankind and our environment.

Notes

Prologue: Visions of Mortality

1 Christians may be struck by the resonant image of Isaac's ascent, carrying the wood on his back like a cross, while his father prepares to sacrifice him.

2 Secondary infertility is defined as infertility after a previous pregnancy or miscarriage. It may be caused by tubal damage – sometimes a result of infection contracted during childbirth. But in Leah's case, the text may suggest that her infertility is due to an impending menopause.

3 Mandrakes are called *dudaim* in Hebrew, a term seemingly connected with the word *dud*, which means 'beloved'. They have violet flowers and a yellow, tomato-like fruit with an intoxicating odour. While I have never seen them in my local supermarket, remarkable medicinal qualities were attributed to them in ancient times, including aphrodisiac properties and the power to stimulate fertility.

4 Gregor Mendel (1822–84) was an Augustinian monk who taught natural science to high-school students in Brno, Moravia. His breeding of pea plants eventually led to the understanding of heredity and the gene (or heredity

factor) as the essential unit of inheritance. His laws of inheritance state that (1) hereditary factors do not mix, but are passed intact; (2) each parent transmits only half of its hereditary factors to each offspring (with certain factors 'dominant' over others); (3) different offspring from the same parents receive different sets of hereditary factors.

Chapter 1: The Beginnings of Life

1 This effect is pulsatile – that is to say, the FSH is released in short bursts at regular intervals. This regular fluctuation is important in producing a response in the target organ, the ovary.

2 The follicle is the cystic container in the ovary in which each egg matures. Follicles start as microscopic objects but, during the weeks leading up to ovulation, some fill with fluid and expand up to about 2 centimetres in diameter. After ovulation, all the follicles which have started to grow during that cycle shrink and, together with the eggs inside them, become minute areas of scar tissue.

3 Twin pregnancy is a risky business. The risk of one or both fetuses dying is high, and twins have an added risk of being abnormal. One in twenty-three twins is born dead and about one in twelve has a serious birth defect.

4 R. Henkel, G. Maass, H. C. Schuppe et al., 'Molecular aspects of declining sperm motility in older men', *Fertility and Sterility* 84: 5 (2005), 1430–7.

5 H. N. Seuanez, 'Chromosomes and spermatozoa of the African great apes', *Journal of Reproduction and Fertility, Supplement*, 28 (1980), 91–104.

6 Once a sperm leaves the male it still has to undergo a number of changes before it can fertilize the egg. The first step in achieving its maturity is called *capacitation*. Capacitation is not fully understood even fifty years after the Cambridge scientists (see page 87) did their pioneering work, but it seems that the sperm's surface and its metabolism change. After capacitation, the sperm are

capable of rapid, effective *hypermotility*, which helps them seek the egg, penetrate and fertilize it. Next, the *acrosome reaction* involves chemical changes to the head of the sperm which help it to pierce the outer coat of the egg. These changes happen when the membranes surrounding the sperm head come into contact with proteins surrounding the egg 'shell'. *Cumulus penetration* is the process whereby the sperm carve their way through the surrounding mass of helper cells (which come from the follicle) to reach the egg. The sperm head produces substances that loosen these cells, dissolving the matrix that holds them together. *Sperm binding* is the final process before the sperm penetrates the egg itself. The sperm stick to the outer shell of the egg (zona pellucida) because there is a further chemical reaction between the proteins that coat its surface.

7 From the Greek *parthenos*, meaning virgin, and *genesis*, meaning birth.

8 T. Kono, Y. Obata, Q. Wu et al., 'Birth of parthenogenetic mice that can develop to adulthood', *Nature* 428: 6985 (2004), 860–4.

9 Ibid.

10 For those who are interested, the more detailed descriptions in my book *Getting Pregnant* (London: Pan Macmillan, 1993) may be helpful.

11 These side-effects are all reversible once a person stops taking the drug.

12 LHRH is the hormone released by the hypothalamus in the brain which stimulates the pituitary gland to produce LH.

13 Y. Sheynkin et al., 'Increase in scrotal temperature in laptop computer users', *Human Reproduction* 29 (2005), 452–5.

14 Or just use a MacBook Pro, in which case they will only get overheating on one side – the left – which is where the processor seems to be situated, judging from the heat this computer emits on my lap.

15 An excellent example of this is seen in the paper by the Austrian group, J. E. Lackner et al., 'Constant decline in sperm concentration in infertile males in an urban population: experience over 18 years', *Fertility and Sterility* 84 (2005), 1657–61. They found that sperm counts had declined rapidly – from around 27 million to 5 million. What they did not emphasize was that sperm motility in the same group of men had risen sixfold. Such studies – and there are many of them – need the greatest care in interpretation.

Chapter 2: IVF: A 'blind assertion of will against our bodily nature'?

1 *Journal of the American Medical Association* 30 (1899), 1428–9.
2 S. L. Schenk, 'Das Säugetierei künstlich befruchtet ausserhalb des Muttertieres', *Mitteilungen aus dem Embryölischen Institut der Kaiserlich-Königlichen Universität in Wien* (1878), 1107.
3 *In vitro* comes from the Latin, meaning 'in glassware'.
4 Spallanzani's friend Malpighi had already attempted artificial insemination of silkworms, but the procedure had failed.
5 W. Heape, 'Preliminary note on the transplantation and growth of mammalian ova within a uterine foster-mother', *Proceedings of the Royal Society of London* 48 (1890), 457–8.
6 It is interesting that Buckley's name does not appear on this classic paper. Nowadays, of course, he would have insisted, if only because of the pressure of the university's Research Assessment Exercise.
7 We now know that the carrying mother, even if not genetically related to the embryo in her uterus, does have some influence on how it develops and some of the characteristics it may exhibit later in life.
8 E. V. Allen and E. A. Doisy, 'An ovarian hormone: preliminary report of its localisation, extraction and par-

tial purification and action in test animals', *Journal of the American Medical Association* 81 (1923), 819–21.

9 Oestriol is the water-soluble oestrogen which used to be tested during pregnancy to try to detect if the placenta was working normally.

10 To be fair, the notion of informed consent and the need for ethical approval were not matters with which many in the medical profession concerned themselves greatly in the 1930s. Hertig and Rock were probably no worse than most of their colleagues in this respect.

11 I. Menken and J. Rock, 'In vitro fertilization and cleavage of human oocytes', *American Journal of Obstetrics and Gynaecology* 55 (1948), 440–51.

12 It is interesting that, even today, sperm capacitation is by no means understood, and there are many aspects of the chemistry that are somewhat mysterious.

13 D. G. Whittingham, 'Fertilization of mouse eggs in vitro', *Nature* 220 (1968), 592–3.

14 M. C. Chang, 'Fertilization of rabbit ova in vitro', *Nature* 184 (1959), 466–7.

15 This was far-sighted. It is now clear that stimulation of ovulation, particularly with high doses of FSH, may produce abnormal follicles containing eggs which are not properly mature. So, on the face of it, Steptoe's strategy was very reasonable, given the knowledge at his disposal.

16 Genesis 38: 2–4. Onan was the younger son of Judah, and his elder brother, Er, died childless. So Onan was required to sleep with Er's wife, Tamar, so that she might have a child – thus carrying on Judah's 'genetic' inheritance. But Onan practised coitus interruptus – either as a contraceptive act, or because he did not want to have a child that would not be regarded legally as his; the Bible does not make his motives clear. For this reason the rabbis of old forbade coitus interruptus, although one dissenting voice – that of Rabbi Eliezer (Talmud: Yevamot 34b) – recommends it as a way of keeping a mother's milk strong during lactation.

17 Leon R. Kass MD PhD is professor at the University of Chicago and Hertog fellow in social thought at the American Enterprise Institute. He was chairman of the President's Council on Bioethics from 2001 to 2005, and had begun his scientific career in research in molecular biology at the National Institutes of Health before (as he says) shifting direction from doing science to thinking about its human meaning.

18 Paul Ramsey, *Fabricated Man: Ethics of Genetic Control* (Harvard, Mass.: Yale University Press, 1970).

19 The Canadian Shulamith Firestone is a feminist writer of considerable repute, the author of *The Dialectics of Sex: A Case for Feminist Revolution*, in which she argued that women should totally control the means of reproduction. Her influence was profound, particularly in the USA.

20 I should point out that, in practice, the death penalty was never enacted for this offence.

21 Tractate *Sanhedrin* 91b. However interesting, the points made in this conversation did not become established in the Jewish legal framework.

22 Pope John Paul II, 'Ad eos qui conventui de biologiae experimentis in Vaticana Civitate habito interfuere', 23 Oct. 1982.

23 'In vitro fertilization and public policy', Catholic Information Services, 1983, p. 8.

24 Aneuploidy is the condition where there is an abnormal number of chromosomes; aneuploidy screening is a technique used to examine individual embryos to check whether the chromosomes show evidence of this abnormality.

25 St Gregory of Nyssa (*c.*330–95), *Adversus Macedonianos*.

26 *Test Tube Babies – a Christian View,* with a foreword by Sir John Peel (London: Becket Publications/Unity Press, 1983).

27 For example, Jerome Lejeune writes in this book: 'If any

exploitation of the early human embryo is intrinsically repugnant, if everyone feels that those conducting experiments must absolutely respect these marvellously young human beings, it is for the scientific reason that a newsman discovered in an intuition of genius, test-tube babies are babies.' This kind of polemical statement does not illuminate.

28 'E fassi un'alma sola, che vive e sente, e se in se rigira': Dante Alighieri, *Purgatorio*, canto 25, 67–75 (*c*.1300).

29 V. Glover and N. M. Fisk, 'Fetal pain: implications for research and practice', *British Journal of Obstetrics and Gynaecology* 106 (1999), 881–6.

Chapter 3: IVF: When, Why and How

1 Kathleen Curran Sweeney, 'The technical child: in vitro fertilization and the personal subject', www.christendom-awake.org (2002).

2 I believe that my assertion here is borne out by fact. Is it not an irony that a number of treatments for infertility, some of which are not evidence-based, are funded by the NHS, while health-care services have refused to purchase IVF for the populations for which they are responsible?

3 This is why those medical practitioners who have advertised their cleverness in achieving a pregnancy in older women are simply boasting. The most infamous of these is probably Dr Antinori from Italy (see chapter 8). There is in fact nothing clever about that treatment at all.

4 There has been more than one study of women not using contraception where the doctors routinely tested for HCG, the pregnancy hormone, around the day the menstrual period was due. These confirm that failed early conception is common. We also know from patients undergoing IVF who are intensively followed that a small, transient rise in pregnancy hormone is very common about 12–14 days after embryo transfer, often giving rise to false hopes that are then dashed – a cruel blow to an infertile couple.

5 Given this low conception rate per month, IVF does very well because it is in practice a single shot at conception. The average pregnancy rate at Hammersmith at present is about 26 per cent – five times better than a single shot (if I may use the term) at sex.

6 The use of the term 'client' seems to me to reduce medicine to a business relationship. To my mind, the people we treat should be seen as patients and given proper status and consideration as such.

7 Obesity is defined as having a Body Mass Index (BMI) above 30. For measurement of BMI (weight plotted against height), visit the American government website www.cdc.gov/nccdphp/dnpa/bmi/index.htm.

8 Claims are always being made that infertility is on the increase. But this is difficult to measure, and most of the claims are dubious. Probably the main reason why it appears to be increasing is that it is now out in the open: it is more talked about, and there are better treatments available. But there are two powerful reasons why we may indeed be becoming less fertile: the advancing age of women at first sexual partnership, and obesity.

9 A. M. H. Hassan and S. R. Killick, 'Negative lifestyle is associated with a significant reduction in fecundity', *Fertility and Sterility* 81 (2004), 384–92.

10 D. M. Campagne, 'Should fertilization treatment start with reducing stress?' *Human Reproduction* 21 (2006), 1651–8.

11 L. Anderheim et al., 'Does psychological stress affect the outcome of in vitro fertilization?', *Human Reproduction* 20 (2005), 2969–75.

12 C. de Klerk et al., 'Effectiveness of a psychosocial counselling intervention for first-time IVF couples: a randomized controlled trial', *Human Reproduction* 20 (2005), 1333–8.

13 Forecasting ovulation is mainly done by measuring the increasing size of the follicles using ultrasound.

Ovulation occurs when they are close to 2 centimetres in diameter.

14 Of course, I am not making light of using the 'morning-after pill' or similar medication, nor of the serious impact of intervention on the emotions of the couples concerned; but open admission of a mistake and delayed effective action is invariably better than finding out by chance that the baby just delivered is not genetically related to its mother.

15 R. Hauser, L. Yogev, G. Paz et al., 'Comparison of efficacy of two techniques for testicular sperm retrieval in nonobstructive azoospermia: multifocal testicular sperm extraction versus multifocal testicular sperm aspiration', *Journal of Andrology* 1 (2006), 28–33.

16 R. H. Asch, L. R. Ellsworth, J. P. Balmaceda and P. C. Wong, 'Birth following gamete intrafallopian transfer', *Lancet* 326 (1985), 8447, 163.

17 R. M. L. Winston and K. Hardy, 'Are we ignoring potential dangers of in vitro fertilization and related treatments?', *Nature Cell Biology* 4 (2002), supplement, 14–18.

18 M. Hansen et al., 'Assisted reproductive technologies and the risk of birth defects: a systematic review', *Human Reproduction* 20 (2005), 328–38.

19 M. Tachataki, R. M. L. Winston and D. M. Taylor, 'Quantitative RT-PCR reveals tuberous sclerosis gene, TSC2, mRNA degradation following cryopreservation in the human preimplantation embryo', *Molecular Human Reproduction* 9 (2003), 593–601.

20 D. J. Barker et al., 'Weight in infancy and death from ischaemic heart disease', *Lancet* 334 (1989), 8663, 577–80.

21 A. S. Chang et al., 'Association between Beckwith-Wiedemann syndrome and assisted reproductive technology: a case series of 19 patients', *Fertility and Sterility* 83 (2005), 349–54.

Chapter 4: Freezing: The Refrigeration Game

1 R. G. Bunge et al., 'Fertilizing capacity of frozen human spermatozoa', *Nature* 172 (1953), 767.

2 A. Trounson and L. Mohr, 'Human pregnancy following cryopreservation, thawing and transfer of an eight-cell embryo', *Nature* 305 (1983), 707–9.

3 Egg-sharing – which is banned in the USA because it is thought to risk being coercive – is where one woman undergoes IVF without charge, but undertakes to give half her eggs to another woman who cannot produce her own. The recipient of the eggs pays for her treatment cycle and that of the donor (who usually enters such an arrangement because she is short of money).

4 By using the term 'good' embryos, I imply only those embryos that are dividing reasonably rapidly and do not show any obvious sign of cellular abnormalities. But, as observed earlier, the scoring of embryos by their visual appearance down a conventional microscope is frequently very misleading.

5 I say it takes about a year advisedly. In our case it took nine months to get the initial animal licence, following which it takes an average of six to nine months to get ethical approval from most local ethics committees to do human work. But before any work can actually start, an application to the HFEA will delay the research for at least four to six months, and often longer.

6 S. J. Pickering, P. R. Braude, M. H. Johnson et al., 'Transient cooling to room temperature can cause irreversible disruption of the meiotic spindle in the human oocyte', *Fertility and Sterility* 54: 1(1999), 102–8.

7 C. Chen, 'Pregnancy after human oocyte cryopreservation', *Lancet* 327 (1986), 8486, 884–6.

8 M. Kuwayama, G. Vajta, O. Kato and S. P. Leibo, 'Highly efficient vitrification method for cryopreservation of human oocytes', *Reproductive Medicine Online* 11 (2005), 300–8.

9 A. Eroglu et al., 'Quantitative microinjection of trehalose into mouse oocytes and zygotes, and its effect on development', *Cryobiology* 46: 2 (2003), 121–34.

10 *Washington Post*, 24 Sept. 2004.

11 K. Oktay et al., 'Embryo development after heterotopic transplantation of cryopreserved ovarian tissue', *Lancet* 363 (2004), 9412, 837–40.

12 This risk was clearly thought to be very low in the case of Jacques Donnez's patient, described earlier.

Chapter 5: Donated Eggs and Sperm: The Most Generous Gift?

1 I think the implication of this satirical story, recounted by the fifteenth-century philosopher Abrabanel, is that the men from Ephraim were roughnecks.

2 Ecclesiasticus, one of the books of the Apocrypha, reads rather like the book of Proverbs. It teaches how man should love wisdom and ethical conduct, and how to behave with modesty in all daily conduct.

3 J. Leeton, A. Trounson and C. Wood, 'The use of donor eggs and embryos in the management of human infertility', *Australian and New Zealand Journal of Obstetrics and Gynaecology* 24 (1984), 265–70.

4 K. K. Ahuja, E. G. Simons, B. Mostyn et al., 'An assessment of the motives and morals of egg share donors: policy of "payments" to egg donors requires a fair review', *Human Reproduction* 13 (1998), 2671–8.

5 M. Y. Thum, A. Gafar, M. Wren et al., 'Does egg-sharing compromise the chance of donors or recipients achieving a live birth?', *Human Reproduction* 18 (2003), 2363–7.

6 This is a fairly standard tactic of the press and some politicians. Science always 'runs fast', but society keeps up surprisingly well. And usually the implementation of scientific knowledge takes vastly longer than expected.

7 There is no central religious authority that decides exactly how Jewish law is applied. Jewish rabbinical law is based on a system of continual refinement and there is

sometimes disagreement among individual rabbis about the acceptability of particular practices. A majority view is important but specific decisions are sometimes made within individual communities about exactly how a particular Jewish law should be implemented.

8 J. David Bleich, *Contemporary Halakhic Problems*, vol. VI (New York: Ktav, 1995). The Halakhah is essentially the body of Jewish law. Rabbi Bleich is one of its foremost modern orthodox exponents, with a very significant international reputation. We debated publicly together in the very early days of in vitro fertilization, when he appeared to me to be very suspicious of many of the modern reproductive technologies, particularly IVF. But since that time, it seems, he has come to feel more at ease with many of the procedures involved.

9 Rabbi Waldenberg's hostility to this technology is certainly not accepted by mainstream Judaism.

10 Talmud: *Yevamot* 97b.

11 'Instruction on respect for human life and its origin and on the dignity of procreation', Catechism of the Catholic Church ('Replies to certain questions of the day'), 1987.

12 J. Cohen et al., 'Pregnancies following the frozen storage of expanding human blastocysts', *Journal of In Vitro Fertilization and Embryo Transfer* 2: 2 (1985), 59–64.

13 J. A. Barrit, C. A. Brenner, H. E. Malter and J. Cohen, 'Mitochondria in human offspring derived from ooplasmic transplantation', *Human Reproduction* 16 (2001), 513–16.

14 Ibid.

15 J. E. Buster, M. Bustillo, I. H. Thorneycroft et al., 'Non-surgical transfer of in vivo fertilized donated ova to five infertile women: report of two pregnancies', *Lancet* 322 (1983), 8343, 223–4.

16 I am not implying that Dr Buster did not get consent for these procedures. But I remain somewhat dubious whether the risk of an unwanted or ectopic pregnancy was fully understood by these women.

17 L. Formigli, G. Formigli and C. Roccio, 'Donation of fertilized uterine ova to infertile women', *Fertility and Sterility* 47 (1987), 162–5.

18 Two studies are particularly valuable: S. Golombok et al., 'Children in lesbian and single-parent households: psychosexual and psychiatric appraisal', *Journal of Child Psychology and Psychiatry and Related Disciplines* 24: 4 (1983), 551–72, and M. Kirkpatrick, C. Smith and R. Roy, 'Lesbian mothers and their children: a comparative survey', *American Journal of Orthopsychiatry* 51: 3 (1981), 545–51.

19 F. Maccallum and S. J. Golombok, 'Children raised in fatherless families from infancy: a follow-up of children of lesbian and single heterosexual mothers at early adolescence', *Child Psychology and Psychiatry* 45 (2004), 1407–19.

20 S. Golombok, V. Jadva, E. Lycett et al., 'Families created by gamete donation: follow-up at age 2', *Human Reproduction* 20 (2005), 286–9.

21 *Daily Telegraph*, 2 Nov. 2003.

22 J. N. Robinson, R. G. Forman, A. M. Clark et al., 'Attitudes of donors and recipients to gamete donation', *Human Reproduction* 6: 2 (1991), 307–9.

23 A. J. Turner and A. Coyle, 'What does it mean to be a donor offspring? The identity experiences of adults conceived by donor insemination and the implications for counselling and therapy', *Human Reproduction* 15: 9 (2000), 2041–51.

24 I. Craft et al., 'Will removal of anonymity influence the recruitment of egg donors? A survey of past donors and recipients', *Reproductive Biomedicine Online* 10: 3 (2005), 325–9.

25 J. E. Scheib, M. Riordan and S. Rubin, 'Choosing identity-release sperm donors: the parents' perspective 13–18 years later', *Human Reproduction* 18 (2003), 1115–27.

26 C. Gottlieb, O. Lalos and F. Lindblad, 'Disclosure of

donor insemination to the child: the impact of Swedish legislation on couples' attitudes', *Human Reproduction* 15: 9 (2000), 2052–6.

27 *Observer*, 30 April 2006.

28 Of course, many other countries have taken similar steps without using the mechanism of a regulatory body similar to the HFEA.

29 Indeed, clinical results of IVF in Britain are a good deal worse than they are in the USA. This is emphasized by a recent survey by Norbert Gleicher et al., 'A formal comparison of the practice of assisted reproduction technologies between Europe and the USA', *Human Reproduction* 21 (2006), 1945–50. Gleicher reports that the chance of successful pregnancy after IVF is around 27% per cycle in the USA and only 17.4% in Europe (including the UK). The reason for the disparity is not clear, though it may in part be associated with the highly dangerous practice of transferring more embryos simultaneously; certainly multiple birth is more common in American programmes, and infant mortality and morbidity are consequently higher. Clearly a child against all the odds should not be achieved at all costs. But it is very unlikely that this 65% improvement is attributable solely to multiple embryo transfer, and European and British workers should be doing more research to improve results.

Chapter 6: Womb to Rent

1 From François Rabelais, *The Lives, Heroic Deeds and Sayings of Gargantua and his Son, Pantagruel*, trans. Sir Thomas Urquart, bk 1.

2 May C. and Elisabeth are not the real names of the women involved in this case.

3 Committee on Human Fertilization and Embryology, *Report of the Committee of Inquiry into Human Fertilization and Embryology*, chaired by Dame Mary Warnock (London: HMSO, July 1984). A minority

opinion on the committee favoured surrogacy in certain circumstances.

4 I am not being particularly generous to Dr Odendal. At the time I remember being dismayed at the arrangement, but mainly because of the multiple embryo transfer, which was frankly dangerous.

5 'Hard', 'Seed', 'Keane', 'Buster': extraordinary how many people associated with human reproduction are so aptly named.

6 Ginea Corea, *The Mother Machine: Reproductive Technologies from Artificial Insemination to Artificial Wombs* (London: Women's Press, 1988).

7 J. Jadva, C. Murray, E. Lycett et al., 'Surrogacy: the experiences of surrogate mothers', *Human Reproduction* 18: 2 (2003), 2196–204.

8 S. Golombok et al., 'Non-genetic and non-gestational parenthood: consequences for parent–child relationships and the psychological well-being of mothers, fathers and children at age 3', *Human Reproduction* (electronic publication in advance of print, 29 March 2006).

9 According to reports published in the *Washington Post*.

Chapter 7: Pre-implantation Diagnosis

1 It has been calculated that the chance of this particular gene mutating is about 1 in 25,000. So the only other credible explanation for Queen Victoria's carrier status is that she was not the daughter of the man who was married to her mother, Edward Duke of Kent. This is the theory suggested by D. M. Potts and W. T. W. Potts in *Queen Victoria's Gene* (Stroud: Sutton Publishing, 1999).

2 Northwick Park Hospital subsequently came under the aegis of Imperial College London, and the Clinical Research Centre was housed at Hammersmith in the 1990s.

3 One of the few examples of a sex-linked disorder that is not so serious is X-linked colour-blindness.

4 R. Saiki et al., 'Analysis of enzymatically amplified beta-globin and HLA-DQ alpha DNA with allele specific oligonucleotide probes', *Nature* 324 (1986), 163–6.

5 Robert Brown, an English botanist, was wont to examine grains of pollen under his microscope. In 1827 he noticed that when these were suspended in fluids such as water, they danced around haphazardly, following a zigzag path. Sixty years later, it was recognized that the smaller the particles suspended in fluid, the more rapid the motion – which also depends on the temperature of the solution: the higher the temperature, the faster the movement. It turns out that any small suspended particle is constantly bombarded by the molecules of the fluid in which it floats. As these collision events are random, so is the resultant movement – which is now called 'Brownian' motion.

6 Gel electrophoresis is a method used to separate molecules on the basis of their physical characteristics such as size or weight. It is used extensively for DNA analysis. The gel is usually a jelly-like matrix in which the molecules, which have been previously stained, are suspended. An electric current is passed through the gel (electrophoresis), and the electrically charged molecules move through it. The speed of their migration depends on their size and shape. After several hours, the gel can be examined and a coloured, stained band will appear at the point to which the molecules being identified have progressed.

There are various ways of refining and improving this process, which involve improving the 'filtering' power of the gel, or changing the electric current.

7 Janet D. is not her real name.

8 This was very definitely woman's work! We decided that if a woman did this part of the procedure there would be much less chance of contaminating our tube with a stray cell from any male, and therefore less chance of getting a false result.

9 The control cells used as a comparison to test the working of the procedure were white blood cells taken from blood samples donated by Alan and myself, and from women working in the lab.

10 Using a negative result like this was far from ideal, but in those days there was no choice. Subsequently we developed a positive method for identifying both the X- and the Y-chromosome, making the procedure more secure. Janet, of course, was well aware of our limitations at the time and happily signed consent – with her husband.

11 Elizabeth E. and Mary C. are not these patients' real names.

12 The clinic was subsequently demolished, its owner, Mr Bira, selling the plot of land on which it stood. Now there are some very beautiful flats on that corner of Avenue Road. On its closure, Raul and I stole the operating theatre sister, Shelagh Simmons, from the clinic to become the nursing director of our fledgling IVF programme.

13 Mark Hughes, quoted in Leslie Mertz, 'Thumbing through the encyclopedia of life', *Wayne Medicine* 2000, published by Wayne State University, USA.

14 This technique was finally perfected by Professor Peter Braude and his team at St Thomas's Hospital (P. N. Scriven, F. O'Mahoney, H. Bickerstaff et al., 'Clinical pregnancy following blastomere biopsy and PGD for a reciprocal translocation carrier: analysis of meiotic outcomes and embryo quality in two IVF cycles', *Prenatal Diagnosis* 20 (2000), 587–92).

15 Hence the name 'Chorea', meaning 'dance', from the Greek word (from which we also get the word 'chorus').

16 André van Steirteghem and Paul Devroey's team in Brussels have been one of the most prolific in the whole field of pre-implantation diagnosis. Their paper on Huntington's Chorea was K. Sermon, M. De Rijcke, W. Lissens et al., 'Pre-implantation genetic diagnosis for

Huntington's disease with exclusion testing', *European Journal of Human Genetics* 10 (2002), 591–8.

17 A. Ao, D. Wells, A. H. Handyside, R. Winston and J. D. Delhanty, 'Pre-implantation genetic diagnosis of inherited cancer: familial adenomatous polyposis coli', *Journal of Assisted Reproduction and Genetics* 15 (1998), 140–4.

18 Moreover, the public consultation – like so many other consultations the HFEA has undertaken – was done via a website: a sure way of getting only the polarized views of a limited number of people who are likely to be completely unrepresentative of the woman or man in the street.

19 Website of Religious Tolerance, based in Ontario (www.religioustolerance.org).

20 Abnormalities in the cell may lead to abnormalities of the DNA – increasing the risk of misdiagnosis.

21 'The case for studying human embryos and their constituent tissues in vitro', in R. G. Edwards and Jean M. Purdy, eds, *Human Conception in Vitro* (London: Academic Press, 1982).

22 Norma is not her real name.

23 There are various forms of Conradi's, and not all are sex-linked. But when it is carried on the X-chromosome, males with it, having no normal X-chromosome, die at birth.

24 Some years ago, I broached this issue with a previous chairman of the HFEA, Dame Ruth Deech (now Baroness Deech). She told me she was not concerned about pressure to give tissues being put on the PGD child by parents. She averred that it would always be possible to ensure that the courts would protect such a child if it felt like refusing. I did not find that answer very reassuring. Having the courts involved is hardly going to make life and family relationships easier for the child.

25 F. Fiorentino et al., 'Short tandem repeats haplotyping of the HLA region in pre-implantation HLA matching',

European Journal of Human Genetics 13: 8 (2005), 953–8.

26 P. Platteau, C. Staessen, A. Michels et al., 'Pre-implantation genetic diagnosis in patients with unexplained recurrent miscarriages', *Fertility and Sterility* 83 (2005), 393–7.

27 M. Twisk, S. Mastenbroeck, M. van Wely et al., 'Pre-implantation genetic screening for abnormal number of chromosomes (aneuploidies) in in vitro fertilization or intracytoplasmic sperm injection' (review), Cochrane Collaboration, *The Cochrane Library*, 2 (Chichester: Wiley, 2006). The Cochrane Collaboration is a highly respected international non-profit organization that provides current information about the effects of health care and publishes the *Cochrane Database of Systematic Reviews*, a resource for evidence-based health care.

Chapter 8: Cloning: A Feat of Clay

1 It may seem unbelievable, but the following two statements are from people contributing to a Russian website which is still running. There are many like them on the internet:

> ANYA MIROVNA 42, CLEANING WOMAN. In a market economy, it is difficult to control employers. And since so many employers happen to be Jewish, it goes without saying that some of them will choose women of child-bearing age in order to get access to newborns. Often a Jew will save a woman employee from having an abortion, thereby performing a service, since abortions are very dangerous to one's health.
>
> VITALY YUDIN, 27, GOVERNMENT WORKER. Do fish swim? Of course Jews drink the blood of infants. They especially like the taste of Slavic blood. The collapse of communism, the rise of Berezovsky, Gusinsky and Abramovich, the 'declining birth-rate'? Hello? These are no coincidences. Once central authority collapsed, these men made their fortunes marketing fresh Russian babies to Israel.

2 Rabbi Loew was regarded as a very fine scholar. His

books still exist in various libraries; as well as writing about many religious issues, ethics and philosophy, he was scientifically literate.

3 There is no evidence at all that Rabbi Loew attempted to make a golem. But if you go to Prague, visit the Altneuschul – one of the very few synagogues the Nazis did not destroy. A brief detour to the graveyard behind this medieval building at dusk is essential. Ignore the huge tomb of Franz Kafka, and just imagine the shambling figure of the golem clumping his way towards you between the crazily pitched and randomly placed gravestones in that gloomy half-darkness.

Golems existed throughout Jewish folklore, back as far as talmudic times. It is said that even the prophet Jeremiah may have created a golem. They are not evil, but have a rather unpleasant tendency to run amok. The elements of Mary Shelley's monster are of course buried in this legend.

4 J. B. Gurdon and V. Uehlinger, '"Fertile" intestine nuclei', *Nature* 210: 42 (1966), 1240–1.

5 S. Budiansky, 'Karl Illmensee: NIH withdraws research grant', *Nature* 309: 5971 (1984), 738.

6 P. M. Zavos and K. Illmensee, 'Possible therapy of male infertility by reproductive cloning: one cloned human 4-cell embryo', *Archives of Andrology* 52: 4 (2006), 243–54.

7 C. B. Fehilly, S. M. Willadsen and E. M. Tucker, 'Interspecific chimaerism between sheep and goat', *Nature* 307 (1984), 634.

8 Probably the most important aspect of this work was the reaction from the animal rights lobby, who argued that a pointless experiment had inflicted suffering on this rather sad-looking animal. There is little doubt that the picture appearing on the front cover of *Nature* did not help laboratory animal science.

9 S. M. Willadsen, 'Nuclear transplantation in sheep embryos', *Nature* 320: 6057 (1986), 63–5.

10 The popular perception is that Frankenstein animated his monster by a bolt of electricity, but this is an image derived primarily from the many films of Mary Shelley's book. The method of the original animation is not clear, but the text suggests that it was chemical.

> It was on a dreary night of November that I beheld the accomplishment of my toils. With an anxiety that almost amounted to agony, I collected the instruments of life around me, that I might infuse a spark of being into the lifeless thing that lay at my feet. It was already one in the morning; the rain pattered dismally against the panes, and my candle was nearly burnt out, when, by the glimmer of the half-extinguished light, I saw the dull yellow eye of the creature open; it breathed hard, and a convulsive motion agitated its limbs. How can I describe my emotions at this catastrophe, or how delineate the wretch whom with such infinite pains and care I had endeavoured to form?

11 S. Antinori, C. Versaci, G. Gholami Hossein, B. Caffa and C. Panci, 'A child is a joy at any age', *Human Reproduction* 8: 10 (1993), 1542.

12 C. Gomez, C. E. Pope and B. L. Dresser, 'Nuclear transfer in cats and its application', *Theriogenology* 66: 1 (2006), 72–81.

13 C. L. Keefer, R. Keyston, A. Lazaris et al., 'Production of cloned goats after nuclear transfer using adult somatic cells', *Biology of Reproduction* 66 (2003), 199–203.

14 F. Xue et al., 'Aberrant patterns of X chromosome inactivation in bovine clones', *Nature Genetics* 31 (2002), 216–20; A. Jouneau et al., 'Developmental abnormalities of NT mouse embryos appear early after implantation', *Development* 133 (2006), 1597–1607; C. B. Herath, 'Developmental aberrations of liver gene expression in bovine fetuses derived from somatic cell nuclear transplantation', *Cloning and Stem Cells* 8: 2 (2006), 79–95.

15 I. Wilmut, A. E. Schnieke, J. McWhir, A. J. Kind and K. H. Campbell, 'Viable offspring derived from fetal and adult mammalian cells', *Nature* 385 (1997), 810–13.

16 L. E. Young, 'Scientific hazards of human reproductive cloning', *Human Fertility* 6: 2 (2003), 59–63.
17 A few somatic cells do not divide – for example, red cells in the circulating blood. Red cells are unusual in that (at least in adults) they do not have a nucleus; and once they have been made from more primitive cells in the bone marrow, they die after about three weeks in circulation.

Chapter 9: Stem Cells: Nature's Magic
 1 There is a much fuller – and lively – account of this dissection, the events that led up to it, and those that followed, in the excellent book by Dr Matthew Cobb of the University of Manchester: *The Egg and Sperm Race* (London and New York: Free Press/Simon & Schuster, 2006).
 2 This is only an approximate figure; I do not know precisely the number of cell types in the body.
 3 M. R. Alison et al., 'Hepatocytes from non-hepatic adult stem cells', *Nature* 406: 6793 (2000), 257.
 4 G. Ferrari et al., 'Muscle regeneration by bone marrow-derived myogenic progenitors', *Science* 279 (1998), 1528–30.
 5 This, sadly, is another example of how worrying parents can be exploited. Commercial enterprises now exist which – for a fee – promise to harvest and store cord blood on the remote chance that such cells might be useful in twenty or more years' time. I suspect that by that time there will far more efficient ways of obtaining stem cells.
 6 J. H. Kim et al., 'Dopamine neurons derived from embryonic stem cells function in an animal model of Parkinson's disease', *Nature* 418: 6893 (2002), 50–6.
 7 Y. Takagi et al., 'Dopaminergic neurons generated from monkey embryonic stem cells function in a Parkinson primate model', *Journal of Clinical Investigation* 115 (2005), 102–10.
 8 A. Atala et al., 'Tissue-engineered autologous bladders

for patients needing cystoplasty', *Lancet* 367 (2006), 1241–6.

9 A. Atala, 'Recent developments in tissue engineering and regenerative medicine', *Current Opinion in Pediatrics* 18: 2 (2006), 167–71.

10 W. S. Hwang et al., 'Patient-specific embryonic stem cells derived from human SCNT blastocysts', *Science* 308: 5729 (2005), 1777–83.

11 I remember one statistic that caught my attention at the time. Hwang's paper stated that an average of between ten and twelve eggs were donated during each cycle of treatment, and that most of these were from women under thirty. The publication specifically stated: 'Although expenses for public transportation and injections administered by medical personnel could have been provided, none of the donors requested this, and therefore no financial reimbursement in any form was paid.' I remember being extremely surprised at the altruism of Korean women and thinking that we would never manage to do this anywhere in Europe or the USA. And then I started thinking rather harder . . .

12 B. C. Lee et al., 'Dogs cloned from adult somatic cells', *Nature* 436: 7051 (2005), 641.

13 Gerry was using perfectly acceptable American English. 'Rambunctious', meaning wild or unruly, was a word coined in the USA in 1854.

14 But why a clone would help, I am not too sure.

15 This is, to my mind, one of the more bizarre aspects of the whole Hwang episode. Among his or her other responsibilities, the senior author, who gets great credit for a publication of this sort, has a duty to check that the experiment has been done in the way set out in the paper, and that ethical issues have been addressed properly.

16 There were no fewer than twenty-five authors on this key paper. One wonders whether the journal editors enquired as to what the role of each author was.

17 Rabbi Moshe Tendler, PhD, the distinguished professor

of biology and Talmud at Yeshiva University, New York, in 'Stem cell research and therapy: a Judeo-biblical perspective', *Ethical Issues in Human Stem Cell Research*, vol. III: *Religious Perspectives*, Sept. 1999. The full text may be downloaded from the National Bioethics Advisory Commission website at http://bioethics.gov/pubs.html.

18 K. Hubner et al., 'Derivation of oocytes from mouse embryonic stem cells', *Science* 300: 5623 (2003), 1251–6.

19 A. T. Clark et al., 'Spontaneous differentiation of germ cells from human embryonic stem cells in vitro', *Human Molecular Genetics* 13 (2004), 727–39.

20 K. Nayernia et al., 'Derivation of male germ cells from bone marrow stem cells', *Laboratory Investigation* 86 (2006), 654–63.

21 K. Nayernia et al., 'In vitro-differentiated embryonic stem cells give rise to male gametes that can generate offspring mice', *Developmental Cell* 11 (2006), 125–32.

Chapter 10: The Trojan Horse

1 Followers of *Star Trek: The Next Generation* may recall a version of this story in the episode 'Darmok' in 1991, where Patrick Stewart, playing the part of Captain Jean-Luc Picard of Star Ship *Enterprise*, boldly going where no man has gone before, is the reincarnation of King Gilgamesh.

2 The idea of a chimaera has fascinated humans since earliest times.

3 Timothy Taylor, *The Prehistory of Sex* (London: Fourth Estate, 1996).

4 Dating of the Venus of Willendorf has been revised several times, with the current estimate at between 24,000 and 22,000 BCE. Two much older figurines, the Venuses of Berekhat Ram and of Tan-Tan, are not culturally related and were primarily formed by natural processes.

5 It is a deliciously androcentric idea to imagine that this entire team meticulously spent five years uncovering just a single phallus, but I understand that they also did much other valuable archaeology over this time.

6 This is similar to the Hebrew biblical account. According to Genesis, before Isaac or Ishmael was conceived Abraham adopted his manservant Eliezer. He complains in Genesis 15: 2 that, 'seeing I go childless, he that shall be the possessor of my house is Eliezer of Damascus'.

7 Translated by Dr H. W. F. Saggs, Professor of Semitic Languages, University College, Cardiff: one of the outstanding Assyriologists of his generation, who died in 2005.

8 Altogether, Set seems to have had the makings of a serial rapist. In another story, the bisexual Set molests his nephew Horus while chatting him up by saying, 'You have a lovely bottom.'

9 The figure of Set will be familiar. He was the god not only of Chaos but also of evil. He is depicted with a long, dog-like snout and pointed ears. The snout may have been that of an aardvark, or possibly a jackal. This nightmare figure generally has a forked tail, but human legs and feet.

10 I have to say this prudishness seemed to continue well into the twentieth century. The book on the subject that I inherited from my grandfather, *Myths and Legends of Ancient Egypt* by Lewis Spence, published in 1915, fails to mention any of the more juicy stories about Set.

11 *The Papyrus Ebers* (translated from the German version by Cyril P. Bryan, with an introduction by Professor J. Elliot Smith (London: G. Bles, [1930]). It is not clear to me how wasp's dung was collected.

12 F. L. Griffith, *The Petrie Papyri: Hieratic Papyri from Kahun and Gurob* (London: Bernard Quaritch, 1898).

13 Soranus, *On Diseases of Women*, bk 1, ch. 19.

14 Other ancient civilizations seem to have had the opposite problem. In the China of the third century BCE, Han Fei

wrote that overstretched resources due to population growth were a cause of war.

15 M. Faerman et al., 'Determining the sex of infanticide victims from the late Roman Era through ancient DNA analysis', *Journal of Archaeological Science* 25 (1998), 861–5.

16 Francis Galton FRS, *Hereditary Genius: An Enquiry into its Laws and Consequences* (London: Macmillan, 1869).

17 In this respect I really did try to do my best for the aristocracy, saying in one speech that the 'noble Lords opposite were more handsome, taller, fitter and possibly more able than those on my side [mostly life peers and members of the Labour Party] of the Chamber'. Needless to say, my argument cut little ice.

18 Dan Kevles, *In the Name of Eugenics* (Berkeley: University of California Press, 1985).

19 US War Department, *Annual Reports*, 1919, quoted by Carl Brigham in *A Study in American Intelligence* (Princeton: Princeton University Press, 1923).

20 In 1949 the Chinese population was 540 million. It has more than doubled in fifty years since then.

21 This was the moment of Laocoön's famous warning: '*Equo ne credite, Teucri. Quidquid id est, timeo Danaos et dona ferentis*' – ' Do not trust the horse, Trojans! Whatever it is, I fear the Greeks even bringing gifts'. Publius Vergilius Maro, *Aeneid*, bk 2.

22 R. Jaenisch and B. Mintz, 'Simian virus 40 DNA sequences in DNA of healthy adult mice derived from preimplantation blastocysts injected with viral DNA', *Proceedings of the National Academy of Sciences of the USA* 71: 4 (1974), 1250–4.

23 J. W. Gordon et al., 'Genetic transformation of mouse embryos by microinjection of purified DNA', *Proceedings of the National Academy of Sciences of the USA* 77: 12 (1980), 7380–4.

24 R. D. Palmiter, R. L. Brinster, R. E. Hammer et al.,

'Dramatic growth of mice that develop from eggs microinjected with metallothionein-growth hormone fusion genes', *Nature* 300: 5893 (1982), 611–15.

25 I have to say that we were entirely honourable about the information and amused only ourselves with the knowledge, not breathing a word to anybody else.

26 L. Naldini et al., 'In vivo gene delivery and stable transduction of non-dividing cells by a lentiviral vector', *Science* 272: 5259 (1996), 263–7.

27 Arthur Schopenhauer, 'Die vermeinte Rechtlosigkeit der Thiere . . . ist geradezu eine emporende Rohheit und Barbarei des Occidents, deren Quelle im Judentum liegt', in *Der Beiden Grundprobleme der Ethik* (Frankfurt, 1841).

28 D. J. Weatherall, 'Thalassaemia and malaria, revisited', *Annals of Tropical Medicine and Parasitology* 91: 7 (1997), 885–90.

29 Lee M. Silver, *Remaking Eden: Cloning and Beyond in a Brave New World* (London: Weidenfeld & Nicolson, 1998).

Glossary

activate The process by which the egg is kick-started into cell division. This normally happens at fertilization, but an egg can be activated by various changes in its chemical or physical environment, leading to parthenogenesis (*see below*).

adenomyosis A uterine disease causing scarring in its muscle tissue by in-growth of the uterine lining, the endometrium. It is associated with painful, irregular periods and infertility.

amniocentesis The procedure that involves aspirating fluid via a needle stuck into the sac surrounding the developing fetus. As the fetus sheds cells from its skin and digestive tract during development, examination of the cells in this fluid can give useful information about the baby's status.

andrology The study of male infertility.

aneuploidy An abnormal number of chromosomes in a cell.

artificial insemination (AI) Injection of semen into the vagina or uterus. The semen may be from the woman's partner, or from a donor.

assisted hatching *See* 'zona drilling' *below*.

beta-thalassaemia A crippling hereditary disease of the red blood cells, causing severe anaemia in children. It is caused by a recessive gene (*see below*). It is particularly common in some Mediterranean countries.

bicornuate uterus During early development, the uterus develops from two separate muscular tubes. Normally in humans these fuse to provide a single body during early fetal life. Sometimes the fusion does not occur, leading to a double uterus. Incomplete fusion results in a bicornuate uterus, and maldevelopment of one side results in a unicornuate uterus. Both types of malformation are associated with varying degrees of infertility, and a higher likelihood of miscarriage or premature labour.

biopsy The removal of a small piece of tissue or a very few cells for diagnostic examination.

blastocyst At approximately five days after fertilization, the human embryo is a ball containing 100–200 cells. A fluid-filled cavity develops in the centre, compressing most of the cells around the periphery of the ball. At this stage the embryo is referred to as a blastocyst.

cervix The cervix is the thick muscular structure at the entrance to the uterus. It has a narrow canal, which contains mucus through which sperm swim to get to the site of fertilization. It opens up (dilates) during labour to allow the baby's head to enter the vagina.

chlamydia A common bacterial organism often found in the vagina which, if it grows out of control, can infect the fallopian tubes and scar them.

chorion villus sampling Biopsy taken from the placental cells surrounding the baby which gives a diagnosis about its status.

chromosomes Tightly coiled structures carrying genes. Humans have 22 paired chromosomes called 'autosomes' – half inherited from the mother and half from the father.

Chromosome 1, the biggest, carries about 3,000 genes; one of the smallest, chromosome 22, has fewer than 300. In addition, there are two sex chromosomes, X and Y. The Y chromosome can be inherited only from the father.

cleaving embryos Fertilized eggs are said to be 'cleaving', i.e. undergoing regular cell division, for the first four days of development.

clomiphene The fertility pill which stimulates the production of follicle stimulating hormone (*see below*).

cortisol A steroid hormone produced by the adrenal glands. Its secretion is increased by stress, raising blood pressure and sugar metabolism; it can be associated with female infertility and tends to suppress the immune system.

cryopreservation The act of preservation by freezing – usually at such low temperatures that there is little molecular activity. At the temperature of liquid nitrogen (–196°C) human cells can be kept without any obvious damage for many years.

cytoplasm The semi-fluid jelly-like substance inside the cell, which surrounds the nucleus and contains its many vital structures such as mitochondria (*see below*).

DNA primer A short sequence of artificially produced DNA which attaches to a region of the DNA to be analysed. It initiates the chain reaction allowing many copies of that region of the genome to be produced for easy analysis.

Down's syndrome Three copies of chromosome 21 cause this defect, which is associated with mental retardation and heart defects. Most Down's syndrome embryos do not survive beyond a few days after fertilization.

ectopic pregnancy A pregnancy outside the uterus, usually implanted in one or the other fallopian tube. If undiagnosed, it is a potentially dangerous condition as it is associated with internal bleeding.

embolization The process of artificially blocking a blood vessel by plugging it. It is sometimes used to treat fibroids.

embryo The definition of 'embryo' is imprecise, but most sources regard the embryo as being the stage of development during the first six to eight weeks. Thereafter, the developing human is usually referred to as a fetus.

embryo transfer The transfer of one or more embryos to the uterine cavity to try to initiate pregnancy.

endocrinology The study of hormones and their diseases.

endometriosis A scarring disease which is caused by pieces of the lining of the uterus, the endometrium, implanting outside the uterus – in the ovaries or peritoneal lining. These can menstruate internally, causing pain.

endometrium *See* endometriosis *above.*

epigenetic effect An inherited attribute caused by a change in gene function, rather than by a change in gene structure. It may occur spontaneously but is often in response to some environmental influence. It may be reversible in future generations.

eugenics Promoting a cleaner, more productive and healthier world by improving the human gene pool. Seems good at first, doesn't it?

fetus *See* embryo *above.*

fibroids Benign muscular tumours arising usually from the uterine wall. If they distort the uterus, they may cause infertility.

fluorescent in situ hybridization (FISH) Attaching a fluorescent dye to a specific sequence of the DNA in a cell, to make a diagnosis.

follicle stimulating hormone (FSH) The hormone, produced by the pituitary gland, which stimulates follicles in the ovary to develop before ovulation.

gene expression This is the complex multi-step process by which a gene, a specific sequence of DNA, produces its effect – usually by making a protein. A gene that is working is said to be expressing, and sometimes this effect can be switched on or off by chemical modifications to the DNA.

germ cell A cell which during development becomes either an egg or a sperm (*cf.* a somatic cell *below*). A mature germ cell may be referred to as a gamete.

gamete intra-fallopian transfer (GIFT) The process of putting a mixture of an unfertilized egg or eggs with some sperm into the fallopian tube, in the hope that fertilization will take place in a natural environment.

Graafian follicles The maturing, enlarging follicle containing an egg.

granulosa cells A group of cells which feed and nurture the egg inside the Graafian follicle.

haemophilia A disease of blood clotting. Haemophilia may be caused by a genetic defect and may therefore be inherited. One common form of haemophilia is X-linked, as the abnormal gene causing it is on the X-chromosome.

Halakhah Jewish law: the term is derived from a word meaning 'the way' or 'the path'.

HFEA The Human Fertilization and Embryology Authority was established in the UK in 1990 to regulate various aspects of human embryology and fertility treatment.

human chorionic gonadotrophin (HGC) Pregnancy hormone produced by placental cells.

human menopausal gonadotrophin (HMG) The pituitary hormones, FSH mixed with LH, excreted in the urine of menopausal women and used to trigger ovulation in infertile women.

hyperstimulation (in full: ovarian hyperstimulation syndrome or OHSS) The potentially dangerous side-effect of fertility drugs. In some women their injection can cause too many follicles to develop. The ovaries become increasingly cystic and there are a series of biochemical changes in the body – fluid accumulation, changes in the blood composition – which in very rare cases can be life-threatening.

hysterectomy Surgical removal of the uterus.

hysterosalpingogram (HSG) The injection of a radio-opaque dye into the uterus and tubes, used to investigate whether the tubes are open and the uterine cavity is normal.

hysteroscopy Telescopic inspection of the inside of the uterus, sometimes done under general anaesthesia.

implantation The process by which an embryo embeds into the lining of the uterus.

imprinting Certain genes only express if they come from one or the other parent. Some imprinted genes are expressed only on a maternally inherited chromosome and silenced on the paternal chromosome; others are expressed only on a paternally inherited chromosome.

inner cell mass The cells in the blastocyst which go on to make the embryo. The outer cell mass will become the placenta and membranes surrounding the baby.

intracytoplasmic sperm injection (ICSI) Injection of a sperm directly into the substance of an egg in an attempt to achieve fertilization.

intrauterine insemination (IUI) Artificial insemination (*see above*) into the uterus.

laparoscope A telescope, usually passed through the navel into the abdomen under anaesthetic, to get a view of the ovaries, tubes and uterus.

LH-releasing hormone (LHRH) The hormone produced by the brain which stimulates the pituitary gland to produce LH (*see below*).

luteinizing hormone (LH) The hormone produced by the pituitary which triggers ovulation.

MESA, MESE, PESA, PESE Various methods for the collection of sperm directly from the testicle in infertile men: microscopic epididymal sperm aspiration or extraction; percutaneous (through the skin) sperm aspiration or extraction.

microsurgery Surgery done under magnification using a microscope.

mitochondria Tiny organelles in the cell cytoplasm which contain small pieces of DNA that control the use of energy.

multiple pregnancy Twins, triplets or more. All multiple pregnancies carry a higher than average risk through pregnancy and even twins are at greater risk than a singleton fetus.

mutation A change in the 'lettering' of the DNA which is often deleterious.

NEAB The National Ethics and Advisory Board in the USA.

neurons Cells which conduct nervous impulses in the brain.

nuclear transfer The process by which a nucleus from one cell is transplanted into an egg that has usually had its nucleus first removed by microsurgery. It is the start of the cloning process.

nucleus The 'headquarters' of the cell, which controls much of its function and contains the DNA and genes.

oestrogen The main female hormone which affects growth, fat deposition and menstruation.

ooplasm The cytoplasm (*see above*) of an egg.

ovulation The process by which a mature egg is released from the Graafian follicle, ready for possible fertilization.

parthenogenesis The process of making an embryo without sperm. In humans, eggs can be activated (*see above*) so that cell division commences; but it is thought that in humans, the resulting embryo is never capable of being viable.

peritonitis Inflammation of the lining of the abdomen, the peritoneum.

Petri dish Contrary to popular belief, test-tube babies are not usually produced in test-tubes but in a plastic or glass dish of the kind invented by one Julius Richard Petri (1852–1921), a German bacteriologist.

pituitary gland The 'master' gland in the body at the base of the brain. It controls other glands including the thyroid, the pancreas, the adrenals and, of course, the ovaries and testes.

pluripotent *See* totipotent *below*.

placenta The baby's afterbirth, which feeds and nurtures it during pregnancy. Sadly, this beautiful and complex organ is usually incinerated after delivery of the infant.

polar body The clump of material extruded from an egg or sperm cell containing a redundant set of chromosomes.

polycystic ovary syndrome (PCO) A complex disorder which is at least partly genetic in origin and is associated with infertility, poor ovulation and cystic ovaries. Many sufferers from PCO have excess body hair and are over-weight.

polymerase chain reaction (PCR) The process by which a single molecule of DNA can be quickly multiplied artificially into many thousands of copies so that there is enough DNA available for easy measurement and analysis.

primary and secondary infertility Primary infertility is infertility in a woman who has never been pregnant.

Secondary infertility is infertility after a birth, ectopic pregnancy or miscarriage.

primitive streak The first definable 'structure' seen on the developing embryo at about fourteen days after fertilization. The primitive streak will eventually become the central nervous system.

progesterone Hormone produced in large quantities by the ovary after ovulation and by the placenta during pregnancy.

recessive gene Each person has two copies of every gene, one from their mother and one from their father, situated on the paired chromosomes. If a genetic trait is recessive, a person needs to inherit both copies of the gene for that trait to be expressed (*see above*). A *carrier* carries only one copy from one or the other parent and so the gene will not be expressed. Thus, both parents have to be carriers of a recessive trait in order for a child to express that trait. If both parents are carriers of a recessive defect, there is a 25% chance that each of their children may show the effect. By contrast, a dominant gene produces its effect if there is only one copy present, from either parent. Thus, there is a 50% chance of a child's being affected if a dominant gene is abnormal.

Rhesus disease A disease of babies during pregnancy which can cause severe anaemia and even death in utero. It is caused by antibodies produced by the mother which attack the fetal red cells.

somatic cell Any differentiated, i.e. specialized, cell – for example a fat cell, muscle cell, or neuron – in the body that is not a germ cell (*see above*).

stem cell A primitive cell capable of becoming a specialized cell of one of several different types after differentiation.

surrogate In reproductive parlance, a woman who is prepared to go through pregnancy and birth on behalf of another woman.

Talmud A massive collection of rabbinic discussions written between 200 and 600 CE about Jewish law, ethics, customs and history. There are two versions, one written in Palestine (the Jerusalem Talmud), and one in Mesopotamia (the Babylonian Talmud); the latter is generally the more influential.

telomere The region of repeated DNA sequences at the end of a chromosome. Chromosomes are incapable of being completely copied to their very end, so that each time a chromosome is replicated during cell division, the very tip of it is lost. The telomeres act as a buffer, preventing loss of genetic information which is needed to sustain the cell's functions. Shortening of the telomeres is associated with ageing.

testes; testicles The male sex organ (or gonad) is the testis. The testicle includes the gonad together with its surrounding membranes, blood vessels and tubing.

testosterone Male hormone produced by cells in the testis.

total fertility rate (TFR) This is a statistical assessment of the average number of children that would be born to a woman living in a particular population. It is an estimate of her fertility over her entire reproductive lifetime (assumed to be 16–49 years old). This synthetic measure gives a general idea of the rise or fall of a given population.

totipotent A stem cell is said to be 'totipotent' if it is capable of developing into any cell type in the body, 'pluripotent' if it can develop into one of many cell types and 'unipotent' if it generally gives rise to only one cell type.

toxaemia of pregnancy A disease causing high maternal blood pressure and fluid retention during pregnancy. It can result in loss of the baby's blood supply and restrict its growth in utero.

transgenic An animal is said to be transgenic if it has had a gene or genes artificially inserted into its genetic structure. Not all transgenes work normally.

trehalose A sugar used as a form of antifreeze during cryopreservation (*see above*).

Turner's syndrome Loss of one of the paired X-chromosomes in a girl. Girls with Turner's suffer with infertility, poor development of the ovaries, and failure to menstruate and ovulate, and are short in stature. On rare occasions, if the X-chromosome is missing in only some of the cells in the body (so-called mosaic), such a girl may be fertile.

ultrasound The use of high-frequency sound as sonar to detect structures inside the abdomen. It is harmless.

unipotent *See* totipotent *above*.

uterus The womb.

varicocele A collection of extra blood vessels in the testicle.

VLA The Voluntary Licensing Authority which historically preceded the formation of the HFEA (*see above*) in 1990.

yolk sac This is the first structure to form during early pregnancy that can be seen on ultrasound. The yolk sac lies outside the embryo and is connected to it by its abdomen. It is the site of where germ cells are produced, as well as very early blood cells.

zygote intrafallopian transfer (ZIFT) The zygote is a fertilized egg that has not yet undergone cell division. This can be transferred to the fallopian tube, as is done during the GIFT (*see above*) procedure. In my view, it has no advantages, but this has not prevented its use commercially.

zona drilling (assisted hatching) A hole is made in the zona pellucida (*see below*) using an acid, a sharp piece of fine glass or a laser, to help the blastocyst to escape – or 'hatch'. It is said that some eggs have a hardened zona and this procedure may be suggested. It is generally of dubious value, but again this does not prevent its being widely used.

zona pellucida The zona pellucida is the protective 'shell' around the mammalian egg. It is made of a glycoprotein which is species-specific. This glycoprotein controls how sperm are attached to the egg, so its specific nature prevents the egg of one species being fertilized by the sperm of another.

Index

Aborigines: infanticide rates 449
abortion 121–2, 125, 415
    and fetal eggs 253–4
    and infection in fallopian
        tubes 61
    psychological injury of 327,
        330
    and surrogate mothers
        309–10
Abraham 13–14, 16
achondroplasia 34
acupuncture 150–1
Addenbrooke's Hospital,
    Cambridge 91
adenomyosis 64, 134
adhesions 60, 64, 134
adoption 89–91, 280, 282
adrenoleukodystrophy 346
Afnan, Masoud 157
age
    and childbirth 43–7, 182,
        188
    and deterioration of eggs 32,
        46–7, 246
    and deterioration of sperm
        32, 34
    and failure rates of IVF 134
Agussi, Shlomi 150
Akkadians 452, 453

alcohol consumption 71, 146
Alexandra, Tsarina 323
Alexei, Prince 323
Alfonso, Prince 324
Ali, Mohammed 417
Alice, Princess 323
Allen, Dr Edgar 83–4
alternative medicine 150–1
Alton, Lord 366–7
amniocentesis 327
Anderheim, Dr L. 148
Anderson, Teresa 319–20
aneuploidy screening 120, 143,
    371–3
Angelman Syndrome 187
Anglican Church: and donor
    insemination 241
animal research 488–91
    see also hares; mice;
    monkeys; rabbits; transgenics
animal activists 491
Anas, George 252–3
anonymity
    of egg donors 273, 274–80
    of sperm donors 243–4,
        271–2, 273, 274–80, 280–3,
        285–6
anosmia 73
antagonists 153

antenatal screening 327
Anthony, Pat 304–5
Antinori, Dr Severino 389–93, 401
Antoninus, Emperor 116
Aquinas, St Thomas 122, 438
Aristophanes: *Peace* 456
Aristotle 122, 409, 457
aromatherapy 151
artificial insemination *see* donor insemination
Asch, Professor Ricardo 171–2, 221, 222
Asian families
egg donation 259–60
surrogacy 259
'assisted hatching' *see* zona drilling
Atala, Anthony 419
Augustus, Emperor 459
Austin, Professor C. R. 87
Australia 95, 112–13, 291
donor recruitment and anonymity 279–80
first IVF baby 96–7
obesity 143–4
reproductive research 69–70, 95, 97, 99, 100–1, 441–2
study of IVF birth defects 181–2
and surrogacy 314
autoplasmic transfer 262
autosomes 352

'Baby M' case 304, 310, 311, 317
Babylon 67, 453
Baird, David 225–6
Barker, David 183, 184
Basil, Bishop of Caesarea 121
Beckwith-Wiedemann Syndrome 186
Ben Sira 235, 236, 256
Bennett, Gaymon 439

Berga, Dr Sarah 149–50
beta-thalassaemia 493–6
BFS *see* British Fertility Society
Bible stories 13–19, 104–5, 235–7
biopsy
cone 64
embryo 338–9
endometrial 56–7
Biosteel 477
birth control *see* contraception/contraceptives
birth defects *see* in vitro fertilization: abnormalities in babies
birth weight, low
and disease 184
see also premature babies
Black, Sir Douglas 110
Blackman, Ann 386
bladder stem cell transplants 419
blastocysts 41, 161–2, 162
freezing 205, 260
transferring 171
Bleich, Rabbi David: *Contemporary Halakhic Problems* 256
Blood, Diane 194–7
Blood, Stephen 194, 197
Bodmer, Sir Walter 360
bone marrow transplants 412, 413
Bonnar, Professor John 93
Bonnet, Charles 37
borage 449
Boston, Charles 466
Bottomley, Virginia 252
Bourn Hall clinic, nr Cambridge 101, 102
bowel cancer genes (*polyposis coli*) 359, 360
Boyson, Rhodes 267
brain, development of the 125

breast cancer 228, 359
Briggs, Robert 378
Brigham, Carl 465
    A Study of Army Intelligence
    465
Brinkley, John Romulus 37–9
Brinster, Ralph 478
British Fertility Society (BFS)
    209, 276
British government 91, 110–11
    legislation on anonymity of
    egg and sperm donors
    273–80
    legislation on surrogacy 298,
    303–4, 314
    sets up Warnock Committee
    107; *see also* Warnock,
    Baroness
    and Unborn Children
    (Protection) Bill 108, 110,
    329, 330
    see also Human Fertilization
    and Embryology Act and
    Authority
British Medical Association:
    Ethics Committee 252–3
British Medical Journal 240–1
Brown, Louise 96, 106, 179
Brown-Sequard, Professor
    Edouard 69, 84
Browne, John McClure 95–6,
    173–4, 326
Buck, Carrie 465
Buckley, Samuel 82
Bunge, Dr R. G. 190
Buserelin 152–3
Bush, President George W./Bush
    administration 213, 392,
    436–7, 439–40
Buster, Dr John 264–5, 332
Bustillo, Maria 265
Bygren, Lars 185

Caesar, Julius 459

Califano, Joseph 113
Cambridge University 81, 87,
    90, 91, 218
Campagne, Daniel 148
Campbell, Keith 381, 393
Canada 290, 312–13
cancer treatments
    and ovarian function 140
    and scarred cervix 64
    and storage of ovarian tissue
    225, 228–9
    and storage of sperm 194,
    214–15, 215, 230
    and storage of sperm stem
    cells 230
    and testicular freezing 230
cancers
    bowel 359, 360
    breast 228, 359
    caused by stem cell injections
    424–5
    and clomiphene 177
    from FSH 177
    and frozen sperm 208, 210
    and gene defects 359–60
    and gene therapy 492
    and ovarian grafts 228–9
    testicular 72
    in transgenic mice 486, 490
    see also Hodgkin's disease;
    leukaemia
capacitation (of sperm) 81, 87,
    502–3 n. 6
Cardiff Assisted Reproduction
    Unit 279
career women 48–9, 217,
    219–20
Carr, Elizabeth 198
Catholic Church, Roman 120,
    127, 128–9, 131
    and adoption of frozen
    embryos 211–12
    and cloning 384
    and donor insemination 241

and egg and sperm donation 258
moment of conception 40
and stem-cell research 439
and surrogacy 305
cells, embryonic 36–7, 41, 376–7
cervix
assessments of 152
and laser treatment 64
mucus 35, 65, 75
and post-coital tests 65
problems and infertility 65, 75
scarred 64, 140
Chandler, Harry 68
Chandrashekran, Anil 485
Chang, Dr M. C. 39, 87, 88–9
chemotherapy 140, 227, 230, 412, 444
chimaeras 411, 480
China
egg donation procedure 262
population 468, 472
Chinese medicine 73–4
chlamydia 61
chorion villus sampling 328
chorionic gonadotrophin 59
Christian Science 128
Christianity 120–2
and animals 348–9
and stem-cell research 439–40
'Virgin' birth 39–40
see also Catholic Church, Roman; Protestant Church
chromosomes
and Down's syndrome 358
and haemophilia 322–3
and polar body biopsy 338
staining 332, 335, 341, 356–8; see also aneuploidy screening
translocations 358

and Turner's syndrome 365
X and Y 78–9
cilia (hair cells) 27, 30, 60
Clark, Dr Amanda 441
Clifford, Charles 202
clomiphene 58–9, 139, 154, 174, 301
and risk of cancer 177
Clonaid 388
cloning 376
and ageing 395–6
dangers of 393–8, 399–401
and endangered species 401
epigenetic effects 395–7
ethical concerns 384–5
and fraudulent claims 379–82, 390, 428–36
of humans 382–7, 394, 397–8, 401–4
of mammals 380–2, 386–7, 393–5, 398
for organ donation 400
therapeutic 425–8, 436
Cobb, Matthew: The Egg and Sperm Race 407, 522 n. 1
Cochrane Collaboration study 372
coffee, drinking 146
Cohen, Jacques 260–2
'coil' 61
complementary medicine 150–1
Conard, Nicholas 451
conception, moment of 36, 40
cone biopsies 64
Conradi's syndrome 367–8
consanguineous relationships 244, 277
contraception/contraceptives
ancient methods 449, 459–67, 458
coil 61
oral ('the Pill') 39, 58, 61
and population growth 468–72

20th-century campaign for 464–5
cord blood 369, 370, 415
Corea, Ginea 310–11
corpus luteum 31
Cotton, Kim 303, 313, 318
counselling 149–50
    and egg donation 251
Coutelle, Charles 336–7, 353, 354
Coyle, Adrian *see* Turner, Amanda
Cromwell IVF and Fertility Centre 249
Cryos International Sperm Bank 243–4, 272
Cryotop 219
Ctesias of Cnida 21
culture media 91, 94, 98, 184–5, 292
cystic fibrosis 337, 352–3, 361, 364
    and pre-implantation diagnosis 338, 353–4, 355, 362
    and gene therapy 492, 493
Cystic Fibrosis Trust 354
cysts, dermoid 140
cytoplasm 378
cytoplasmic transfer 260–1

Daily Mail 253, 305
Daily Telegraph 348
Dante Alighieri: *The Divine Comedy* 123, 125
Darwin, Charles 127–8, 461, 462
Davis, Junior Lewis and Mary Sue 197–203, 231
Dawes, Geoffrey 110
Deech, Baroness Ruth 518 n. 24
Delhanty, Dr Joy 357, 358, 360
della Cortes, Rosana 389–90
Deni Indians 449
dermoid cysts 140

'designer' babies 368–70
Devonshire, Georgiana, Duchess of 52
Devroey, Paul 196, 372
diabetes/diabetics 184, 332, 476
diet: and chemical changes in genes 185–6
dimethyl sulphoxide 223
DNA 354
    and FISH 356–7, 358
    modifying 263, 491–2
    and single-gene diseases 331–2, 336, 341, 356, 363
    testing 334–5, 339–40
    and viruses 474–5, 475–6, 447–8, 482–5
dogs, experiments with 80–1
Doisy, Dr Edward 82–8
Dolly the sheep 381, 386, 393–4, 398, 426, 427
'Dominic Egg' 85
Donald, Professor Ian 56, 325
Donaldson, Dame Mary 109–10, 298
Donnez, Dr Jacques 227–8
donor-conceived children 266–9, 282–3
donor insemination/artificial insemination 74, 239–45
    and anonymity of donors 244, 271, 273, 273–80, 280–3, 285–6
    early examples 239–40
    fraudulent practices 270–1
    and illegitimacy of offspring 241, 242
    and lesbian and single mothers 267–9
    in myth and legend 237–8
    religious attitudes to 241
    success rates 246
dopamine production 417, 425
Doron, Chief Rabbi Yehoshua Bakshi 232

Douglas, Justice William 467–8
Down's syndrome 248, 358, 365–6
drugs
    fertility 58–60
    and low-quality sperm 71
    steroids 73
Duchenne muscular dystrophy 245, 331, 362
Dunstan, Professor Gordon 121
Dussik, Dr Karl Theodore 56
dwarfism 34
dystrophin gene 363

Ebers papyrus 454, 455
ectopic pregnancy 28, 61–2
    after IVF 95, 177–8, 210
Eddy, Carl 221–2, 223–4, 225
Edwards, Robert 87, 90–5, 96, 98, 105, 234, 260, 333, 366
egg banks 217
egg donation 245–55, 426
    age limit of donors 246
    and anonymity of donors 247, 273, 273–80
    and autoplasmic transfers 262
    and counselling 246–7, 251
    and cytoplasmic transfers 260–1
    egg giving 249
    egg sharing 204, 249, 278, 510 n. 3
    fetal eggs 251–4
    motivation of donors 247, 250, 266, 284
    in overseas clinics 288–9
    and parental rights 255–6
    payment of donors 248–9, 283–4
    religious views 255–9
    screening of donors 246
    shortage of donors 248
    success rates 246

eggs, human 23–4
    and ageing 31–2, 47
    collecting 85, 93, 98, 133, 154, 156–7, 177
    fertilization of 25, 31, 34–7
    freezing 218–19
    maturation of 26–7, 36, 94, 154
    and nuclear transfer 378
    zona drilling ('assisted hatching') 170, 179, 260
    see also egg donation; fertilization; ovulation
Egypt, ancient 67, 453–6
Ehrlich, Paul: The Population Bomb 469–70
ejaculate 32, 33
    split 74
Elder, Professor Murdo 300
electrofusion 480
embryo freezing 87, 205
    and adoption of embryos 211–13
    benefits of 204
    and husbands' rights 197–203, 213–15, 232–4
    and IVF success rates 206
    risks 182–3, 207–11
    UK legislation 211, 213, 214
    and wastage of embryos 205, 205–6, 211–12
embryos, animal 480
embryos, human
    cells and cell division 37, 41, 324, 377–8
    culture media for 94, 98, 184–5, 292
    diagnosing sex of 331
    freezing see embryo freezing
    genetic screening 213, 291, 295; see also pre-implantation genetic diagnosis
    implantation in uterus 25, 41–2, 57

religious views on 114–23
scientific and medical views on 123–30
'surplus' 117–18, 123, 263–6, 366, 437–9
transferring to uterus 81, 161–3, 205–6
endometriosis 64, 93, 140
ovarian 60, 64
endometrium 41
biopsies of 56–7
Epic of Gilgamesh 447
epididymis 32
epigenetic effects 396–7
ethics committees 300
eugenics 191–2, 384, 459, 460–8, 497
European Congress on Human Reproduction and Embryology 473
European Court of Human Rights 213–14
European Human Rights Act 275–6
Evans, Natalie 213–15, 231

Faerman, Dr Marina 460
faith-healing 150–1
fallopian tubes 27–35, 95
and adhesions 60
blocked 60, 95, 138, 216
cilia 28–31, 60
damaged by infection 60–1
diagnosing problems 62, 62–3
dilated 134
scarring in 60, 64
surgery on 62, 87
see also ectopic pregnancy
Family Law Reform Act 241
feminism/feminists 38, 106, 464–5
Ferreira-Jorge, Karen 304
fertility

female 46–9, 450–1, 452–3
male 67, 447, 451, 453–4
'fertility tourism' 286–9
fertilization 23, 31, 36–7, 133
see also in vitro fertilization
fetu
eggs and ovaries in 251–4
stem cells in 415–16, 417–18
fibroids 63, 134
Fiorentino, Dr F. 370
Firestone, Shulamith 106–7
FISH (fluorescent in situ hybridization) 356–8, 371
fish, transgenic 477, 487
Fishel, Dr Simon 165–6
Fisk, Professor Nicholas 126
Fletcher, Dr John 253
foetus *see* fetus
follicles, ovarian 25, 83, 502 n. 2
follicle stimulating hormone (FSH) 25, 26, 55–60, 58, 59, 151, 394, 476
and clomiphene 58, 154, 301
injections 59, 139, 153–4, 228, 247, 302
in men 72–3
see also ovarian hyperstimulation syndrome
follicular puncture 60
Formigli, Drs G. and L. 265
Foundation on Economic Trends 384
Fox, Michael J. 418
France 113
ban on commercial surrogacy 312–13
conception rates 135–6
first IVF baby 103
study of embryo freezing 207–8
work on ovarian grafting 225–6
Franks, Stephen 24
Frederick William, Prince 323

freezing
    eggs 217–20
    embryos *see* embryo freezing
    ovaries 222–3
    sperm 190, 192–3, 243–4
    testes 229–31
Friedler, Dr Shevach 150
frogs' eggs, artificial fertilization of 80, 81
FSH *see* follicle stimulating hormone
Fukuyama, Francis 497, 499

Galgenburg Venus 450
Galton, Sir Francis 461
gamete intra-fallopian transfer (GIFT) 171–2, 172–4
Gandhi, Indira 471
Gandhi, Sanjay 471
gay couples: and surrogacy 314
    see also lesbian couples
'Geep' (goat/sheep) 381
genes
    and effects of grandparents' diet 185–6
    imprinted 40, 396–7
    see also gene therapy; genetic disorders; genetic engineering/modification; transgenics
Genesis 13–19, 117
gene therapy 476–7, 492
    germ-line 493–9
genetic disorders 142–3, 331–2
    diagnosing *see* pre-implantation genetic diagnosis
    late onset 358–61
    and older men's sperm 32, 34
    single-gene 331–2, 352
genetic engineering/modification 477, 496–9
    see also transgenics

germ cells 24, 415, 442, 443–4
germ-line therapy *see* gene therapy
Germany 113, 473
    ban on commercial surrogacy 312
    stem-cell research 441
GIFT *see* gamete intra-fallopian transfer
gland extracts 68–9
Glover, Professor Vivette 126
glycerol 190
golem legends 375–6
Golombok, Professor Susan: studies 317–19
gonadotrophin therapy 292
gonorrhoea 61
Gonzalez, Luisa 319–20
Gordon, Harry 327
Gordon, Jon 478–9, 480
Gosden, Dr Roger 225–6, 251–2, 253
Gould, Stephen Jay 448
Graaf, Regnier de 407–9
    The Generative Organs of Men 408
    The Generative Organs of Women 408
Graafian follicles 25
Graham, Dr James 52–3
Graham, Robert K. 191–2
    The Future of Man 191
granulosa cells 26, 31
Greeks, ancient 21, 22, 40, 456–7, 458, 474
Gregory of Nyssa, St 121
Grisez, Germain 384
Gurdon, John 378

haemophilia 322–3, 492
    screening for 323, 335
Hagar 16, 306, 315
hair cells *see* cilia

Hall, Jerry 383, 386
Hamilton, Lady Emma (née Lyon) 52
Hammersmith Hospital 29, 56, 173
  HFEA licence 294–5
  IVF clinic 62–3, 91, 101, 103, 126, 149, 157, 158, 160, 163, 206, 208–10, 325
  IVF surrogacy attempts 298–303
  pre-implantation diagnosis 324–30
  zona drilling 170, 260
Handel, Bill 308–9
Handyside, Alan 324, 325, 328, 332–3, 334, 337, 338–40, 343, 345, 348, 350, 354, 355, 357, 481
Hansen, Dr Michele 181
Harappa, Pakistan 453–4
Hard, Dr Addison 239–40
Hardy, Kate 24, 332–3, 334, 344, 357, 363, 423
hares, artificial fertilization of 82–3
'Harvard Egg' 85
Harvey, William 24
Hassan, Mohammed, and Killick, Stephen: study 146
HCG see human chorionic gonadotrophin
Heape, Walter 81–2
herbal treatments 150–1
Herodotus 454
Hertig, Arthur 85, 86
HFEA see Human Fertilization and Embryology Authority
Hillier, Stephen 101, 102–3, 156, 157–8, 264, 325
Hindu legends 237–8
Hippocrates 456
HIV (human immunodeficiency virus) 193–4, 482, 492

HMG see human menopausal gonadotrophin
Hodgkin's disease 226, 227, 229, 444
Hogan, Brigid 415
Hohle Fels Cave: stone penis 451
Holmes, Oliver Wendell 465–6
Homer: Odyssey 21, 474
Horlock, Carole 314–15, 320
Horlock, Stephen 314
hormones see follicle stimulating hormone; luteinizing hormone; oestrogen; progesterone
Horne, Professor Johannes van 406, 408
Hornett, Amanda 132
Hornett, Beth 132
Hovatta, Outi 224, 230, 445
HSGs see hysterosalpingograms
Hubner, Karin 440, 442
Hughes, Dr Mark 354
human chorionic gonadotrophin (HCG) 41, 154
Human Fertilization and Embryology Act (1990) 112, 267, 273, 275, 278, 291, 349
Human Fertilization and Embryology Authority (HFEA) 110, 112, 267, 289–95
  anonymity debate 273, 275, 277, 278
  and Blood case 195–7
  and donor insemination 242–3, 243
  and egg donors 243, 248, 248–9, 251
  and embryo freezing 206, 208–10, 210, 213
  and embryo screening 361, 366, 369–70

Human Fertilization (*cont.*)
  and embryo transfers 160, 161, 290
  insistence on counselling 149–50
  lack of data collection 186, 206, 292–3
  licences 230, 294–5, 366, 369–70, 445–6
  The Patients' Guide to Infertility 293
  rulings on payments 283–4
human menopausal gonadotrophin (HMG) 58
Human Reproduction (journal) 65, 262, 281
Hungarian IVF clinics 287
Hunter, Dr John 239
Huntington's Chorea 359
husbands' rights 163, 194–7, 200–3, 213–15, 232–4
Huxley, Aldous: *Brave New World* 84
Hwang Woo-Suk, Dr 428–36
hydrosalpinx 134
hyperstimulation 59
hypospadias 168, 239
hypothalamus 53–4, 59
hysterosalpingograms (HSGs) 62–3

ICSI *see* intra-cytoplasmic sperm microinjection
Iliescu, Adriana 43–6
Illmensee, Karl 379–80, 391
imprinted genes 40, 187, 396–7
Inanna 453
India 40, 237–8
  mass sterilization campaign 470–1
infanticide 449, 460, 472
infections 60–1
infertility, female 50–1, 51–2, 136–8
  and alternative and complementary medicine 150–1
  due to cervical damage 51, 140–1
  due to damaged fallopian tubes 51, 138–9
  due to ovulation problems 51, 53–60, 139
  due to premature menopause 139–40
  due to uterine abnormalities 51, 57, 63–5, 134
  and intra-cytoplasmic sperm microinjection 165–8
  and obesity 143–5
  of older women 46–7
  and radiation 147
  secondary 501 n. 2
  and smoking 146–7
  and stress 148–50
  and tea drinking 146–7
  testing for causes 50, 52, 56–7
  and underweight 145–6
  'unexplained' 141
infertility, male 65–6, 142
  and alcohol consumption 146
  and decline in sperm count 70
  early 'cures' 67–9
  and low-quality sperm 71
  and MESA 169–70
  and PESA 170
  and radiation 147
  and TESA/TESE 168–9
  and testicular temperature 65–6
  see also intra-cytoplasmic sperm microinjection
insemination
  IVF 159
  see also donor insemination
insulin 476
  resistance to 54

intra-cytoplasmic sperm
  microinjection (ICSI) 159,
  165–8, 179
  and abnormalities 182,
  187–8, 218–19
  and testicular freezing 229–30
intra-uterine devices 61
intra-uterine insemination (IUI)
  65, 74–5
in vitro fertilization (IVF) 50–1,
  51, 57–8
  and abnormalities in babies
  181–8, 209, 218–19, 292–3
  and after-care 165
  and aneuploidy screening
  371–3
  as best treatment 136–43
  blastocyst transfer 171
  causes of failure 132–5
  and cruelty to fetus 125–6
  early experiments 85–90
  expense of 75, 129, 131–2,
  153–4
  first live births 96, 98–9, 103
  and freezing of embryos 203–6
  gamete intra-fallopian
  transfer (GIFT) 171–4
  and hysterosalpingograms
  62–3
  and legislation 108–13; see
  also Human Fertilization and
  Embryology Authority
  and low birth weight 183–4
  and miscarriage 47, 165
  moral and ethical
  implications 105–8, 113–14,
  129–30
  and multiple pregnancies 17,
  161, 171, 178–9
  'natural-cycle' 154
  opposition to 106, 113–14,
  127, 128–9, 131, 132; see
  also religious views below
  at overseas clinics 286–9
  pioneers of 590–5
  and premature births 179–80
  preparing for 143–51
  religious views on 114–23,
  127, 128, 130
  risks to child 179–88
  risks to patient 175–9
  scientific and medical views
  123–7, 130
  success rates 76, 134–5
  and surplus embryos 123–7,
  204, 205, 263–6, 366, 437–9
  and surrogacy 297, 298–305,
  312–13
  ten-step programme 150–65
  'unnaturalness' of 127–8
  and zona drilling ('assisted
  hatching') 170
  zygote intra-fallopian
  transfer (ZIFT) 175
Ishtar 453
Isis 453
Islam
  and adoption 119
  and egg and embryo freezing
  231–2
  and IVF 119, 130, 257–8
  and stem-cell research 438
  and surrogacy 119, 307
Israel 113, 293–4
Italy 391–2
IUI see intra-uterine
  insemination
IVF see in vitro fertilization

Jackson, Dr Margaret 241–2
Jackson, Mark 241
Jacob 15, 17–18
Jacobson, Dr Cecil 270–1
Jaenisch, Dr Rudolf 475–6
Jains 40
Jakobovits, Chief Rabbi Dr
  Immanuel (Lord) 115, 118,
  306–7

Jehovah's Witnesses 216
Jeremiah, Prophet 235, 236
Jewish Chronicle 463
Jews 235–7; *see also* Judaism
   blood libel 374–5
   golem legends 375–6
   see also Bible stories;
   Judaism
John Paul II, Pope 119–20
Johns Hopkins University,
   Baltimore 186, 387–8
Johnson v. Calvert 313
Johnston, Ian 96
Johnston, Howard 214–15
Jones, Dr Georgeanna Seegar
   198
Jones, Dr Howard 198
Jones, Steve 463
Josephus 460
Journal of the American
   Medical Association 77
Judah the Prince 116
Judaism
   and abortion 437–8
   and egg donation 254, 255–7
   and freezing eggs and
   embryos 232–4
   and IVF 114–19, 130
   and masturbation 232
   and ovarian transplants 254
   and stem-cell research
   437–8
   and surrogacy 306–7
   and treatment of animals
   489–90

Kahun papyrus 455
Kallmann's syndrome 72–3
Kass, Leon 106, 497, 499
Kathasaritsagar 237
Keane, Noel 307–8
Kemeter, Peter 301
Kevles, Dan: *In the Name of*
   Eugenics 462

Khameini, Ayatollah Ali
   Hussein 257
Killick, Stephen *see* Hassan,
   Mohammed
King, Dr Irving 202
King, Thomas 378
Klerk, Dr C. de 149
Kono, Dr Tomohiro 40
Lancet 56, 69
laparoscope/laparoscopy 57, 60,
   62, 92–3, 101–2, 155, 156–7
lead: and low sperm count 71
Leah 17
Ledger, Professor William 75
Leeuwenhoek, Antonie van 239
Lejeune, Dr Jerome 201
Leopold, Prince 332, 333
lesbian couples 45, 193, 243,
   267–8, 269
Lesch-Nyhan syndrome 342,
   355
leukaemia 229, 369, 412, 444
Leventhal, Dr Michael 54
Levin, Ira: *The Boys from Brazil*
   379, 384, 398
Levin, Dr Richard L. 310
LH *see* luteinizing hormone
Life magazine 39
Lester, Lord 196
Lockwood, Dr Gillian 216
Loeb, Jacques 37–8
Loew ben Bezalel, Chief Rabbi
   Yehuda 375
Lopata, Alex 96
luteinizing hormone (LH) 26,
   58, 59–60, 72–3, 155
   testing levels 55, 103

MacGregor, William 91
McKay, Dr Ron 418
Mackay of Clashfern, Lord 112
McLaren, Anne 94, 110
McVeigh, Enda 149
Maddox, John 348

Mahabharata 238
Mahavira 40
Maimonides, Moses 489
Making Babies (BBC series) 284
malaria 495–6
Malpighi, Marcello 409
Malthus, Thomas 469
Margara, Raul 145, 350–2
Mariescu, Dr Bogdan 44, 46
marijuana: and sperm counts 71
masturbation 104–5, 232
Mattei, Dr Jean-François 384
media 38, 107–8, 209, 302, 384, 386–7
Medical Research Council (MRC) 94, 109, 428
Medical World 239, 240
Melbourne University 96
  see also Monash University
Mendel, Gregor 19, 501–2 n. 4
Mendiola, Professor Michael 439
menopause 58, 135
  premature 139–40
menstrual cycles 26, 47, 54, 175
MESA *see* microscopic epididymal sperm aspiration
Mesopotamians 452–3
mice, experiments on 39, 40, 87, 98, 207–8, 251, 331, 332, 475–6, 477–8, 478–9, 484, 486
microscopes, invention of 409–11
microscopic epididymal sperm aspiration (MESA) 169–70
microsurgery 62, 74, 140, 350
Millport Marine Biology Station, Scotland 81
minerals 151
miscarriage 17, 415
  after IVF treatment 47, 164–5
  and aneuploidy screening 373
  caused by obesity 143–5
  and Down's syndrome pregnancies 365
  and infection in fallopian tubes 61
  and polycystic ovary syndrome 54
  and uterine problems 63–4
mitochondria 261
Mitterrand, François 384
Modell, Bernadette 328
Mohr, Linda 198
Monash University, Melbourne 95, 97, 98, 99
Monk, Dr Marilyn 385
monkeys, experiments with 222
Montegazza (Italian scientist) 189
Moreno, Enrique 319
mortality rates 49–50
MRC *see* Medical Research Council
Muggleton-Harris, Dr Audrey 261
Mullis, Dr Kary B. 333–4
multiple pregnancies 16, 17, 37, 139
  and blastocyst transfer 171
  and egg donations 262
  and embryo transfer 161, 262
  and ovarian hyperstimulation 59, 139
  risks 178–9, 319–20
mumps 71
muscular dystrophy 245, 331, 362, 492
Muslims *see* Islam
myelin 420

Nahmani, Dani and Ruth 232–3
Naldini, Luigi 482
National Institute for Clinical Excellence (NICE) 154
'natural' and 'unnatural' 127

Nature 88, 333–4, 348, 380, 381, 430, 433, 481
Nayernia, Karim 441
Neesen, Dr 77
New England Journal of Medicine 84–5, 355
New York Times 384
newspapers *see* media
NICE *see* National Institute for Clinical Excellence
Nicholas II, Tsar 323
Nielsen, Heine 272
nuclear power stations 147
nuclear transfer 378
  and cloning 381–2, 389, 480
  generating stem cells by *see* somatic-cell nuclear transfer
nun's urine 58

obesity 75, 134, 143–5
Observer 288
occupations: and low sperm counts 71
Odendal, Dr Hein 305
oestrogen 26, 70, 83, 96, 206
oestrus cycles 26
Oeztuerk-Winter, Feride 485
OHSS *see* ovarian hyperstimulation syndrome
Oinoe, myth of 22
Oktay, Dr 228–9
older women 47–8, 49
  deaths in childbirth 49–50
Oldham General Hospital 90, 92
oligodendrocytes 420
Onan/onanism 104–5
Onania 104
one-parent families *see* single women
ooplasm 260, 261
ooplasmic transfer *see* cytoplasmic transfer

orgasm
  female 35
  male 34
Osiris 453
Osservatore Romano, L' 384
ovarian hyperstimulation syndrome (OHSS) 59, 176–7, 247, 288
ovaries 24–5
  cancer 117, 229
  egg production 24, 25
  endometriosis 60
  follicles 25, 83, 502 n. 2
  freezing 222–5
  and infertility 54
  stimulation 152, 153–4; *see also* ovarian hyperstimulation syndrome
  suppression 152–3
  use of tissue 220, 225–9
  see also polycystic ovary syndrome
ovulation 26–7, 35
  problems with 51, 53–60, 139
  testing for 55
Owens, Michael 158

Packham, Dina 347
Palmer, Raoul 93
Palmiter, Dr Richard 476
Pancoast, Dr William 240
parental rights 255, 263, 273–4
  and surrogacy 313
  see also husbands' rights
Parkinson's disease 416–18, 424
parthenogenesis 37–40, 159, 429, 440
PCO *see* polycystic ovary syndrome
PCR *see* polymerase chain reaction
PCTs *see* post-coital tests
Pearson, Professor Karl 463

Penketh, Richard 327–8, 331, 332, 341, 356
percutaneous epididymal sperm aspiration (PESA) 170
Peretz ben Elijah, Rabbi 236
periods, irregular 54
    after IVF 176
Perry, Emily Louise 216–17
Perry, Helen 216–17, 219
PESA see percutaneous epididymal sperm aspiration
Petrie, William Matthew Flinders 455
PGD see pre-implantation genetic diagnosis
Phillips, Dr Perry 361, 362
pigs: and xenotransplantation 485–6
Pincus, Gregory 39, 86
pituitary gland 25, 26, 53, 55, 58, 59, 70, 152, 154, 228
Pius XII, Pope 119, 241
polar body biopsy 337–9
Polge, Chris 190
Polybius 485
polycystic ovary syndrome (PCO) 54, 55, 60, 144, 332
polymerase chain reaction (PCR) 334, 338–40, 343, 344–5, 346, 354, 356, 362–3, 481, 483, 484
Poon, Ellen 485
population growth 468–70
    and sterilization 470–1
post-coital tests (PCTs) 65
Powell, Enoch: Unborn Children (Protection) Bill 108, 110, 329, 331, 349
Prader-Willi Syndrome 187
pregnancy tests 41
pre-implantation genetic diagnosis (PGD) 142–3, 295, 329

aneuploidy screening 120, 143, 371–3
    blastocyst biopsies 338
    of cancer genes 360–1
    and chromosome staining 332, 335, 341, 356–8, 371
    commercialization of 364
    and 'designer' babies 368–70
    and Down's syndrome 365–6
    and embryo biopsies 337
    ethical considerations 344, 365–8
    future improvements 364–5
    limitations 361–5
    and PCR 333–5, 339–40, 343, 345, 346, 363, 370
    and polar body biopsies 377–9
    and sex-linked diseases 322–3, 331, 333–5, 342–9, 365, 366–8
    and sexing embryos 344–5, 358, see chromosome staining above
    and single-gene diseases 336–7, 352–6, 358–61, 361, 362
    for tissue transplants 369–70
premature births 180
    and disease in later life 181–2, 183–4
'primitive streak' 42
progesterone 31, 55, 57, 164–5, 206
prostaglandins 35
prostate gland 33
Protestants 119, 120–1, 439
    see also Anglican Church
Purdey, Jean 95
Putin, Vladimir 473
'pygmies'/Pygmaioi 22–3

rabbits, experiments on 28–31, 39, 79, 81–2, 87, 88, 95, 222, 223–4

Rabelais, François: *Pantagruel* 296–7
Rachel 17
racism 465
radiation, background 147
radiotherapy 204, 412, 444
Raelians 388–9
Rainbow Flag Health Services 272
Ramsay, Paul 106
Rasputin, Gregory 323
Readhead, Carol 230, 443, 444–5, 480–5
Rebecca 15, 16, 17
record-keeping 159–60
Reeve, Christopher 420
reflexology 150–1
Reisman, Geoffrey 421
religious views 89, 436–9
  see Christianity; Islam; Judaism
Repository for Germinal Choice, California 191
reproductive process, human 31–7, 41–3
retinoblastoma genes 359–60
rheumatoid arthritis: and gene therapy 492
Rifkin, Jeremy 384
Rock, John 85–7
Roe v. *Wade* 113
Roh Sung-Il, Dr 433
Roman Catholic Church *see* Catholic Church
Romans, ancient 116, 457–60
Rose, Joanna 275–6
Rothschild, Lord 81
Royal College of Obstetricians and Gynaecologists 92, 101–2, 109
Royal Women's Hospital, Melbourne 98–9
Ruddle, Frank 498
Salle, Bruno 226

salmon, transgenic 477
Sanger, Margaret 464
Sarah xi, xiii, 306, 315
Sardinia: thalassaemia 494–6
Schatten, Dr Gerald 430–1, 433, 437
Schenk, Dr S. L. 77–9
Schill, Professor 32
Schopenhauer, Arthur 488
Science 430, 431, 433, 434
SCNT *see* somatic cell nuclear transfer
screening *see* pre-implantation genetic diagnosis
sea-urchins 37, 40, 81
semen/seminal fluid 33, 35
  see also sperm
Semenax 83
Set 67, 453
sexual intercourse
  and infection 60–1
  and pregnancy rates 135–6
  and timing ovulation 55
sexually transmitted diseases 60–1
Sheffield Centre for Reproductive Medicine and Fertility 279
Shenykin, Dr Yefim 65
Sherbahn, Richard 373
Sherman, Dr 190
Shiva (Hindu god) 238, 454
Shockley, William 192
Silber, Dr Sherman J. 32
Silver, Lee: *Remaking Eden* 496–7, 498
Simmons, Shelagh 517 n. 12
single women 193, 243, 268
Skinner, Dennis 108
Skinner, Jack 466–7
Slovenian IVF clinics 287
Smith, Shelly 272
smoking 71, 146–7
'Snowflakes Program' 212–13

somatic-cell gene therapy 492
somatic cell nuclear transfer
    (SCNT) 426–8
Soranus of Ephesus 457–8
Sorenson court case 242
South Africa: surrogacy case
    304–5
Spallanzani, Lazaro 80–1, 189,
    238–9
Spemann, Hans 377–8
sperm (atozoa) 31–3, 34
    absence of 74, 168, 230–1
    antibodies to 73
    capacitation of 81, 87, 88,
    502 n. 6
    collection for IVF treatment
    158
    examining and screening 72,
    158–9, 193–4
    freezing 190, 192–3, 243
    genetically modifying 482
    and intra-cytoplasmic sperm
    microinjection 165–8
    low count 70, 169
    motility of 32, 72
    of older men 32, 34
    and orgasm 34
    stem cells 230–1, 444–6
    see also donor insemination;
    fertilization; husbands'
    rights
sperm banks 190–1, 192–3,
    242–3
spermatogonia 443
starflower 449
Stein, Dr Irving 54
Steirteghem, André van 196
stem cells 324, 326, 376, 391–2,
    409–14, 443–6
    from aborted fetuses 418
    and differentiation 424,
    439–40, 442
    embryonic v. adult 422–5
    multipotent 414

and nuclear transfer 389, 425–8
    pluripotent 414
    religious attitudes to 436–9
    sources 412–13, 414, 422,
    443–4
    sperm 230–1, 444–6
    therapeutic uses 417–18
    totipotent 414
Steno, Nicolas 405, 406, 407,
    409
Steptoe, Patrick 90, 92–3, 96–7,
    98, 100–3, 105–6, 179, 234,
    260
sterilization
    in China 472
    in India 471
    in USA 465–8
Stern, Betsy and Bill 304, 307,
    311
steroid drugs 73
Stillman, Bob 382–4, 385
Stoddart, Ron 211–12
Stopes, Marie 464
    The Control of Parenthood
    464
stork myths 2–3
stress 130
    and infertility 55, 148–50
Sumerians 452, 453
Sumner, Francis B. 38
Sunday Times 314
Superhuman (BBC TV series)
    261
surrogacy 16, 107
    and baby brokers 307–11
    care for the bearing mother
    308–9
    'full' 298
    IVF 297–305
    parental rights 312–13
    'partial' 297–8, 311, 313
    and problems with babies 309
    religious views on 305, 306–7
    and telling children 317–19

Sutcliffe, Dr Alastair 187
Swammerdam, Jan 406–7, 409
Sweeney, Kathleen Curran 131
Sylvius, Dr Franciscus 407
Szombathy, Josef 449

Tachataki, Maria 182, 208
Tacitus: *Histories* 460
Takagi, Dr Yasushi 418
Talansky, Beth 479
Talmud 115
Tawaret 454
tea consumption 146
technology, views on 127–8
Telethon Institute for Child
    Health Research, Perth,
    Australia 181
television series, author's
    Making Babies 284
    Superhuman 261
telomeres 395
temperature testing 55
Temple of Health, Adelphi
    Theatre 52–3
Tendler, Rabbi Moshe 437
Tertullian 121
TESA *see* testicular sperm
    aspiration
TESE *see* testicular sperm
    extraction
*Test Tube Babies – a Christian
    View* 122
Testart, Jacques 103–4
testes
    cancer 72
    and decline in sperm count
    70
    freezing 229–31
    gland extracts from 69,
    83–4
    shrunken 72–3
    temperature of 65–6, 66
testicular sperm aspiration
    (TESA) 168, 169

testicular sperm extraction
    (TESE) 169
testosterone 69–70, 73
TFRs *see* 'total fertility rates'
thalassaemia 493–6
Thatcher, Margaret 306, 349
Thum, Dr M. Y. 250
Thurnham, Peter 111–12
Time magazine 386
tissue transplants 369–70
'total fertility rates' (TFRs)
    472–3
Touirat, Ouarda 226
toxaemia of pregnancy 178
transgenics 443, 476–86, 498
    and ethical concerns 486–91
    patenting animals 488
    religious views 488–90
trehalose 219
triplets 37, 178–9
Trobriand Islanders 39
Trojan Horse 474–5
Trounson, Dr Alan 97, 99–100,
    198–9, 245, 260, 262, 301, 442
TSC-2 (gene) 182
tuberculosis: and adhesions 64
Tuddenham, Ted 335
Turner, Amanda, and Coyle,
    Adrian: study 274–5
Turner's syndrome 365
twins 25–6, 37, 178, 309

UK DonorLink 277
UK Stem Cell Bank 428
ultrasound scans 56, 59, 103
    for embryo transfer 163
UN International Conference on
    Population and Development
    471
Unborn Children (Protection)
    Bill 108–9, 110–11, 329, 331
Underground travel 147
underweight: and infertility
    145–6

United States of America
   abortion 61, 113
   donor insemination 239–40, 242
   egg donors 248
   embryo research 113
   fertility clinics and sperm banks 86, 190–1, 192–3, 266–7, 272
   National Ethics and Advisory Board (NEAB) 113
   obesity 143–5
   stem-cell research 391–2, 439
   sterilization laws 465–8
   surrogacy 304, 307–11, 313, 315, 320–1
University College, London 187, 455, 463
urethra 33
urine 58, 84
uterus 35, 63
   adenomyosis 64
   adhesions 64
   contractions 35
   endometriosis 64
   fibroids 63
   implantation of embryo in 25, 41–3, 57
   malformations 62, 134
   scarring 64
   transferring embryo to 81, 161–3, 205–6
Utian, Wulf 227

vagina: and sperm 34
variocele 71
vas deferens 33
vasectomy 243
Venus figurines 449–51
Verlinsky, Yuri 337
Vibraye, Marquis Paul de 450
Victoria, Queen 322, 323
'virgin' births 39

viruses 474–6, 482–3
Voluntary Licensing Authority (VLA) 109, 110

Wachtel, Erica 326
Waldenberg, Rabbi Eliezer 256
Walpole, Horace 53
Warnock, Baroness Mary 275
   Committee of Inquiry into Human Fertilization and Embryology 107, 120, 297, 303
Watson, James 493
Weatherall, Sir David 495
Weinberger, Caspar 113
Weismann, August 377, 386
Weiss, Rabbi Benjamin 254
Whitehead, Mary Beth 304, 311–12
Whittingham, Dr David 87, 90, 94, 199, 203
WHO see World Health Organization
Widdecombe, Ann 110–11
Wigley, Dafydd 329–30
Willadsen, Steen 380–2, 386
Willendorf, Venus of 449
Williamson, Bob 335–6, 337, 354
Wilmut, Ian 381, 393
Wilton, Dr Leeanda 385
Winston, Joel 351
Winston, Lira 350
Winston, Tanya 351
Winston-Fox, Ruth 90
Wood, Dr Carl 95, 96, 100
Woolf, Lord Justice 196
World Health Organization (WHO) 470
Wright, Professor Nick 413
Wright, Stuart Pearson 285–6
Wuermeling, Professor Hans-Bernhard 384–5

X-chromosomes *see* chromosomes

X-rays 147, 207

Y-chromosomes *see* chromosomes

Zavos, Dr Panos 380, 389, 390–3, 401

Zideman, David 157, 158

ZIFT *see* zygote intra-fallopian transfer

zona drilling ('assisted hatching') 170, 179, 260

zona pellucida 218, 333

Zoroaster 40

zygote intra-fallopian transfer (ZIFT) 175